Apoptosis in Health and Disease
Clinical and Therapeutic Aspects

The process of programmed cell death or apoptosis has, in the past decade, been shown to be centrally involved in the pathogenesis of the significant majority of human illnesses and injury states. The cellular attrition observed in most degenerative conditions is apoptotic in nature; conversely, a failure of apoptosis has been proposed to underlie many forms of cancer. The central role of apoptosis in human disease clearly brings with it clinical promise: for example, the strong possibility exists that attenuation of apoptotic death will significantly modulate the severity of degenerative disorders. Similarly, conditions such as cancer, autoimmune disease, psoriasis, and endometriosis, in which aberrant cellular proliferation is observed, may benefit from enhanced rates of apoptosis. This book surveys the underlying molecular mechanisms of apoptosis, investigates its role in degenerative and other diseases, and evaluates potential therapies that will permit appropriate activation or inhibition of apoptosis in disease and injury states.

Martin Holcik, Ph.D. is a scientist at the Apoptosis Research Centre of the Children's Hospital of Eastern Ontario. He is also a Canadian Institutes of Health Research New Investigator.

Eric C. LaCasse, Ph.D. is a scientist and Group Leader at Ægera Oncology Inc., a biotech company founded on the initial discovery of the cellular inhibitors of apoptosis.

Alex E. MacKenzie, M.D., Ph.D. is a molecular geneticist and an attending pediatrician at the Children's Hospital of Eastern Ontario. He was the leader of the spinal muscular atrophy positional cloning effort which, in 1995, resulted in the isolation of the NAIP gene.

Robert G. Korneluk, Ph.D., F.R.S.C. is a well-recognized human molecular geneticist with an established track record in cloning inherited human disease genes, notably the gene for myotonic dystrophy in 1990. He is a director of Apoptosis Research Center.

Apoptosis in Health and Disease

Clinical and Therapeutic Aspects

Edited by

Dr. Martin Holcik
University of Ottawa

Dr. Eric C. LaCasse
Ægera Oncology Inc.

Professor Alex E. MacKenzie
Children's Hospital of Eastern Ontario

and

Professor Robert G. Korneluk
University of Ottawa

CAMBRIDGE
UNIVERSITY PRESS

CAMBRIDGE UNIVERSITY PRESS
Cambridge, New York, Melbourne, Madrid, Cape Town, Singapore,
São Paulo, Delhi, Dubai, Tokyo, Mexico City

Cambridge University Press
The Edinburgh Building, Cambridge CB2 8RU, UK

Published in the United States of America by Cambridge University Press, New York

www.cambridge.org
Information on this title: www.cambridge.org/9780521159449

© Cambridge University Press 2005

First published 2005
First paperback edition 2010

A catalogue record for this publication is available from the British Library

Library of Congress Cataloguing in Publication Data
Apoptosis in health and disease: clinical and therapeutic aspects / [edited by] Martin
Holcik . . . [et al.].
  p. cm.
Includes bibliographical references and index.
ISBN 0 521 52956 5 (paperback)
1. Apoptosis. 2. Pathology, Cellular. I. Holcik, Martin, 1967–
[DNLM: 1. Apoptosis – physiology. 2. Apoptosis – drug effects. QH 671 A643793 2005]
QH671.A6576 2005
571.9'36 – dc22 2004051844

ISBN 978-0-521-52956-3 Hardback
ISBN 978-0-521-15944-9 Paperback

Additional resources for this publication at www.cambridge.org/9780521159449

Contents

Contributors

Martin R. Bennett
British Heart Foundation Professor of
Cardiovascular Sciences, Unit of
Cardiovascular Medicine, Addenbrooke's
Centre for Clinical Investigation, Level 6,
Box 110, Addenbrooke's Hospital, Hills
Road, Cambridge CB2 2QQ, UK

R. Chris Bleackley
Department of Biochemistry, University
of Alberta, Faculty of Medicine and
Dentistry, Edmonton, Alberta T6G 2H7,
Canada

Armin Ensser
Institut für Klinische und Molekulare
Virologie, Friedrich-Alexander-Universität
Erlangen-Nürnberg, Schlossgarten 4,
D-91054 Erlangen, Germany

Gerard Evan
Cancer Research Institute, Box 0875,
University of California San Francisco,
San Francisco, CA 94143-0875,
USA

Helmut Fickenscher
Abteilung Virologie, Hygiene-Institut,
Ruprecht-Karls-Universität Heidelberg, Im
Neuenheimer Feld 324, D-69120
Heidelberg, Germany

Bernhard Fleckenstein
Institut für Klinische und Molekulare
Virologie, Friedrich-Alexander-Universität
Erlangen-Nürnberg, Schlossgarten 4,
D-91054 Erlangen, Germany

Martin Holcik
Department of Pediatrics, University of
Ottawa and Apoptosis Research Centre,
Children's Hospital of Eastern Ontario
Research Institute, 401 Smyth Road,
Ottawa, ON, Canada, K1H 8L1

Robert G. Korneluk
Departments of Pediatrics, and
Biochemistry, Microbiology and
Immunology, University of Ottawa and
Apoptosis Research Centre, Children's
Hospital of Eastern Ontario Research
Institute, 401 Smyth Road, Ottawa, ON,
Canada, K1H 8L1

Eric C. LaCasse
Ægera Oncology Inc., Room 306, Children's
Hospital of Eastern Ontario, 401 Smyth
Road, Ottawa, ON, Canada, K1H 8L1

Zhiping Liu
Department of Pathology, Johns Hopkins
University School of Medicine, Baltimore,
Maryland, USA

Alex E. MacKenzie
Department of Pediatrics, University of Ottawa and Apoptosis Research Centre, Children's Hospital of Eastern Ontario Research Institute, 401 Smyth Road, Ottawa, ON, Canada, K1H 8L1

Lee J. Martin
Department of Pathology, Division of Neuropathology, Neuroscience, and Neurology, Johns Hopkins University School of Medicine, Baltimore, Maryland, USA

Linda Z. Penn
Department of Medical Biophysics, University of Toronto, Ontario Cancer Institute, Princess Margaret Hospital, 610 University Avenue, Toronto, Ontario, Canada, M5G 2M9

Donald L. Price
Department of Pathology, Division of Neuropathology, Neuroscience, and Neurology, Johns Hopkins University School of Medicine, Baltimore, Maryland, USA

Hyung Don Ryoo
Howard Hughes Medical Institute, The Rockefeller University, 1230 York Avenue, Box 252, New York, NY 10021, USA

Pere Santamaria
Department of Microbiology and Infectious Diseases and Julia McFarlane Diabetes Research Centre, Faculty of Medicine, The University of Calgary, 3330 Hospital Drive NW, Calgary, Canada, T2N 4N1

Erinn L. Soucie
Department of Medical Biophysics, University of Toronto, Ontario Cancer Institute, Princess Margaret Hospital, 610 University Avenue, Toronto, Ontario, Canada, M5G 2M9

Hermann Steller
Howard Hughes Medical Institute, The Rockefeller University, 1230 York Avenue, Box 252, New York, NY 10021, USA

Juan Troncoso
Department of Pathology, Division of Neuropathology, Neuroscience, and Neurology, Johns Hopkins University School of Medicine, Baltimore, Maryland, USA

Preface

Multicellular organisms are remarkable in their complexity, comprised of an almost unimaginably diverse array of molecular mechanisms enabling the propagation, differentiation, and maintenance of their component cells. But just as there exist precisely programmed mechanisms to generate and maintain the cellular constituents of metazoans, so too there exist elaborate means of ending the lives of cells. Homeostasis is achieved at the cost of a certain intolerance; cells have defined life spans, limited ability to deal with physical, chemical, electrical, thermal, or biologic stress, and a narrow scope of acceptable behavior, deviation from which rapidly results in death. Thus, cells which have served their purpose in development, have reached the end of their natural life span post-development, have sustained an injury, or in some way have become dysregulated conduct a rapid self-disassembly and then die efficiently, committing suicide. This programmed cell death or apoptosis is a natural ongoing process, necessary for life. As logical as this state of affairs may appear in hindsight, a full appreciation has only come over the last two decades.

Although, as might be expected, the number of molecular pathways enacting this cellular attrition pales in comparison with that required to generate a cell, they are nonetheless remarkably complex. The past decade has witnessed the delineation of many, and likely most, of the central molecular mechanisms and the constituent parts by which apoptosis occurs. This gratifying molecular progress leads directly to the question of therapeutic relevance. Just as "normal" programmed cell death is essential for life, dysregulated programmed cell death is observed in the significant majority of diseases and injury states. The ubiquity of such dysregulation invokes the obvious question: does the pharmacologic or biologic modulation of this process impact disease severity? In the following pages, internationally recognized experts attempt to address this central and still largely unanswered question showing where we are in the pursuit of apoptosis modulation-based therapies.

This book attempts to summarize some of the key apoptosis findings and how they apply to medicine. This general subject is vast and cannot be justly covered in a single chapter or book. In addition, we apologize to our many colleagues whose work could not be cited for lack of space.

Apoptosis in health, disease, and therapy: overview and methodology

Eric C. LaCasse[1], Martin Holcik[2], Robert G. Korneluk[3]
and Alex E. MacKenzie[2]

[1] Ægera Oncology Inc.
[2] Department of Pediatrics, University of Ottawa, Apoptosis Research Centre and the Solange Gauthier Karsh Molecular Genetics Laboratory, Children's Hospital of Eastern Ontario Research Institute
[3] Departments of Pediatrics, and Biochemistry, Microbiology, and Immunology, University of Ottawa Apoptosis Research Centre and the Solange Gauthier Karsh Molecular Genetics Laboratory, Children's Hospital of Eastern Ontario Research Institute

1.1 Introduction: life cannot exist without cellular death

Apoptosis, or programmed cell death, is the mechanism by which most cells die both physiologically and pathologically. The realization in the mid 1980s that cells die by an active, genetically defined process changed not only our views on cellular life but led to a whole new discipline of biologic study with significant implications for medicine (Thompson, 1995; Robertson *et al.*, 2002). Apoptosis research has advanced our understanding of a basic cellular process, shed insight into many diseases, and is poised to affect the future practice of medicine by the introduction of therapies targeting this cell death process.

In this book, the term "apoptosis" is used synonymously, for right or wrong, with programmed cell death (PCD). While PCD may be a more appropriate term, encompassing all forms of active physiological cell death, apoptosis, which is defined morphologically and biochemically, is used here for historical purposes (Lockshin and Zakeri, 2002; Melino, 2002; Sloviter, 2002). The original "anatomical" characteristics of apoptosis were noted in the nineteenth century (reviewed in Clarke and Clarke, 1996; Rich *et al.*, 1999). However, it was not until publications in 1951 and in the 1960s described developmental cell death or "shrinkage necrosis" that the PCD concept was recognized, reintroduced, and formalized (Lockshin and Zakeri, 2001; Kerr, 2002; Vaux, 2002). The term "apoptosis" was coined in 1972, referring to this morphologically defined form of cell death (Kerr *et al.*, 1972). However, a broader recognition of the

Apoptosis in Health and Disease: Clinical and Therapeutic Aspects, ed. Martin Holcik, Alex E. MacKenzie, Robert G. Korneluk, and Eric C. LaCasse. Published by Cambridge University Press. © Cambridge University Press 2004.

centrality and import of PCD did not occur until seminal studies performed by Horvitz and colleagues in the early 1980s demonstrated the genetic underpinnings of this process in the nematode, *Caenorhabditis elegans* (see Chapter 2; Ellis and Horvitz, 1986). The impact of these genetic discoveries was recently recognized by the awarding of the 2002 Nobel Prize in Physiology or Medicine to Sidney Brenner, John E. Sulston, and H. Robert Horvitz (Benitez-Bribiesca, 2003).

1.1.1   Apoptosis characteristics

Apoptosis is the physiological mechanism by which cells die, characterized by the features listed in Table 1.1, and is distinct from the accidental form of cell death called necrosis which represents an "extreme" in the continuum of cell death. Although initial work sought to characterize cell death as strictly apoptotic or necrotic, many cell death events display both apoptotic and necrotic features. Thus, necrosis represents one extreme of a continuum, with classical apoptosis at the other. In certain cell types and tissues, such as neurons, cell death usually proceeds with unique characteristics that generated long debate with respect to mechanism. Many researchers now recognize that all features of apoptosis need not be present. In necrosis, there is a loss of cellular plasma membrane integrity accompanied by an inflammatory response, while in apoptosis the plasma membrane integrity is preserved and the cells are discretely disassembled and phagocytosed without an inflammatory response. Necrosis occurs in severe circumstances, such as cases of sudden transfer of energy (kinetic, electrical, thermal), frostbite, or exposure to certain toxins. These can rupture the plasma membrane and/or severely compromise the cells' energy-producing respiratory process. However, the normal physiological route to cell death is, under most circumstances, apoptosis. Only under extreme conditions, when the cell cannot properly execute its apoptotic program, does the cell die, by default, by necrosis, thus blurring the distinction between these two forms of cell death. While necrosis can be considered pathophysiological, it is a medically undesired form of cell death because the ensuing inflammatory response causes secondary cell death and tissue damage. For example, the acute use of high-dose steroid therapy to suppress inflammation in spinal cord injury has led to improvements in neurologic outcomes by suppressing some of this secondary cell death. Arguments for a physiologic, and potentially beneficial, role of necrosis have also been made (Proskuryakov *et al.*, 2003).

1.1.2   Apoptosis pathways

There exist multiple cellular pathways triggering apoptosis, two of which, the extrinsic and intrinsic pathways, are the best studied. Various other pathways exist, and

Table 1.1 *Characteristics distinguishing apoptosis from necrosis*

Defining features	Apoptosis	Necrosis
Physiologic/pathologic features		
Cellular role	Usually normal	Abnormal, accidental
Process	Active, energy dependent	Passive, results from lack/loss of energy
Distribution	Dispersed, affects individual cells	Contiguous, simultaneous, and massive affects in damaged tissue areas
Triggers	100s of physiologic and noxious stimuli	Sudden transfer of energy, specific toxins, or ATP depletion
Induction	Slow (hours), stochastic	Rapid (seconds, minutes)
Tissue inflammation	Absent	Present
Cell removal	Rapid and discrete	Slow
Morphologic features		
Cellular membranes	Integrity preserved, blebbing of intact plasma membrane	Loss of integrity, with spilling of cell constituents
Cell volume	Decreased, as well as the formation of small, fragmented "apoptotic bodies" or inclusions	Increased
Organelle structure	Late preservation, with exception of nuclear condensation and fragmentation	Swelling of nucleus and other organelles
Chromatin	Discrete, organized condensation, margination and fragmentation (e.g. pyknotic nuclei)	Pattern conserved
Biochemical and molecular features		
Mitochondrial permeability transition	Moderate	Severe
Mitochondrial membrane potential (delta psi-m)	Transient loss	Permanent loss
Requirement for ATP	Yes	No
Membrane phospholipid asymmetry	Exteriorization of phosphatidylserine from inner to outer leaflet of plasma membrane	Unchanged
Cell pH	Acidification	Unchanged
DNA cleavage	Initial specific large cleavage products of 300, then 50, kbp, followed by internucleosomal cleavage leading to DNA ladder pattern of 180 bp unit repeats	Random DNA cleavage
Caspase dependence	Yes	No

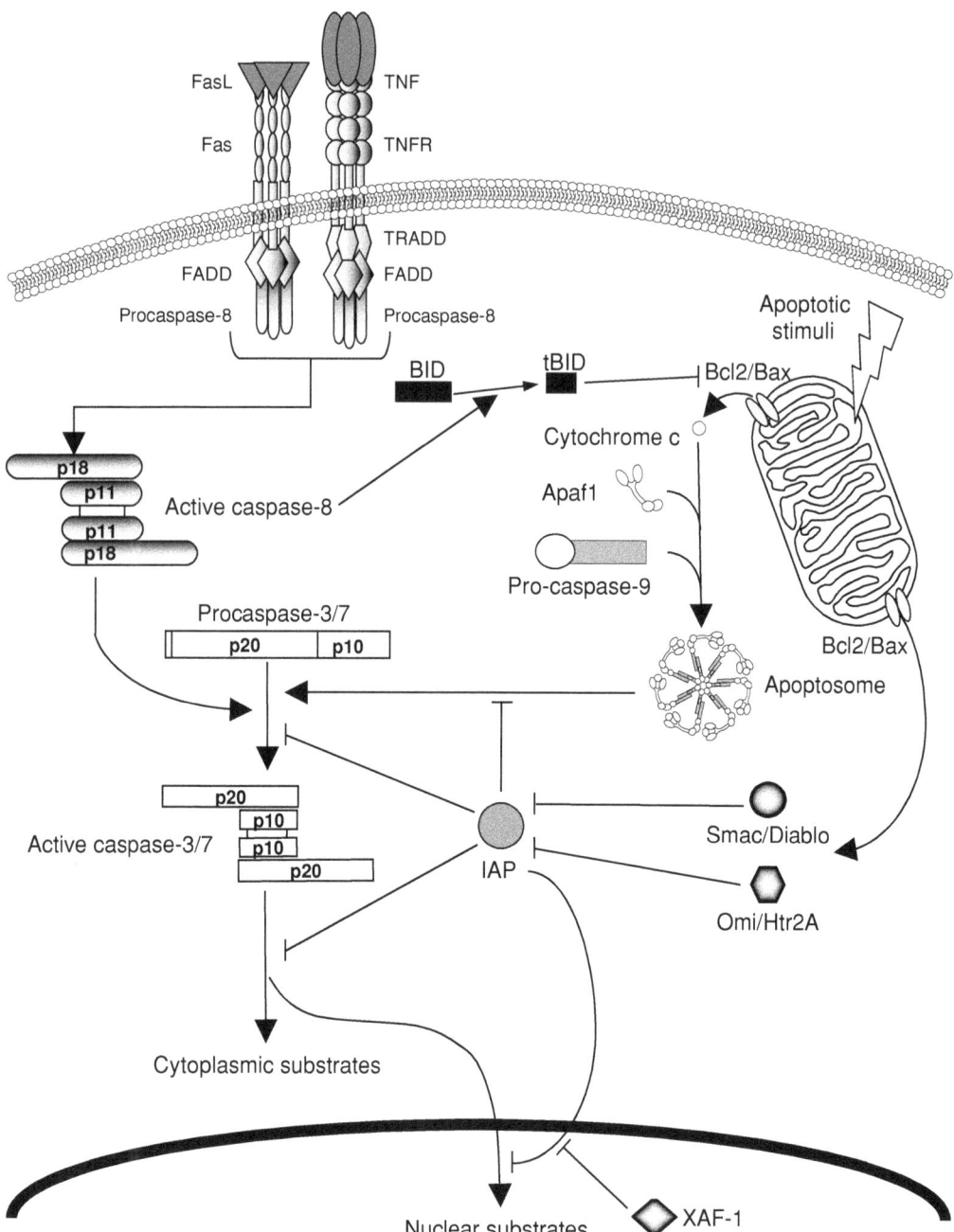

Figure 1.1   *Intrinsic and extrinsic apoptotic pathways in mammalian cells.* Intracellular stress results in the activation of the mitochondrial, or intrinsic, pathway which leads to cytochrome c release, apoptosome formation, and caspase activation. Extracellular ligand binding to death receptors triggers the extrinsic pathways that can either directly result in the activation of the caspases, or requires further amplification through the mitochondrial pathway dependent

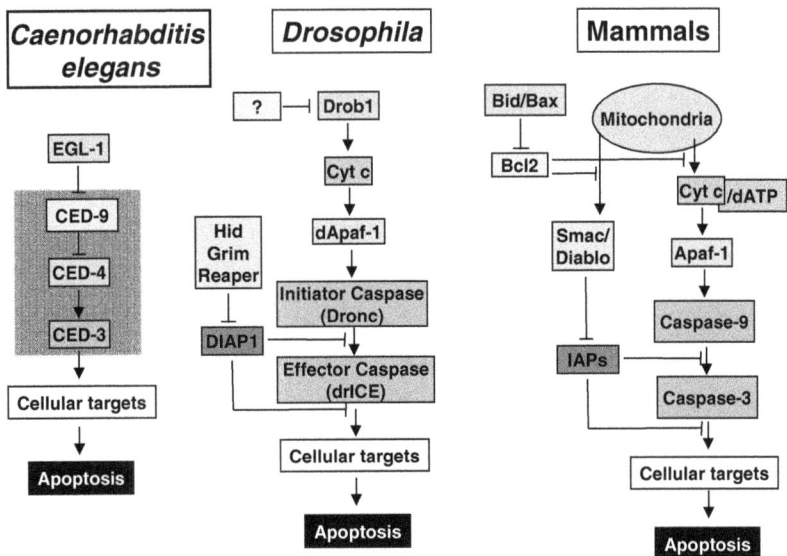

Figure 1.2   *Core elements of the apoptotic pathways conserved across several phyla.* The core apoptotic machinery is shown for *C. elegans, D. melanogaster*, and for mammals. The original core of death genes identified in *C. elegans* (ced-3, ced-4, and ced-9) are shown boxed in gray (Ellis and Horvitz, 1986; Hengartner *et al.*, 1992). The shading scheme for each gene identifies orthologs and paralogs across the phyla. For additional information see chapter 2, as well as references by Meier *et al.*, 2000, Aravind *et al.*, 2001, and Lawen, 2003.

some are discussed in a review by Ferri and Kroemer (2001). The extrinsic pathway involves plasma membrane receptors of the tumor necrosis factor receptor (TNFR) superfamily that recognize extracellular death-inducing ligands, while the intrinsic pathway utilizes the coordinated control and release of apoptogenic factors from the mitochondria which "sense" many death-inducing stimuli through regulatory factors of the bcl2 family (Figure 1.1). All known death pathways culminate in the activation of a proteolytic cascade involving a family of proteases, the caspases (Figures 1.1 and 1.2). The caspases are cysteinyl-containing active center proteases with specificity for protein cleavage after aspartyl residues (Thornberry and Lazebnik, 1998; Earnshaw *et al.*, 1999; Nicholson, 1999). Thus, the term *caspases* for cysteinyl-containing aspartate-specific proteinase. The caspases are responsible

on the cell type. Both apoptotic signaling pathways converge at the level of effector caspases, such as caspase-3 and -7. Multiple control points exist along these pathways, controlling either the release of cytochrome c and other apoptogenic factors from the mitochondria or by regulating the caspase inhibitors, the inhibitors of apoptosis (IAPs), through their antagonists or through other regulatory mechanisms.

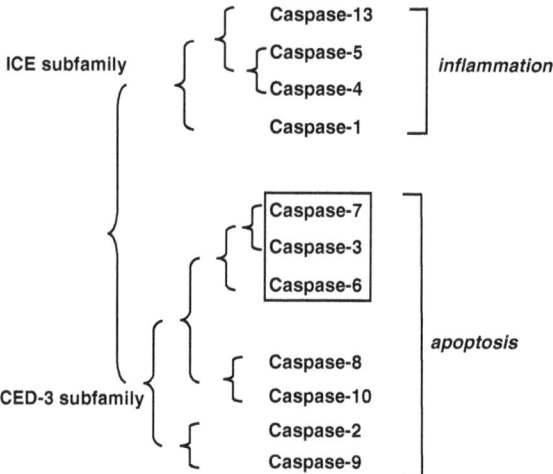

Figure 1.3   *The caspase family.* Phylogenetic analysis segregates the human caspases into two major subfamilies, one based on caspase-1 previously referred to as ICE, for interleukin-converting enzyme, and the other based on similarities to the *Caenorhabditis elegans* cell death gene, *ced-3*. Further classification of the caspases is possible: into those that mediate cytokine maturation that are involved in inflammation, those with a short pro-domain involved in the effector phase of apoptosis (shown boxed), and those with a long pro-domain and involved in the initiator phase of apoptosis (not boxed). Note evolutionary distances are not accurately represented in this dendrogram. Based on Nicholson, 1999.

for many of the hallmarks of apoptosis listed in Table 1.1, through their cleavage of specific polypeptide substrates (Fischer *et al.*, 2003). The caspases – e.g. caspase-2, -3, -6, -7, -8, -9, -10 (and 12 in the mouse) – are not the only proteases involved in PCD; calpains have been shown to have a role in some instances, and there is the possibility of others as yet undiscovered. However, the caspases are the prime effectors of apoptosis, and their activation produces a catalytic cascade leading to further caspase activation, ultimately committing the cell to death by cleavage of structural proteins, activation of nucleases, as well as the resultant inactivation of survival, repair, or anti-apoptotic factors and activation of pro-death factors. Note that not all caspases are involved in apoptosis and that certain caspases are involved in the processing of pro-inflammatory cytokines, e.g. caspase-1, -4, -5, and -11 (Figure 1.3). In addition, certain caspases may play highly specific roles in normal cell differentiation and maturation, such as the enucleation of red blood cells, formation of lens cells of the eye, or platelet formation from megakaryocytes, or in blast phase activation of lymphocytes, apparently without inducing apoptosis (see Chapter 2; Ishizaki *et al.*, 1998; Weil *et al.*, 1998; Alam *et al.*, 1999; Kennedy *et al.*, 1999; Zeuner *et al.*, 1999; De Botton *et al.*, 2002; Newton and Strasser, 2003; Olson *et al.*, 2003; Perfettini and Kroemer, 2003; Philchenkov, 2003).

1.1.3 Conservation and diversification of the apoptosis machinery

Caspases exist within the cell as inactive zymogens and are activated primarily by two distinct mechanisms both involving protein–protein interactions within large complexes and proximity-induced processing of the caspases (Chen *et al.*, 2002; Boatright *et al.*, 2003; Chang *et al.*, 2003). One mechanism involves the generation of the DISC, the death-inducing signaling complex, which is formed by the trimerization of plasma membrane death receptors upon binding of death ligands, such as TNFα or TNFα-related apoptosis-inducing ligand (TRAIL). Receptor trimerization subsequently triggers the recruitment of adaptor molecules and initiator or apical caspases, such as caspase-8 or -10, to the cytoplasmic portion of the death receptors. This induced proximity then allows for activation of the apical caspase, which can then in turn activate downstream effector caspases. The second mechanism involves the release of apoptogenic factors from the mitochondria, particularly cytochrome c, which results in the formation of the apoptosome, a large helical complex comprised of apaf-1, cytochrome c, and caspase-9 (Acehan *et al.*, 2002; Shi, 2002). This complex formation induces a conformational shift and allows the proximity-induced activation of the apical caspase-9, which can then cleave downstream effector caspases, such as caspase-3 and -7. The two death pathways do not function in isolation with evidence of both cross-talk and feed-forward loops. Additional, less well-understood pathways and mechanisms for caspase activation are also found (Nakagawa *et al.*, 2000; Sperandio *et al.*, 2000; Ferri and Kroemer, 2001; Forcet *et al.*, 2001; Troy *et al.*, 2001; Rao *et al.*, 2002; Read *et al.*, 2002; Chandra and Tang, 2003). A third characterized pathway of caspase activation that is specific to cytotoxic T cell-mediated cell death is reviewed in Chapter 6. This form of cell death results from the introduction of the serine proteases, known as granzymes, into the cytoplasm which can then either directly activate the caspase cascade (by cleavage of specific caspases or bid) or bypass the caspases (by cleaving their substrates directly) (Barry and Bleackley, 2002; Lieberman, 2003).

The cell is armed with an elaborate self-destruct mechanism, comprised of inactive zymogens that can be activated by numerous stresses or triggers, but which remain tightly controlled (e.g. Fussenegger *et al.*, 2000; Bortner and Cidlowski, 2002). For example, the cell contains endogenous caspase inhibitors, the inhibitors of apoptosis (IAPs) (Miller, 1999; Verhagen *et al.*, 2001; Salvesen and Duckett, 2002; Shi, 2002; Stennicke *et al.*, 2002), as well as Bcl-2 family regulators of the mitochondrial release of apoptogenic factors (Gross *et al.*, 1999; Wang, 2001; Newmeyer and Ferguson-Miller, 2003; Scorrano and Korsmeyer, 2003; Tsujimoto, 2003), and modulators of DISC signaling, such as FLIP (Thome and Tschopp, 2001; Peter and Krammer, 2003). All these factors are controlled by several different mechanisms, including transcriptional and post-transcriptional regulation, translational control, phosphorylation, proteolysis, and ubiquitin-mediated protein degradation. In addition, the caspase-inhibiting IAPs are antagonized by mitochondrial factors (i.e. SMAC/DIABLO – Du *et al.*, 2000, Verhagen *et al.*, 2000 – and OMI/HTra2 – Suzuki *et al.*, 2001), cytoplasmic proteins (i.e. eRF3 – Hegde *et al.*, 2003), and nuclear factors (i.e. XAF-1 – Liston *et al.*, 2001), while the bcl-2 anti-apoptotic proteins are antagonized by the BH3-only and Bax-like members of the bcl-2 family (Figure 1.1). Many of these protein interactions have been solved structurally (Fesik, 2000; Shi, 2001). Thus, within the cell, there exists a highly regulated and structured machinery to induce or suppress cell death. The identity and interplay of factors controlling this ongoing internal conflict form the basis of an enormous amount of basic, pre-clinical, and clinical research.

Multicellular organisms are remarkable in their complexity. They are characterized by an almost unimaginable diverse array of molecular mechanisms which enable the cellular propagation, differentiation, and maintenance which underpin eukaryotic life forms. But just as there exist wide-ranging, precisely programmed mechanisms to generate the constituent cellular components, so too there exist elaborate means of ending the lives of cells. The eukaryotic milieu is of necessity an intolerant one – cells have a narrow scope of appropriate behavior, deviation from which rapidly results in cellular death. Thus, cells which have served their purpose in development, cells which have reached the end of their natural life span post-development, and cells which have sustained an injury or have in some way become dysregulated die efficiently; they commit suicide and are disassembled. Although the series of mechanisms enacting this cellular attrition pale in comparison with those required to generate a cell, as might be expected, they are nonetheless remarkably complex. As logical as this state of affairs may appear to be in hindsight, a full appreciation has only come over the last 15 years.

Included in the many factors involved in apoptosis control are protein motifs such as bcl2 homology regions 1–4 (BH1, BH2, BH3, and BH4) and baculovirus IAP repeats (BIR domains), as well as CARD, DD, DED, NACHT, and PYRIN domains or motifs (see abbreviations and references in Table 1.2). These motifs or domains are not exclusive to apoptosis; many are found in proteins that have roles in immunity, in particular reflecting the co-evolution of the two processes (see Section 1.2 and Chapters 6 and 7; Bertin and DiStefano, 2000; Koonin and Aravind, 2000; Pawlowski *et al.*, 2001; Staub *et al.*, 2001; Bouchier-Hayes and Martin, 2002; Gumucio *et al.*, 2002; Martinon *et al.*, 2002; Chamaillard *et al.*, 2003; Creagh *et al.*, 2003). The number of genes known to contain these motifs has grown with the completion of the human genome project, with more than 100 genes identified to date (Table 1.2). Among the many other factors involved in apoptosis are those that appear to exist as "orphans" (cytochrome c) or as small family clusters (p53, p63, and p73), only one member of which may play a predominant role in apoptosis control.

Table 1.2 *Death gene motifs and numbers of human genes*

Gene family based on homology or motif [InterPro accession number]	Number of genes identified* (at 06/2003)	References (some key references listed at bottom)
Caspase [IPR001309, IPR002138]	11	Lamkanfi *et al.*, 2002
Bcl-2 (BH domains, includes BH3-only members) [IPR000712, IPR003093, IPR002475]	25	
IAP (BIR-containing) [IPR001370]	8	Salvesen and Duckett, 2002
TNFR (TNF receptor family) [IPR001368]	29	Aggarwal, 2003
TNF ligands [IPR001875]	18	Aggarwal, 2003
DED (death effector domain) [IPR001875]	12	Tibbetts *et al.*, 2003
DD (death domain) [IPR000488]	33	
CARD (caspase recruitment domain) [IPR001315]	25	Hofmann *et al.*, 1997; Bouchier-Hayes and Martin, 2002
NACHT [IPR007111]	20	Koonin and Aravind, 2000
NOD (nucleotide-binding oligomerization domain)	24	Inohara and Nunez, 2003
NALPS (a "caterpiller" sub-domain; includes PYPAF)	14	Tschopp *et al.*, 2003; Harton *et al.*, 2002
PAAD (includes PYRIN or PYD, and DAPIN) [IPR004020]	19	Bertin and DiStefano, 2000; Fairbrother *et al.*, 2001; Pawlowski *et al.*, 2001; Staub *et al.*, 2001

IAP, inhibitor-of-apoptosis; BH, bcl-2 homology domain; BIR, baculovirus IAP repeat; CATERPILLER, CARD/Transcription enhancer/R(purine)-binding/pyrin/lots of leucine repeats; NACHT, NAIP/CIITA/HET-E/TP1; NOD, nucleotide-binding oligomerization domain; PAAD, Pyrin/AIM/ASC/DD-like; PYD, pyrin domain; PYPAF, pyrin-containing APAF1-like; TNF, tumor necrosis factor. *These categories are based on protein families, or specific motifs, and are not mutually exclusive, with some proteins possessing two or more different motifs. Not all the proteins containing these motifs are involved in apoptosis; many are involved in immunity and inflammation and other functions as well. The number of genes is a conservative estimate based on published reports. The true number of genes is likely to be different, and for the most part higher than those stated, based on some bioinformatic analysis (see below for examples and caveats) and due to conservative estimates of the predicted total number of genes. The EnsMART/EnsemblMART data mining tool (Clamp *et al.*, 2003) at Ensembl (www.ensemble.org) was used to create a non-redundant list of human gene products that contained specific domains as described by InterPro (www.ebi.ac.uk/interpro/), a database for protein families, domains, and functional sites (Mulder *et al.*, 2003). Lists of genes, sequences, and descriptions can be output from any of the nine species in EnsMART with a variety of filter options from the predicted genes of the complete genomes in an easy three-step process on the Ensembl website. For example, an analysis of the human genome for the death domain (*DD* motif) identified 31 genes compared with the 33 listed here, while the baculovirus *IAP* repeat (BIR motif) identified 11 genes compared with the eight listed here. The three additional IAP genes are hILP3, a likely XIAP expressed pseudogene, and two NAIP-like sequences which also represent expressed pseudogenes (Xu *et al.*, 2002). Key reviews and references include: Koonin and Aravind, 2002; Reed, 2002; Doctor *et al.*, 2003; Liu *et al.*, 2003; Reed *et al.*, 2003.

Genetic ablation of genes involved in apoptosis, mainly in mice, has demonstrated the necessity of cell death in the shaping of metazoan life, such as the genes involved in blastocyst cavitation (Joza *et al.*, 2001).

In hindsight, it seems logical that the death of a cell should be as important as its birth, thus completing the circle of life. In fact, proliferation and apoptosis are intimately linked, as errors during cell cycle progression will result in a default pathway to apoptosis. Questions still remain as to when particular apoptotic processes evolved and if unicellular organisms die by some form of PCD (Ameisen, 2002; Koonin and Aravind, 2002; Huettenbrenner *et al.*, 2003). Although some of the death motif-containing genes exist in yeast, they are likely to be involved in functions other then cell death (e.g. BIR domain-containing proteins that play a role in cytokinesis rather than apoptosis control; Miller, 1999; Silke and Vaux, 2001). While caspases are a central part of apoptosis in metazoans, no counterpart has yet to be identified in plants, but metacaspases or other proteases may fulfil this role (Ameisen, 2002; Koonin and Aravind, 2002; Hoeberichts and Woltering, 2003). Nevertheless, it is becoming clear that even plant cells can undergo apoptosis and that many of the genes involved in apoptosis control in mammals also work in transgenic plants, with dramatic results (Dickman *et al.*, 2001).

The sequencing of several metazoan genomes has brought a greater appreciation of the true complexity involved in the control of apoptosis, with many newly identified genes sub-serving more specialized and frequently redundant roles, increasing in parallel with the complexity of the organism (Table 1.3 and Chapter 2). This redundancy has confounded the analysis of many murine strains with ablated apoptotic loci, with compensatory changes often attenuating the predicted outcome (e.g. Holcik *et al.*, 2000; Zheng *et al.*, 2000; Harlin *et al.*, 2001).

The importance of several of these apoptosis gene families is demonstrated by the subversion of these pathways by viruses (see Chapter 7). Viruses use various strategies, most likely appropriated from the host cell, to suppress (or induce in some cases) the normal cellular response to viral infection, such as the induction of host cell apoptosis or the shut down of protein synthesis. Many examples of viral proteins targeting p53, the mitochondria (e.g. bcl-2 homologs), the TNFR family, caspases, NFkB (nuclear factor κB), IFN (interferon), and protein kinase R (PKR) exist, illustrating the key role these nodes play in the "apoptosis" pathway. Viruses have taught us much about the fundamental aspects of cellular physiology.

1.2   Apoptosis: roles in health

Apoptosis is a normal physiologic process beginning with the deletion, fusion, and sculpting of structures during embryogenesis such as in the removal of cells

Table 1.3 *Conservation and expansion of apoptosis genes in model organisms and humans*

Gene family and information	Worm (*Caenorhabditis elegans*)	Fly (*Drosophila melanogaster*)	Mouse (*Mus musculus*)	Man (*Homo sapiens*)
Size of genome (megabase pairs)	100	128	2497	3232
Predicted no. of genes	19 921	13 525	24 968	24 261
Caspases	3	7	10	11
Bcl-2 (pro- and anti-apoptotic)	2	2	27	25
Apaf-1/ced-4	1	1	1	1
TP53	1	1	3	3
IAP (caspase inhibitor)	2	4	7 (11–13*)	8
IAP-antagonists (RHG motif)	? (likely none exist)	5	3 (likely more exist)	3 (likely more exist)

RHG, reaper-HID-grim. *Expansion of NAIP locus in mice produces 5–7 functional copies of the NAIP gene depending on the mouse strain, which are included in the total number in parentheses (Yaraghi *et al.*, 1998; Endrizzi *et al.*, 2000; Growney and Dietrich, 2000). Key reviews and references include: Aravind *et al.*, 2001; Venter *et al.*, 2001; Baehrecke, 2002; Reed *et al.*, 2003.

comprising the interdigital webbing (see Chapter 2; Baehrecke, 2002; Mirkes, 2002), and maintaining tissue homeostasis through the elimination of defective, infected, or auto-reactive cells. It has been estimated that as many as a trillion cells die every 2 weeks in adult humans, enabling the continued turnover of tissue and removal of potentially harmful cells. Certain processes, such as B- or T-cell maturation or spermatogenesis, engender large waves of apoptosis during development to produce a small number of functional mature cells (Print and Lakoski-Loveland, 2000; Rathmell and Thompson, 2002; Marsden and Strasser, 2003). Elsewhere, apoptosis is less conspicuous, proceeding asynchronously, with apoptotic bodies quickly engulfed and digested, leaving little trace. Apoptosis is also central to remodeling in the adult, such as the follicular atresia of the post-ovulatory follicle and mammary gland involution post-weaning, to name a couple of examples.

1.3 Apoptosis: roles in disease

The death of cells in human disease, whether the cause or effect of the condition, occurs, in most cases, by a process indistinguishable from "physiologic" apoptosis. This has been demonstrated by different means, and for many, many different pathologies. Necrosis may also often be observed in disease because of the severity of the insult that can disrupt cellular integrity by shear forces or poison a cell such that it can no longer maintain its energetic and osmotic balance. The release of cellular contents from necrotic cells leads to an inflammatory response at the site of injury which can cause secondary cell death by apoptosis in the surrounding tissue. Additional apoptotic death occurs in some injury states due to loss of trophic support or cell contact, as is seen following nerve transection injuries.

Therapies based on the modulation of apoptosis can be broadly classified as those that induce apoptotic death, chiefly but not exclusively for cancer, and those that inhibit cell death. In the latter case, pathologic apoptosis clearly occurs as a result of a cellular stress whether intra- or extra-cellular in origin. Still largely unanswered is the fundamental question: will the effective suppression of apoptosis in a given pathologic process result in cells which are functionally competent or will the same process triggering the apoptosis render the cells dysfunctional or even non-functional, possibly resulting in their death either by necrosis or by some other non-apoptotic process? This question is particularly germane to conditions characterized by neuronal apoptosis (e.g. traumatic brain injury, stroke, spinal cord injuries, or in chronic disorders like Alzheimer's disease, Parkinson's disease, multiple sclerosis, HIV/AIDS, or amyotrophic lateral sclerosis) because the regenerative capacity of this differentiated class of cells is limited at best. The 'saved' neuron would still have to function properly for any benefit to be seen. Indeed, this appears to be possible in proof of principle studies in animal models of neurodegenerative diseases and in cell culture experiments. The over-expression of caspase inhibitors by gene therapy or transgenic approaches, as well as the use of cell-permeable small molecule caspase inhibitors, have clearly demonstrated that it is possible to save neurons and to derive functional benefits in animal models of acute and chronic disorders (Robertson *et al.*, 2000; Sapolsky, 2003). In some cases, the interventions only delay cell death or show no benefit at all, and this is most likely related to the severity and duration of the insult. In addition, other deleterious cellular processes may have been engaged, such as calpain activation, which may not have been inhibited by the therapeutic approach tested (Han *et al.*, 2002). Furthermore, the poor cellular permeability of small molecule caspase inhibitors has led to many disappointing results (Bilsland and Harper, 2002). Overall, the many highly promising animal studies suggest that we can limit cell death in injury and disease to only those cells that are too far gone. If intervention is early enough then one can save cells that will ultimately recover their full function. The approach, then, is one of emergency medicine at a cellular level – that is, to treat the life-threatening sequelae (block caspases), stabilize the patient's vitals (the cell's mitochondrial respiration), and provide support (e.g. trophic factors) and time for recovery and return of normal biologic parameters (e.g. re-establishment of synaptic connections in the case of neurons). Strategies attempting these anti-apoptotic approaches will be discussed below in Section 1.4.

While "pathogenic" apoptosis is seen in neurodegenerative disorders (see Chapter 4; Yuan and Yankner, 2000; Friedlander, 2003), diabetes (see Chapter 6; Mathis *et al.*, 2001), myocardial infarctions (see Chapter 5), and many other diseases, it is the failure of cells (cancerous clones, auto-reactive T cells) to undergo apoptosis that underlies cancer (see Chapter 3) and proliferative disorders (see Chapter 6; Grodzicky and Elkon, 2002). Resistance to apoptosis is considered a hallmark of cancer, a property that allows for the development of additional malignant traits in the cancer cell (Hanahan and Weinberg, 2000; Green and Evan, 2002; Johnstone *et al.*, 2002). The observation in 1988 that the Bcl-2 oncogene, originally identified as a prominent translocation in follicular lymphoma (Cleary *et al.*, 1986; Tsujimoto and Croce, 1986), was identified not as a growth stimulatory gene but as an apoptosis-suppressing gene (Vaux *et al.*, 1988), thus defined, for the first time, apoptosis suppression as an oncogenic process. Examples of mutations of virtually all genes involved in apoptosis have been reported in the literature. This includes deletion and gene silencing of initiator and effector caspases, over-expression of caspase inhibitors, and loss of expression of pro-apoptotic Bcl2 family members. Indeed, many of the most common mutations that underlie cancer in humans directly subvert the apoptotic process (i.e. tp53).

Table 1.4 lists apoptosis genes directly involved in disease, and those that are merely implicated in disease. Genes directly involved in disease are identified through causative mutations, deletions, gene rearrangements (translocations), amplifications, or loss-of-heterozygosity (LOH). It should be noted that this direct evidence for apoptosis involvement in disease applies to only a small percentage of monogenic disorders, some with rare incidence or prevalence. However, these examples provide greater significance in terms of identifying apoptosis genes as causes of disease. As a case in point, Li–Fraumeni syndrome is a rare familial disease involving the p53 tumor suppressor and specific soft tissue sarcomas, but the important role that sporadic p53 mutations play in the majority of cancers is well known.

Also included in Table 1.4 are genes implicated in disease through associations, alterations in expression, and/or correlations that do not necessarily discriminate between cause or effect. These genes have been implicated in a large number of diseases, which are mostly polygenic or multifactorial in nature, and reflect the majority of diseases with significant morbidity and mortality. The burden of proof for direct gene involvement in these cases is more difficult to attain. For this reason,

Table 1.4 *Apoptosis genes and disease associations*

Disease or syndrome (abbreviations listed below)	Genes directly involved (through mutations, deletions, translocations, LOH, or amplifications)*	MIM no. (OMIM database)	References
ADULT	P63	103285	van Bokhoven and McKeon, 2002
ALPS (type 0 or Ia)	Fas/CD95/TNFRSF6	601859	Rieux-Laucat *et al.*, 2003
ALPS (type Ib)	Fas ligand	601859	Rieux-Laucat *et al.*, 2003
ALPS (type II)	Caspase-10	603909	Rieux-Laucat *et al.*, 2003
Blau syndrome	NOD-2/CARD15	186580	
BLS	CIITA	209920	
cancer	tp53 (*LOH*)		
Cancer (esophageal, cervical, lung)	cIAP1 and/or cIAP2 (*A*)		Imoto *et al.*, 2001, 2002; Dai *et al.*, 2003
CINCA/NOMID	NALP3/cryopyrin/ PYPAF1/CIAS1	607115	
Crohn's disease	NOD-2/CARD15	266600	
EEC	P63	604292	van Bokhoven and McKeon, 2002
Familial cylindromatosis	CYLD	132700; 123850; 313100; 605041	Bignell *et al.*, 2000; Brummelkamp *et al.*, 2003; Kovalenko *et al.*, 2003; Trompiuki *et al.*, 2003
FCAS/FCU	NALP3/ cryopyrin/PYPAF1/CIAS1	120100	
FHL	perforin	603553	
FMF	Pyrin/marenostrin	249100	
Follicular lymphoma	Bcl-2 (*T*)	151430	Cleary *et al.*, 1986; Tsujimoto and Croce, 1986
Hay–Wells syndrome (AEC)	P63	106260	van Bokhoven and McKeon, 2002
HED/EDA	EDA1, EDAR, EDARADD	305100, 224900	Smahi *et al.*, 2002
Incontinentia pigmenti	NEMO/ IKKgamma	308300	Smahi *et al.*, 2002
Li–Fraumeni syndrome	tp53	151623	Malkin *et al.*, 1990; Srivastava *et al.*, 1990
LMS	P63	603543	van Bokhoven and McKeon, 2002

Table 1.4 (*cont.*)

Disease or syndrome (abbreviations listed below)	Genes directly involved (through mutations, deletions, translocations, LOH, or amplifications)*	MIM no. (OMIM database)	References
MALT lymphoma	Bcl-10 (*T*)	603517	Willis *et al.*, 1999; Zhang *et al.*, 1999
MALT lymphoma	cIAP2 and MALT1/ paracaspase (*T*)	604860	Dierlamm *et al.*, 1999; Dierlamm *et al.*, 2002; Liu *et al.*, 2002
MALT lymphoma	MALT1/paracaspase (*T*)		Murga-Penas *et al.*, 2003; Streubel *et al.*, 2003
MWS	NALP3/cryopyrin/PYPAF1/ CIAS1	191900	
NHL	MALT1/paracaspase (*A*)	604860	Sanchez-Izquierdo *et al.*, 2003
NHL and CLL	Bcl-3		
Primary macroglobulinemia	cIAP2 and MALT1 (*T*)		Hirase *et al.*, 2000
SHFM	P63	605289	van Bokhoven and McKeon, 2002
SMA (type 1)	NAIP (*D*)	253300	Roy *et al.*, 1995; Thompson *et al.*, 1995
TRAPS	TNFRSF1A/TNFR1	142680	

Disease or syndrome (abbreviations listed below)	Genes implicated** (through altered expression/ association)	MIM no. (OMIM database)	References
Advanced cancer, multiple myeloma	Alterations in NFkB pathway/ activity		Karin *et al.*, 2002; Orlowski and Baldwin, 2002
Cancer (many)	Survivin, XIAP, bcl-2, bcl-Xl, A1, cIAP2		LaCasse *et al.*, 1998; Tamm *et al.*, 2000; Ramaswamy *et al.*, 2001; Altieri, 2003; Li, 2003; Ramaswamy *et al.*, 2003
Cancer (many or various specific forms)	Bax, apaf-1, caspase-8, CD95, DR4, DR5		
Cancer (many)	XAF-1 (*LOH*)		Fong *et al.*, 2000; Liston *et al.*, 2001

(*cont.*)

Table 1.4 (*cont.*)

Disease or syndrome (abbreviations listed below)	Genes implicated** (through altered, expression/association)	MIM no. (OMIM database)	References
Colon cancer	Bax		Rampino *et al.*, 1997
Colorectal cancer	cIAP2		Mori *et al.*, 2003
Diffuse large cell lymphoma (NHL) and Hodgkin's disease	cIAP2, A1, NFkB pathway		Alizadeh *et al.*, 2000; Davis *et al.*, 2001; Hinz *et al.*, 2001; Ramaswamy *et al.*, 2001
Legionnaire's disease	NAIP5/ birc1e (in mice; no evidence of human involvement yet)	600355	Diez *et al.*, 2003; Wright *et al.*, 2003
Melanoma, renal cell carcinoma	Livin		
Multiple sclerosis	XIAP, cIAP1, cIAP2, survivin	126200	Sharief and Semra, 2001a, b, 2002; Sharief *et al.*, 2002a, b; Semra *et al.*, 2002
Mycosis fungoides/ CTCL	cIAP2	254400	Tracey *et al.*, 2003
Ovarian cancer	XIAP		Ramaswamy *et al.*, 2001
Pancreatic cancer	cIAP2		Ramaswamy *et al.*, 2001
Pancreatic cancer	Survivin (*A*)		Mahlamaki *et al.*, 2002
Rheumatoid arthritis	NFkB	164011	

ADULT, acro-dermato-ungual-lacrimal-tooth; AEC, ankyloblepharon ectodermal dysplasia, clefting; ALPS, autoimmune lymphoproliferative syndrome (Canale–Smith syndrome); BLS, bare lymphocyte syndrome; CINCA, chronic infantile neurologic cutaneous articular syndrome; CLL, chronic lymphocytic leukemia; CTCL, cutaneous T-cell lymphoma; EDA, anhidrotic ectodermal dysplasia; EEC, ectrodactyly, ectodermal dysplasia, clefting; FCAS, familial cold autoinflammatory syndrome; FCU, familial cold urticaria; FHL, familial hemophagocytic lymphohistiocytosis; FMF, familial Mediterranean fever; HED, hypohidrotic ectodermal dysplasia; LMS, limb–mammary syndrome; MALT, mucosa-associated lymphoid tissue; MWS, Muckle–Wells syndrome; NHL, non-Hodgkin's lymphoma; NOMID, neonatal-onset multisystem inflammatory disease; OMIM, On-line Mendelian Inheritance in Man™; SHFM, split hand/foot malformation; SMA, spinal muscular atrophy; TRAPS, TNFRSF1A-associated periodic syndrome; *T*, translocation, *A*, amplification, *D*, deletion, LOH, loss-of-heterozygosity; *, strong or direct link; **, causal relationship not conclusively proven. Key reviews and references include: Mullauer *et al.*, 2001; Tamm *et al.*, 2001; Zornig *et al.*, 2001; Igney and Krammer, 2002; Johnstone *et al.*, 2002; McDermott, 2002; Nathan, 2002; Inohara and Nunez, 2003; Tschopp *et al.*, 2003; Zhivotovsky and Orrenius, 2003.

these genes are simply listed as "implicated" and not as causative, as this may require further validation to demonstrate convincingly a direct or causative role. For several of the genes listed in Table 1.4, there exists a corresponding model in the mouse, either through natural mutations (e.g. *gld* and *lpr* mice) or through gene knock-out or transgenic manipulations, all of which can aid in demonstrating the role of apoptosis in disease (Gil-Gomez and Brady, 1998; Ranger *et al.*, 2001; Joza *et al.*, 2002; Wajant *et al.*, 2003; Zeiss, 2003).

1.4   Therapeutic strategies targeting apoptosis

The pivotal role apoptosis plays in a cell's demise, be it a primary cause or a secondary effect in a disease process, makes it a most attractive "drug target" for the development of novel therapeutics. Many different drug and gene therapy approaches targeting apoptosis are underway validating the "target." The "drug" approaches can be either traditional small molecules, or antisense oligonucleotides (ASOs), or biologics such as monoclonal antibodies and recombinant proteins. More experimental approaches are underway using gene therapy or oncolytic viruses. Several of the compounds that are under development that specifically target the apoptosis machinery or derive from screening approaches reliant on specific apoptosis markers or screening platforms are listed in Table 1.5. Many of these compounds may never become drugs due to the many inherent problems in drug development (unexpected toxicities, poor pharmacokinetics, poor stability, poor oral bioavailability, etc.). However, they represent important first steps in demonstrating the validity of this therapeutic approach. These compounds may serve as gold standards in screening approaches or as precursor molecules for more effective drugs, all aimed at developing marketable drugs targeting key apoptosis regulatory molecules that will impact on health care and society.

The drug and therapeutic approaches can be divided into six clearly identifiable categories: ASOs, caspase inhibitors (small molecules), other small molecules, biologics, gene therapy (including oncolytic viruses), and, finally, diagnostics. The potential application of modulating apoptosis in the production and manufacturing of biologics and other medicaments in the agricultural area or in veterinary medicine is not discussed. ASOs are discussed first because they will most likely be the first "apoptosis" drugs on the market. The bcl-2 antisense drug, Genasense™, developed by Genta is nearing completion of several pivotal phase III cancer trials, and has been partnered with Aventis in the second largest financial deal for a single drug (Dorey, 2002). Caspase inhibitors hold much promise for disorders such as sepsis, liver disease, acute neurological disorders, and myocardial infarctions, but have to overcome drug delivery (pharmacokinetic) problems. Biologics, such as monoclonal antibodies targeting TNFα and recombinant proteins such as

Table 1.5 *Therapeutics targeting apoptosis*

Class (*)	Company or institution (partner, collaborator, acquisition, or parent)	Drug(s)/therapeutic(s)	Description of agent or strategic approach	References or website address
1	Aegera (Hybridon)	AEG35156/ GEM640	XIAP antisense	www.aegera.com
1	Genta (Aventis)	Genasense™/ G3139/ augmerosen/ oblimersen sodium	Bcl-2 antisense	www.genta.com Jansen *et al.*, 2000; Pirollo *et al.*, 2003
1	Hybridon	GEM240	MDM2/HDM2 antisense (increased p53)	www.hybridon.com Chen *et al.*, 1998
1	Isis (Eli Lilly)	ISIS-23722/LY2181308	Survivin antisense	www.isispharm.com Carter *et al.*, 2003
1	Isis (Elan)	Orasense™/ISIS-104838	TNFα antisense	www.isispharm.com
1	Isis	ISIS-16009; ISIS-15999	Bcl-Xl antisense	www.isispharm.com Hayward *et al.*, 2003
1	Isis	ISIS-22023	Fas antisense	www.isispharm.com
1	Novartis	NCH-4401	HDM2 antisense	www.novartis.com Geiger *et al.*, 2000
1	Oncogenex (Isis)	OGX-011/ISIS-112989	Clusterin/TRPM2 antisense	www.oncogenex.ca
1	University of Zurich (Novartis)	4625, 3005	Bcl-2 and bcl-Xl antisense (bispecific)	Olie *et al.*, 2002
2	AstraZeneca	Anilinoquinazolines, AQZ-3, AQZ-6	Caspase inhibitors	www.astrazeneca. us.com Scott *et al.*, 2003
2	BASF	Aryloxymethyl and acylomethyl ketones	Caspase 1 inhibitor	www.basf.com Brady, 1998
2	Exegenics (Cytoclonal)	?	Caspase inhibitor	www.exegenicsinc. com
2	GSK	?	Caspase inhibitor	www.gsk.com
2	Idun	IDN-1965, IDN-6556, IDN-5370, IDN-6734, IDN-8066, IDN-7314, acyl dipeptides	Caspase inhibitors	www.idun.com Linton *et al.*, 2002; Gujral *et al.*, 2003; Natori *et al.*, 2003; Wencker *et al.*, 2003
2	Maxim (Cytovia)	MX-1013/CV-1013/Z-VD-fmk	Caspase inhibitor	www.maxim.com Yang *et al.*, 2003
2	Merck (Merck–Frosst)	L-826,791/M-791, L-826,920/M-920, MF-286, MF-867	Caspase inhibitors	www.merck.com Hotchkiss *et al.*, 2000; Han *et al.*, 2002
2	Merck (Merck–Frosst)	Nicotinyl aspartyl ketone	Caspase-3 inhibitor	www.merck.com Isabel *et al.*, 2003
2	Pfizer	MMPSI	Caspase inhibitor	www.pfizer.com Chapman *et al.*, 2002

Table 1.5 (*cont.*)

Class (*)	Company or institution (partner, collaborator, acquisition, or parent)	Drug(s)/therapeutic(s)	Description of agent or strategic approach	References or website address
2	Sunesis Pharmaceuticals (Merck)	Compound 7 and others	Caspase inhibitor	www.sunesis.com Erlanson et al., 2003
2	Texas Biotechnology	TBC-4521	Caspase inhibitor	www.tbc.com
2	Vertex (Aventis)	Pralnacasan/VX-740/ HMR-3480	Caspase-1 and -4 inhibitor	www.vpharm.com
2	Vertex	VX-799	Caspase inhibitor	www.vpharm.com
2	Yamanouchi	Compound 27c	Caspase 1 inhibitor	www.yamanouchi. com Okamoto et al., 1999
3	Abbott	PETCM	Caspase-3 activator	www.abbott.com Jiang et al., 2003
3	Abbott (Idun)	A-385358, A-317267	Bcl-Xl/Bcl-2 antagonists	www.abbott.com
3	Abbott (Idun)	?	Smac mimics	www.abbott.com
3	Aegera	?	IAP antagonists and regulators	www.aegera.com
3	AstraZeneca	Thiophenecarboxamides	IKK inhibitor	www.astrazeneca.com
3	Aventis (Millenium)	PS-1145	IKK inhibitor	www.aventis.com
3	Aventis	RU360	Mitochondrial calcium uniporter	www.aventis.com Vanderluit et al., 2003
3	Biomeasure	?	BH3 mimics	
3	BMS	BMS-345541	IKK inhibitor	McIntyre et al., 2003
3	Burnham Institute	TPI-1396-34	IAP-caspase disruptor	www.burnham-inst.org
3	Celgene-Signal (Serono)	SPC0023579, SPC839/AS602868	IKK inhibitors	www.celgene.com
3	Celgene-Signal	SP600125	JNK inhibitor	www.celgene.com Scapin et al., 2003
3	Cephalon	CEP-1347/KT7515	MLK inhibitor	Roux et al., 2002
3	Cephalon	CEP-11004	JNK inhibitor	www.cephalon.com
3	GeminX	GX-01	Bcl2-BAP31 disruptor	www.geminx.com
3	Genomics Inst. of the Novartis Research Foundation (Scripps)	TWX006, TWX024, TWX041	IAP-caspase disruptors	Wu et al., 2003
3	Georgetown Univ. (NCI)	NSC365400/ compound 6, NSC7233, NSC252041, NSC140067, NSC357777	BH3 mimic	Enyedy et al., 2001
3	Harvard	BH3I-1, BH3I-2, BH3I-1SCH3	BH3 mimics	Degterev et al., 2001

(*cont.*)

Table 1.5 (*cont.*)

Class (*)	Company or institution (partner, collaborator, acquisition, or parent)	Drug(s)/therapeutic(s)	Description of agent or strategic approach	References or website address
3	Harvard	Minocycline	Neuroprotectant, unknown mechanism	Chen *et al.*, 2000; Zhu *et al.*, 2002
3	Igen (NCI)	?	MDM2 E3 ligase inhibitors	www.igen.com
3	Thomas Jefferson University	HA14-1	Bcl2-BH3 disruptor	Wang *et al.*, 2000
3	Thomas Jefferson University	8-MOP/8-methoxypsoralen	Smac binder	
3	Karolinska	PRIMA-1	P53 stabilizer	Bykov *et al.*, 2002
3	Karolinska	RITA	P53-MDM2 disruptor	
3	Kyowa-Hakko Kogyo	Tetrocarcin-A	Bcl-2 inhibitor	www.kyowa.co.jp
3	Maxim (Shire)	MX-2105, MX-2060, MX-77356	Caspase activators	www.maxim.com
3	Maxim	Substituted N-phenyl nicotinamides	Apoptosis inducers	www.maxim.com Cai *et al.*, 2003
3	Millenium	Velcade™/PS-341/Bortezomib	Proteasome inhibitor/ NFkB "inhibitor"	
3	Morphochem	?	BclX disruptor/ BAD mimic	www.morphochem.com
3	NCI	Ellipticine	P53 activator	Peng *et al.*, 2003
3	NCI (Princeton)	NSC 321206	Smac mimic	Glover *et al.*, 2003
3	Novartis	NCPIKK004, NVPIKK005	IKK inhibitor	www.novartis.com
3	Novartis	?	P53-MDM2 disruptors	www.novartis.com
3	Novartis	NIM811, PKF220-384/SDZ220-384	PTP inhibitors (CsA derivatives)	www.novartis.com Waldmeier *et al.*, 2002
3	Pfizer	CP-31398, CP-257042	P53 stabilizer	www.pfizer.com Foster *et al.*, 1999
3	Pharmacia (Pfizer)	SC-514	IKK-2 inhibitor	Kishore *et al.*, 2003
3	Roche	Ro106-9920	NFkB inhibitor	www.roche.com Swinney *et al.*, 2002
3	Roche	Ro68-3400	VDAC inhibitor	www.roche.com Cesura *et al.*, 2003
3	Structural Bioinformatics	?	Bcl-2 inhibitors	
3	Sunesis	Compound 2	Apoptosis inducer	Nguyen and Wells, 2003
3	Yale	?	BH3 mimic	
3	University of Illinois at Chicago	Pifithrin/ PFT	P53 inhibitor	Komarov *et al.*, 1999
3	?	Z-1-117 (pifithrin analog)	P53 inhibitor	Duan *et al.*, 2002
3	Wyeth	Amifostine (WR1065)	Cytoprotective agent	
3	University of Florida	UCF-101	Omi inhibitor	Cilenti *et al.*, 2003

Table 1.5 (*cont.*)

Class (*)	Company or institution (partner, collaborator, acquisition, or parent)	Drug(s)/therapeutic(s)	Description of agent or strategic approach	References or website address
3	University of Washington	Antimycin A$_3$	Bcl-2 disruptor	Tzung *et al.*, 2001
3	?	Apoptolidin, isoapoptolidin	Fo-F1 ATPase inhibitor	Salomon *et al.*, 2000
3	Various	Thalidomide and derivatives	NFkB inhibitors	Orlowski and Baldwin, 2002
3	Various	Natural products and derivatives, e.g. DHMEQ, jesterone dimer	NFkB inhibitors	Haefner, 2002; Liang *et al.*, 2003
3	Various	Ligands of peripheral benzodiazepine receptor, and other mitochondrially targeted drugs (e.g. CsA, bongrekic acid, atractyloside)	Mitochondrially targeted drugs	Debatin *et al.*, 2002; Watts and Kline, 2003
3	Various	Various small molecules	P53-MDM2 disruptors	Chene, 2003
3	Institute of Molecular and Cell Biology (Singapore)	A5, A11 (tetrabromo-fluorescein)		Tan *et al.*, 2003
4	Apoxis	MegaFasL	Trimerized FasL	www.apoxis.com
4	Amgen	Kineret^{TM}/Anakinra	IL-1R antagonist MoAb	www.amgen.com
4	Amgen	rhOPG	osteoprotegerin	Capparelli *et al.*, 2003
4	Biogen	Avonex^{TM}	IFNβ	www.biogen.com
4	Berlex (Schering)	Betaseron^{TM}	IFNβ	www.berlex.com
4	Centocor	Remicade^{TM}/Infliximab	TNFα MoAb	www.centocor.com
4	Centocor (Santen)	?	Fas/APO-1 Ab	www.centocor.com Coney *et al.*, 1994
4	Eli Lilly	Xigris^{TM}/Drotrecogin alpha/rh APC	Recombinant human activated protein C	www.lilly.com Joyce *et al.*, 2001; Cheng *et al*, 2003
4	EMEA	?	TNFα	www.emea.eu.int
4	Genentech (Immunex/Amgen)	TRAIL/Apo2L		www.gene.com
4	HGS (Cambridge Antibody Technology Group)	TI-1, TI-2	DR4/ TRAILR1-MoAb/ scFv	
4	Hoffmann–La Roche (Aventis)	Pegasys	Pegylated interferon α2a	
4	Immunex (Amgen)	Enbrel^{TM}/Etanercept	soluble TNFRII–IgG1 fusion	
4	Immunex (Amgen)	?	Caspase-3 linked to Herceptin	www.immunex.com
4	Protherics	CytoTAb^{TM}	TNFα antibody	www.protherics.com

(*cont.*)

Table 1.5 (*cont.*)

Class (*)	Company or institution (partner, collaborator, acquisition, or parent)	Drug(s)/therapeutic(s)	Description of agent or strategic approach	References or website address
4	Regeneron (Novartis)	IL-1 TRAP	IL-1 antagonist	Gabay, 2003; Dinarello, 2003
4	Sankyo	TRA-8	DR5 MoAb	www.sankypharma.com
4	Schering-Plough	PEG-Intron	Pegylated interferon $\alpha2b$	www.sch-plough.com
4	Xencor	DN-TNF	Dominant–negative TNF variants	www.xencor.com Steed *et al.*, 2003
5	Aegera	?	XIAP gene therapy	www.aegera.com
5	Ariad	Argent/AP-1903	Gene therapy and modulator	www.ariad.com
5	Calydon	CV706	PSA selective oncolytic adenovirus	DeWeese *et al.*, 2001
5	Entremed	ICP10 PK (HSV2)	Herpes simplex virus-2	Perkins, 2002
5	GeminX	E4orf4	Adenoviral death protein gene therapy	www.geminx.com Robert *et al.*, 2002
5	GenVec	TNFerade	TNFα producing adenovirus	Rasmussen *et al.*, 2002
5	GTI (Novartis)	OAV001, OAS403	Oncolytic adenovirus	Jakubczak *et al.*, 2003
5	Introgen (Canji)	RPR/INGN201	Adenovirus p53 gene therapy	www.introgen.com
5	MDA (University of South Carolina)	?	Bax adenovirus	Gu *et al.*, 2000
5	Oncolytics Biotech	Reolysin™	Oncolytic reovirus	www.oncolyticsbiotech.com
5	Onyx	ONYX-015/ dl1520/ CI-1042/CG7870	Oncolytic adenovirus for p53 mutant or null cells	www.onyxpharma-ceuticals.com Makower *et al.*, 2003
5	Onyx	Onyx-411	Oncolytic adenovirus targeting Rb pathway	www.onyxpharma-ceuticals.com Johnson *et al.*, 2002
5	Wellstat Biologics (Pro-virus)	PV701	Oncolytic NDV	www.wellstat.com Pecora *et al.*, 2002
5	Wellstat Biologics (University of Ottawa)	PV327	Oncolytic VSV	www.wellstat.com Stojdl *et al.*, 2000
5	Introgen	ING241	Adeno-Mda7/ IL-24	www.introgen.com Fisher *et al.*, 2003
5	Regulon (Intelligence)	LipoHER	Caspase-9 gene therapy for breast cancer	www.regulon.org
5	Regulon (Intelligence)	LipoPSA	Caspase-9 gene therapy for prostate cancer	www.regulon.org
5	Selective Genetics	?	Bcl-2 gene therapy	www.selectivegenetics.com

Table 1.5 (*cont.*)

Class (*)	Company or institution (partner, collaborator, acquisition, or parent)	Drug(s)/therapeutic(s)	Description of agent or strategic approach	References or website address
5	Leiden University	Apoptin™	Adenovirus, or parvovirus, apoptin gene therapy	Olijslagers *et al.*, 2001; van der Eb *et al.*, 2002
5	Schering-Plough	SCH58500	Adenovirus p53 gene therapy	www.sch-plough.com
5	SUNY	PV1 (RIPO)	Intergeneric poliovirus	Gromeier *et al.*, 2000
5	Targeted Genetics	?	Adenovirus E1A oncolytic gene therapy	www.targen.com
6	Exact Sciences	?	Diagnostics	www.exactsciences.com
6	Fujirebio Diagnostics (Abbott)	?	Survivin diagnostics for cancer	www.fdi.com
6	Theseus Imaging Corp.	Apomate™	Tc-99m rh-Annexin V imaging	www.theseusimaging.com Belhocine *et al.*, 2002

*****Class designations:** 1, antisense oligonucleotides; 2, caspase inhibitors-small molecule; 3, other small molecules; 4, biologics; 5, gene therapy and oncolytic viruses; 6, diagnostics. ? = not named/unknown. **Abbreviations:** BMS, Bristol-Myers-Squibb; DR4, death receptor 4; DR5, death receptor 5; EMEA, European Agency for the Evaluation of Medicinal Products; GSK, GlaxoSmithKline; GTI, Gene Therapy Inc.; HDM2, human MDM2 ortholog; HGS, Human Genome Science; IFN, interferon; IKK, IkB kinase; IgG, immunoglobulin; JNK, c-jun N-terminal kinase; MDA, M. D. Anderson; MDM2, murine double minute 2; MLK, mixed-lineage kinase; MoAb, monoclonal antibody; NCI, National Cancer Institute; NDV, Newcastle disease virus; NFkB, Nuclear factor kappa B; scFv, single chain immunoglobulin variable fragment; SUNY, State University of New York; TRAIL, TNFα-related apoptosis-inducing ligand; TRPM-2, testosterone-repressed prostatic message of 2 kilobases; VSV, vesicular stomatitis virus; XIAP, X-linked IAP. Data are based on a combination of published reports (reviews or specific examples are cited in table or below), published patent applications, meeting reports, press releases, and other internet-derived information sources (website addresses provided for corporate sites). Some key reviews and references: Bullock and Fersht, 2001; Huang and Oliff, 2001; Baell and Huang, 2002; Debatin *et al.*, 2002; Ferreira *et al.*, 2002; Haefner, 2002; Hawkins *et al.*, 2002; Huang, 2002; Jansen and Zangemeister-Wittke, 2002; Johnstone *et al.*, 2002; Kiechle and Zhang, 2002; Reed, 2002; Roshak *et al.*, 2002; Zhang, 2002; Aggarwal, 2003; Andreakos, 2003; Beauparlant and Shore, 2003; Brekke and Sandlie, 2003; Bykov *et al.*, 2003; French and Tschopp, 2003; Los *et al.*, 2003; Makin and Dive, 2003; Murphy *et al.*, 2003; Nygren and Larsson, 2003; Reed, 2003; Shabbits *et al.*, 2003; Shangary and Johnson, 2003; Vila and Przedborski, 2003; Waldmeier, 2003.

Note the inclusion or not of a "product" or Company in this table does not constitute an endorsement or rejection. Furthermore, the authors make no guarantees as to the validity or accuracy of the statements or data included herein. This table is simply meant to illustrate the tremendous amount of pre-clinical and clinical effort devoted to therapeutics targeting apoptosis control that may be, one day, of use to clinicians and other health-care providers.

interferon β, are already on the market but are mainly used as immunomodulatory drugs, and are included here for the sake of completeness. More direct applications targeting apoptosis directly, such as through the use of TRAIL as a cancer therapy, await further development and the clearing of toxicity hurdles. Gene therapy and oncolytic viruses have to overcome even greater developmental hurdles, inherent to the manufacturing of live viruses for therapeutics, as well as overcoming some recent setbacks in gene therapy clinical trials and other limitations (Bell *et al.*, 2003). Oncolytic viruses were included as "apoptosis modulatory drugs" because they make use of apoptotic pathways and responses inherent to viral infection to kill cancer cells. Diagnostics are barely mentioned here, but they too will impact on medicine in the future as we gain further understanding as to how the apoptosis genes are involved in disease initiation or progression. The compounds in Table 1.5 are being developed for many purposes, i.e. anti-neoplastics, neuroprotectants, and immune modulators, and appear theoretically to have unlimited applications as PCD plays a role in almost every pathology. These compounds hold much promise to affecting outcomes for a variety of acute and chronic disorders, and we expect to see the results of such endeavors in the upcoming years. Very few of these compounds are in advanced stages of clinical development. With the first successes, we hope to see many more compounds developed and applied in clinical practice in the not too distant future.

1.5 Methods in apoptosis research

1.5.1 An overview

Apoptosis methods have evolved and changed rapidly over the last two decades owing to the increasing practical knowledge base and greater understanding of the genes and mechanisms involved. Listed in Table 1.6 are several general methodologies used in the detection of PCD. Included in the list are some outdated methods, listed for historical reasons, as they define or measure certain of the characteristics of PCD defined in Table 1.1. Many more specific, and potentially useful, molecular markers and techniques now exist but only a few examples are listed under "other markers and techniques" at the end of the table. These "other" assays are either of a highly specific nature or of a more exploratory nature and may not be fully validated.

The analysis or study of specific molecules involved in PCD and how they affect PCD are not listed in Table 1.6. Traditional methods for gene function analysis include over-expression studies, the use of dominant–negative constructs, antisense or RNAi approaches (reviewed in LaCasse, in press), generation of knock-out or transgenic mice, as well as the measurement of endogenous messenger RNA or protein levels. In addition, the identification of apoptotic cells expressing a DNA

Table 1.6 *General methods for apoptosis detection*

Characteristic measured or observed	Method	Comment or reference
Gross cellular morphology	Histochemical stains and light microscopy (LM)	
	Phase contrast, or Nomarski, LM	Loss of adherence to substrate and neighboring cells, and rounding of cells with increased light birefringence is used for PCD detection of cells in culture
	EM	Tedious but unequivocal
Cell volume decrease	Forward light scatter (flow cytometry) or LSC	Quantitative
Chromatin condensation	DNA-binding dyes (EF, confocal microscopy)	e.g. DAPI, Hoescht dyes
	Single-stranded DNA detection (IHC, ELISA)	Changes in chromatin structure measured after formamide-induced denaturation; Frankfurt and Krishan, 2001a,b
	EM	e.g. Sakahira *et al.*, 1999
Lysosome integrity or lysosomal activity	Enzyme histochemistry	Helpful in discriminating autophagy from apoptosis
DNA fragmentation and content	Laddering of low molecular weight DNA (agarose GE)	Defining hallmark, not often used as routine test anymore due to poor sensitivity and introduction of more sensitive molecular markers
	High molecular weight DNA fragments (pulsed-field GE)	Oberhammer *et al.*, 1993
	Comet assay (SCGE)	Does not discriminate PCD from genotoxic DNA damage
	TUNEL or ISEL staining (IHC)	Detection of DNA strand breaks by end labeling. Does not distinguish from breaks induced during necrosis. Hornsby and Didenko, 2002; Loo, 2002; Walker *et al.*, 2002
	Hypodiploid DNA content (flow cytometry, LSC)	Quantitative detection of DNA loss
	Free histone/nucleosome detection (ELISA)	Quantitative detection of DNA fragmentation
	Cell-free systems with isolated nuclei	Clarke, 2002
	JAM assay	Quantitative analysis of fragmented ^{3}H-DNA. Prone to false positives. Bohm *et al.*, 1998; Hoves *et al.*, 2003
Plasma membrane integrity or blebbing	Dye exclusion tests (LM, EF, flow cytometry, LSC)	e.g. trypan blue, eosin yellow, propidium iodide, 7-AAD. e.g. Lecoeur *et al.*, 2002

(*cont.*)

Table 1.6 (*cont.*)

Characteristic measured or observed	Method	Comment or reference
	Light side scatter (flow cytometry)	Diffraction is a measure of cell granularity/membrane blebbing
Phosphatidylserine (PS exposure)	PS detection by Annexin V labeling (IHC, flow cytometry, LSC, gamma camera, CT, PET, MRI)	Often combined with dye exclusion test to differentiate PCD from necrosis. Note that in tissue culture experiments, PCD may proceed to secondary necrosis due to lack of phagocytic cells. In vivo detection being validated with 99Tc-labeled Annexin V (Belhocine *et al.*, 2002; van de Wiele, 2003), [124]I-labelled annexin V (Glaser *et al.*, 2003a, b), or SPIO-labelled annexin V (Zhao *et al.*, 2001; Brauer, 2003; Hakumaki and Brindle, 2003)
	PS detection by PSS-380 fluorescence	Koulov *et al.*, 2003
Mitochondrial activity	ADP/ATP ratio (luciferin assay)	Used to discriminate PCD from necrosis
	Mitochondrial dehydrogenase activity (reduction of TTZ salts)	Viability/metabolic assay, not necessarily reflective of PCD
Mitochondrial permeability (MMP)	Release of cytochrome c, SMAC, into cytosol (western, ELISA)	e.g. Ott *et al.*, 2002; Arnoult *et al.*, 2003; Waterhouse and Trapani, 2003
Mitochondrial transmembrane potential	Assessed with voltage-sensitive dyes (EF, confocal microscopy, flow cytometry, LSC)	
Caspase activity	Enzyme assays, may involve IP or IC ELISA, as well as flow cytometry or LSC (e.g. FLICA™). May also employ FRET assays with caspase cleavable reporters (e.g. Tawa *et al.*, 2001). Optical imaging approaches being developed for in vivo detection.	Use of specific synthetic colorimetric or fluorimetric substrates that may be membrane permeable or not. Assay may use recombinant proteins, crude cell lysates, or immuno-affinity purified complexes or components, e.g. the apoptosome
	Detection of cleaved polypeptides (western)	e.g. PARP, ICAD, BID
	Detection of neo-epitopes exposed by cleavage of caspases themselves or their substrates (western, IHC, ELISA, flow cytometry, LSC)	e.g. caspase-3/CM1 antibody, cytokeratin 18/M30 antibody
	^{35}S-metabolic labeling of caspases (IP and PAGE/autoradiography)	

Table 1.6 (*cont.*)

Characteristic measured or observed	Method	Comment or reference
Other markers and techniques	Loss of Y1 RNA (RT-PCR)	Relatively new test; Asselbergs and Widmer, 2003
	Change in Bax/bcl-2 ratio (western)	
	Ubiquination, cleavage, or loss of XIAP protein (western)	
	Induction of clusterin (RPA or microarray)	
	Stabilization of p53 protein or transcriptional induction of p53 targets	
	Increased ubiquination of proteins (western, IHC)	Seldom used, may make a comeback
	Protein synthesis shut down	Tedious; a measure of ATP depletion or disassembly of the translation machinery by caspases
	Transglutaminase activation	Underestimation of apoptosis; Grabarek *et al.*, 2002
	Alterations in nuclear structure (high frequency ultrasound)	In vivo imaging due to increased scatter; Czarnota *et al.*, 1999
	Alterations in cellular free water (MRI)	In vivo imaging; Brauer, 2003
	Altered sub-cellular localization of proteins or formation of higher molecular complexes, e.g. Bax translocation to mitochondria (western, confocal microscopy), or shift in sedimentation coefficients of specific proteins (centrifugation)	Tedious

7-AAD, 7-aminoactinomycin D; ADP, adenosine diphosphate; ATP, adenosine triphosphate; CT, computed tomography; DAPI, 4′,6-diamidino-2-phenylindole; ELISA, enzyme-linked immuno-sorbent assay; EM, electron microscopy; EF, epifluorescence microscopy; FLICA™, fluorochrome-labeled inhibitor of caspase; FRET, fluorescence resonance energy transfer; GE, gel electrophoresis; HC, histochemistry; IHC, immunohistochemistry; IP, immunoprecipitation; IC, immunocapture; ISEL, in situ end labeling; LM, light microscopy; LSC, laser scanning cytometry; MMP, mitochondrial membrane permeability; MRI, magnetic resonance imaging; PAGE, polyacrylamide gel electrophoresis; PET, positron emission tomography; PS, phosphatidyl serine; PTP, permeability transition pore; RPA, Rnase protection assay; RT-PCR, reverse transcriptase polymerase chain reaction; SCGE, single cell gel electrophoresis; TTZ, tetrazolium; TUNEL, TdT-mediated dUTP nick end-labeling. Key reviews and references include: Liming *et al.*, 1999; Saraste, 1999; Castedo *et al.*, 2002; Kiechle and Zhang, 2002; Kohler *et al.*, 2002; Zakeri and Lockshin, 2002; Bai *et al.*, 2003; Otsuki and Shibata, 2003; Schultz and Harrington, 2003. Additional technical reports and methods comparisons include: Goldstein *et al.*, 2000a, b; Stadelmann and Lassmann, 2000; Micoud *et al.*, 2001; Drag-Zalesinska *et al.*, 2002; Kylarova *et al.*, 2002; Duan *et al.*, 2003; Garrity *et al.*, 2003; Pozarowski *et al.*, 2003; Yasuhara *et al.*, 2003.

vector of interest can be done by co-expressing a marker such as green fluorescent protein (GFP) (Vezina *et al.*, 2001). An example of the discovery of the cellular IAPs through medical genetic and genomic approaches is listed below as a "case study" in methods applied to PCD research.

1.5.2   Medical genetics, genomics, and molecular biology – a case study in the discovery of the caspase inhibitors, the inhibitors-of-apoptosis (IAPs)

A positional cloning effort was undertaken to identify the causative gene of type 1 spinal muscular atrophy (SMA, or acute Werdnig–Hoffman disease) through traditional linkage analysis of affected individuals. Analysis of candidate genes at the identified locus revealed a gene that was deleted in approximately 67% of affected individuals but not in unaffected individuals. Homology searches revealed that this novel human gene contains sequence motifs called baculovirus IAP repeats (BIRs) previously only reported in baculovirus genes called IAPs (inhibitors-of-apoptosis) as they were shown to suppress host insect cell apoptosis (Miller, 1999). Thus, the first cellular IAP was cloned, and called NAIP for <u>n</u>euronal <u>a</u>poptosis <u>i</u>nhibitory <u>p</u>rotein (now identified by HUGO gene symbol, *birc1*; Roy *et al.*, 1995). Postulating that more mammalian IAPs exist in the human genome, as well as that of mouse, searches of expressed sequence tag (EST) and genomic sequence databases were undertaken to identify other BIR-containing genes. Primers were designed on putative hits, and used to amplify and screen complementary DNA libraries to clone other human or mouse IAPs. With this approach, several other IAPs were discovered, and shown to suppress PCD in tissue culture over-expression studies (Liston *et al.*, 1996), and extended to complex in vivo animal models of disease using viral gene delivery (Xu *et al.*, 1997; Eberhardt *et al.*, 2000; Kugler *et al.*, 2000; Perrelet *et al.*, 2000; Crocker *et al.*, 2001; McKinnon *et al.*, 2002; Perrelet *et al.*, 2002; Straten *et al.*, 2002; Petrin *et al.*, 2003). IAP knock-out and IAP transgenic mice demonstrated further apoptosis modulatory roles for these genes (Holcik *et al.*, 2000; Conte *et al.*, 2001; Crocker *et al.*, 2003; Trapp *et al.*, 2003). Antisense approaches were also undertaken to validate these genes (Holcik *et al.*, 2000; Sasaki *et al.*, 2000; Hu *et al.*, 2003). As one can clearly see, multiple methodologies (more than can be listed here) are used in the discovery, study, and validation of apoptosis-related molecules. No-one can predict how, when, and where the next interesting discovery of an apoptosis-related molecule will be made. Many surprises still await in this ever increasing field of endeavor which will impact on the practice of medicine.

1.6   Conclusion

The fact that cell death is a highly controlled and genetically programmed process, as important as mitosis is to cell birth, should not be unexpected. However, this realization took a long time to occur (Melino, 2002; Vaux, 2002). Cells simply do not decay as one might assume, and their orderly removal, dismantling, and recycling by apoptosis is needed to maintain proper homeostasis in complex organisms, to fight infections, to shape the immune response, and to sculpt embryonic structures, to name but a few examples (see Chapters 2, 6, and 7). Indeed, life in complex organisms would not exist without cell death to remove unwanted cells that have completed their support function or which have become DNA damaged or infected.

The advent of therapeutics aimed at the apoptotic process will likely herald in a new era of medicine based on rational molecular approaches. These future promising therapies have been supported in part by the genomics revolution and the great strides in apoptosis understanding garnered over the last two decades.

ACKNOWLEDGEMENTS

Thanks to Stephen Baird for bioinformatics assistance, Sandy Boehmer for help with manuscript preparation, Dr. Wai Gin Fong for preparation of Figure 1.1, and Dr. Peter Liston for comments. Cell death work in the authors' laboratories is supported by grants from the Canadian Institutes of Health Research (MH, RGK, and AEM), Howard Hughes Medical Institute (RGK), National Cancer Institute of Canada (RGK), Canadian Genetic Disease Network (RGK and AEM), Canada Foundation for Innovation, Ontario Innovation Trust, and the Ontario Research and Development Challenge Fund (MH, RGK, and AEM).

REFERENCES

Acehan, D., Jiang, X., Morgan, D. G., Heuser, J. E., Wang, X., and Akey, C. W. (2002). Three-dimensional structure of the apoptosome: implications for assembly, procaspase-9 binding, and activation. *Mol. Cell*, **9**, 423–32.

Aggarwal, B. B. (2003). Signalling pathways of the TNF superfamily: a double-edged sword. *Nat. Rev. Immunol.*, **3**, 745–56.

Alam, A., Cohen, L. Y., Aouad, S., and Sekaly, R. P. (1999). Early activation of caspases during T lymphocyte stimulation results in selective substrate cleavage in nonapoptotic cells. *J. Exp. Med.*, **190**, 1879–90.

Alizadeh, A. A., Eisen, M. B., Davis, R. E. *et al.* (2000). Distinct types of diffuse large B-cell lymphoma identified by gene expression profiling. *Nature*, **403**, 503–11.

Altieri, D. C. (2003). Validating survivin as a cancer therapeutic target. *Nat. Rev. Cancer*, **3**, 46–54.

Ameisen, J. C. (2002). On the origin, evolution, and nature of programmed cell death: a timeline of four billion years. *Cell Death Differ.*, **9**, 367–93.

Andreakos, E. (2003). Targeting cytokines in autoimmunity: new approaches, new promise. *Expert. Opin. Biol. Ther.*, **3**, 435–47.

Aravind, L., Dixit, V. M., and Koonin, E. V. (2001). Apoptotic molecular machinery: vastly increased complexity in vertebrates revealed by genome comparisons. *Science*, **291**, 1279–84.

Arnoult, D., Gaume, B., Karbowski, M., Sharpe, J. C., Cecconi, F., and Youle, R. J. (2003). Mitochondrial release of AIF and EndoG requires caspase activation downstream of Bax/Bak-mediated permeabilization. *Embo. J.*, **22**, 4385–99.

Asselbergs, F. A. and Widmer, R. (2003). Rapid detection of apoptosis through real-time reverse transcriptase polymerase chain reaction measurement of the small cytoplasmic RNA Y1. *Anal. Biochem.*, **318**, 221–9.

Baehrecke, E. H. (2002). How death shapes life during development. *Nat. Rev. Mol. Cell Biol.*, **3**, 779–87.

Baell, J. B. and Huang, D. C. (2002). Prospects for targeting the Bcl-2 family of proteins to develop novel cytotoxic drugs. *Biochem. Pharmacol.*, **64**, 851–63.

Bai, L., Wang, J., Yin, X.-M., and Dong, Z. (2003). Analysis of apoptosis-basic principles and procedures. In *Essentials of Apoptosis: A Guide for Basic and Clinical Research*, ed. X.-M. Yin and Z. Dong. Totowa, NJ: Humana Press Inc., pp. 239–51.

Barry, M. and Bleackley, R. C. (2002). Cytotoxic T lymphocytes: all roads lead to death. *Nat. Rev. Immunol.*, **2**, 401–9.

Beauparlant, P. and Shore, G. C. (2003). Therapeutic activation of caspases in cancer: a question of selectivity. *Curr. Opin. Drug Discov. Devel.*, **6**, 179–87.

Belhocine, T., Steinmetz, N., Hustinx, R. *et al.* (2002). Increased uptake of the apoptosis-imaging agent (99m)Tc recombinant human Annexin V in human tumors after one course of chemotherapy as a predictor of tumor response and patient prognosis. *Clin. Cancer Res.*, **8**, 2766–74.

Bell, J. C., Lichty, B., and Stojdl, D. (2003). Getting oncolytic virus therapies off the ground. *Cancer Cell*, **4**, 7–11.

Benitez-Bribiesca, L. (2003). The two pathways of apoptosis: one led to Stockholm, the other led home. *Arch. Med. Res.*, **34**, 1–2.

Bertin, J. and DiStefano, P. S. (2000). The PYRIN domain: a novel motif found in apoptosis and inflammation proteins. *Cell Death Differ.*, **7**, 1273–4.

Bignell, G. R., Warren, W., Seal, S. *et al.* (2000). Identification of the familial cylindromatosis tumour-suppressor gene. *Nat. Genet.*, **25**, 160–5.

Bilsland, J. and Harper, S. (2002). Caspases and neuroprotection. *Curr. Opin. Investig. Drugs*, **3**, 1745–52.

Boatright, K. M., Renatus, M., Scott, F. L. *et al.* (2003). A unified model for apical caspase activation. *Mol. Cell*, **11**, 529–41.

Bohm, C., Hanski, M. L., Gratchev, A. *et al.* (1998). A modification of the JAM test is necessary for a correct determination of apoptosis induced by FasL+ adherent tumor cells. *J. Immunol. Methods*, **217**, 71–8.

Bortner, C. D. and Cidlowski, J. A. (2002). Cellular mechanisms for the repression of apoptosis. *Annu. Rev. Pharmacol. Toxicol.*, **42**, 259–81.

Bouchier-Hayes, L. and Martin, S. J. (2002). CARD games in apoptosis and immunity. *EMBO. Rep.*, **3**, 616–21.

Brady, K. D. (1998). Bimodal inhibition of caspase-1 by aryloxymethyl and acyloxymethyl ketones. *Biochemistry*, **37**, 8508–15.

Brauer, M. (2003). In vivo monitoring of apoptosis. *Prog. Neuropsychopharmacol. Biol. Psychiatry*, **27**, 323–31.

Brekke, O. H. and Sandlie, I. (2003). Therapeutic antibodies for human diseases at the dawn of the twenty-first century. *Nat. Rev. Drug Discov.*, **2**, 52–62.

Brummelkamp, T. R., Nijman, S. M., Dirac, A. M., and Bernards, R. (2003). Loss of the cylindromatosis tumour suppressor inhibits apoptosis by activating NF-kappaB. *Nature*, **424**, 797–801.

Bullock, A. N. and Fersht, A. R. (2001). Rescuing the function of mutant p53. *Nat. Rev. Cancer*, **1**, 68–76.

Bykov, V. J., Issaeva, N., Shilov, A. *et al.* (2002). Restoration of the tumor suppressor function to mutant p53 by a low-molecular-weight compound. *Nat. Med.*, **8**, 282–8.

Bykov, V. J., Selivanova, G., and Wiman, K. G. (2003). Small molecules that reactivate mutant p53. *Eur. J. Cancer*, **39**, 1828–34.

Cai, S. X., Nguyen, B., Jia, S. *et al.* (2003). Discovery of substituted N-phenyl nicotinamides as potent inducers of apoptosis using a cell- and caspase-based high throughput screening assay. *J. Med. Chem.*, **46**, 2474–81.

Capparelli, C., Morony, S., Warmington, K. *et al.* (2003). Sustained antiresorptive effects after a single treatment with human recombinant osteoprotegerin (OPG): a pharmacodynamic and pharmacokinetic analysis in rats. *J. Bone Miner. Res.*, **18**, 852–8.

Carter, B. Z., Wang, R. Y., Schober, W. D., Milella, M., Chism, D., and Andreeff, M. (2003). Targeting survivin expression induces cell proliferation defect and subsequent cell death involving mitochondrial pathway in myeloid leukemic cells. *Cell Cycle*, **2**, 488–93.

Castedo, M., Ferri, K., Roumier, T., Metivier, D., Zamzami, N., and Kroemer, G. (2002). Quantitation of mitochondrial alterations associated with apoptosis. *J. Immunol. Methods*, **265**, 39–47.

Cesura, A. M., Pinard, E., Schubenel, R. *et al.* (2003). The voltage-dependent anion channel is the target for a new class of inhibitors of the mitochondrial permeability transition pore. *J. Biol. Chem.*, **278**, 49812–18.

Chamaillard, M., Girardin, S. E., Viala, J., and Philpott, D. J. (2003). Nods, nalps and Naip: intracellular regulators of bacterial-induced inflammation. *Cell Microbiol.*, **5**, 581–92.

Chandra, D. and Tang, D. G. (2003). Mitochondrially localized active caspase-9 and caspase-3 result mostly from translocation from the cytosol and partly from caspase-mediated activation in the organelle. Lack of evidence for Apaf-1-mediated procaspase-9 activation in the mitochondria. *J. Biol. Chem.*, **278**, 17408–20.

Chang, D. W., Ditsworth, D., Liu, H., Srinivasula, S. M., Alnemri, E. S., and Yang, X. (2003). Oligomerization is a general mechanism for the activation of apoptosis initiator and inflammatory procaspases. *J. Biol. Chem.*, **278**, 16466–9.

Chapman, J. G., Magee, W. P., Stukenbrok, H. A., Beckius, G. E., Milici, A. J., and Tracey, W. R. (2002). A novel nonpeptidic caspase-3/7 inhibitor, (S)-(+)-5-[1-(2-methoxymethylpyrrolidinyl)sulfonyl]isatin, reduces myocardial ischemic injury. *Eur. J. Pharmacol.*, **456**, 59–68.

Chen, L., Agrawal, S., Zhou, W., Zhang, R., and Chen, J. (1998). Synergistic activation of p53 by inhibition of MDM2 expression and DNA damage. *Proc. Natl. Acad. Sci. USA*, **95**, 195–200.

Chen, M., Ona, V. O., Li, M. *et al.* (2000). Minocycline inhibits caspase-1 and caspase-3 expression and delays mortality in a transgenic mouse model of Huntington disease. *Nat. Med.*, **6**, 797–801.

Chen, M., Orozco, A., Spencer, D. M., and Wang, J. (2002). Activation of initiator caspases through a stable dimeric intermediate. *J. Biol. Chem.*, **277**, 50761–7.

Chene, P. (2003). Inhibiting the p53-MDM2 interaction: an important target for cancer therapy. *Nat. Rev. Cancer*, **3**, 102–9.

Cheng, T., Liu, D., Griffin, J. H. *et al.* (2003). Activated protein C blocks p53-mediated apoptosis in ischemic human brain endothelium and is neuroprotective. *Nat. Med.*, **9**, 338–42.

Cilenti, L., Lee, Y., Hess, S. *et al.* (2003). Characterization of a novel and specific inhibitor for the pro-apoptotic protease Omi/HtrA2. *J. Biol. Chem.*, **278**, 11489–94.

Clamp, M., Andrews, D., Barker, D. *et al.* (2003). Ensembl 2002: accommodating comparative genomics. *Nucleic Acids Res.*, **31**, 38–42.

Clarke, P. G. and Clarke, S. (1996). Nineteenth century research on naturally occurring cell death and related phenomena. *Anat. Embryol. (Berl.)*, **193**, 81–99.

Clarke, P. R. (2002). Apoptosis: lessons from cell-free systems. In *Apoptosis: The Molecular Biology of Programmed Cell Death*, ed. M. D. Jacobson and N. McCarthy. Oxford: Oxford University Press, pp. 176–99.

Cleary, M. L., Smith, S. D., and Sklar, J. (1986). Cloning and structural analysis of cDNAs for bcl-2 and a hybrid bcl-2/immunoglobulin transcript resulting from the t(14;18) translocation. *Cell*, **47**, 19–28.

Coney, L. R., Daniel, P. T., Sanborn, D. *et al.* (1994). Apoptotic cell death induced by a mouse-human anti-APO-1 chimeric antibody leads to tumor regression. *Int. J. Cancer*, **58**, 562–7.

Conte, D., Liston, P., Wong, J. W., Wright, K. E., and Korneluk, R. G. (2001). Thymocyte-targeted overexpression of xiap transgene disrupts T lymphoid apoptosis and maturation. *Proc. Natl. Acad. Sci. USA*, **98**, 5049–54.

Creagh, E. M., Conroy, H., and Martin, S. J. (2003). Caspase-activation pathways in apoptosis and immunity. *Immunol. Rev.*, **193**, 10–21.

Crocker, S. J., Liston, P., Anisman, H. *et al.* (2003). Attenuation of MPTP-induced neurotoxicity and behavioural impairment in NSE-XIAP transgenic mice. *Neurobiol. Dis.*, **12**, 150–61.

Crocker, S. J., Wigle, N., Liston, P. *et al.* (2001). NAIP protects the nigrostriatal dopamine pathway in an intrastriatal 6-OHDA rat model of Parkinson's disease. *Eur. J. Neurosci.*, **14**, 391–400.

Czarnota, G. J., Kolios, M. C., Abraham, J. *et al.* (1999). Ultrasound imaging of apoptosis: high-resolution non-invasive monitoring of programmed cell death in vitro, in situ and in vivo. *Br. J. Cancer*, **81**, 520–7.

Dai, Z., Zhu, W. G., Morrison, C. D. *et al.* (2003). A comprehensive search for DNA amplification in lung cancer identifies inhibitors of apoptosis cIAP1 and cIAP2 as candidate oncogenes. *Hum. Mol. Genet.*, **12**, 791–801.

Davis, R. E., Brown, K. D., Siebenlist, U., and Staudt, L. M. (2001). Constitutive nuclear factor kappaB activity is required for survival of activated B cell-like diffuse large B cell lymphoma cells. *J. Exp. Med.*, **194**, 1861–74.

Debatin, K. M., Poncet, D., and Kroemer, G. (2002). Chemotherapy: targeting the mitochondrial cell death pathway. *Oncogene*, **21**, 8786–803.

De Botton, S., Sabri, S., Daugas, E. *et al.* (2002). Platelet formation is the consequence of caspase activation within megakaryocytes. *Blood*, **100**, 1310–17.

Degterev, A., Lugovskoy, A., Cardone, M. *et al.* (2001). Identification of small-molecule inhibitors of interaction between the BH3 domain and Bcl-xL. *Nat. Cell Biol.*, **3**, 173–82.

DeWeese, T. L., van der Poel, H., Li, S. *et al.* (2001). A phase I trial of CV706, a replication-competent, PSA selective oncolytic adenovirus, for the treatment of locally recurrent prostate cancer following radiation therapy. *Cancer Res.*, **61**, 7464–72.

Dickman, M. B., Park, Y. K., Oltersdorf, T., Li, W., Clemente, T., and French, R. (2001). Abrogation of disease development in plants expressing animal antiapoptotic genes. *Proc. Natl. Acad. Sci. USA*, **98**, 6957–62.

Dierlamm, J., Baens, M., Wlodarska, I. *et al.* (1999). The apoptosis inhibitor gene API2 and a novel 18q gene, MLT, are recurrently rearranged in the t(11;18)(q21;q21) associated with mucosa-associated lymphoid tissue lymphomas. *Blood*, **93**, 3601–9.

Dierlamm, J., Murga Penas, E. M., Daibata, M. *et al.* (2002). The novel t(11;12;18)(q21;q13;q21) represents a variant translocation of the t(11;18)(q21;q21) associated with MALT-type lymphoma. *Leukemia*, **16**, 1863–4.

Diez, E., Lee, S. H., Gauthier, S. *et al.* (2003). Birc1e is the gene within the Lgn1 locus associated with resistance to *Legionella pneumophila*. *Nat. Genet.*, **33**, 55–60.

Dinarello, C. A. (2003). Setting the cytokine trap for autoimmunity. *Nat. Med.*, **9**, 20–2.

Doctor, K. S., Reed, J. C., Godzik, A., and Bourne, P. E. (2003). The apoptosis database. *Cell Death Differ.*, **10**, 621–33.

Dorey, E. (2002). Genta strikes bumper deal with Aventis. *Nat. Biotechnol.*, **20**, 533–4.

Drag-Zalesinska, M., Wysocka, T., Dumanska, M., Jagoda, E., and Zabel, M. (2002). Comparison of techniques permitting to detect apoptosis in situ. *Folia. Histochem. Cytobiol.*, **40**, 125–6.

Du, C., Fang, M., Li, Y., Li, L., and Wang, X. (2000). Smac, a mitochondrial protein that promotes cytochrome c-dependent caspase activation by eliminating IAP inhibition. *Cell*, **102**, 33–42.

Duan, W., Zhu, X., Ladenheim, B. *et al.* (2002). p53 inhibitors preserve dopamine neurons and motor function in experimental parkinsonism. *Ann. Neurol.*, **52**, 597–606.

Duan, W. R., Garner, D. S., Williams, S. D., Funckes-Shippy, C. L., Spath, I. S., and Blomme, E. A. (2003). Comparison of immunohistochemistry for activated caspase-3 and cleaved cytokeratin 18 with the TUNEL method for quantification of apoptosis in histological sections of PC-3 subcutaneous xenografts. *J. Pathol.*, **199**, 221–8.

Earnshaw, W. C., Martins, L. M., and Kaufmann, S. H. (1999). Mammalian caspases: structure, activation, substrates, and functions during apoptosis. *Annu. Rev. Biochem.*, **68**, 383–424.

Eberhardt, O., Coelln, R. V., Kugler, S. *et al.* (2000). Protection by synergistic effects of adenovirus-mediated X-chromosome-linked inhibitor of apoptosis and glial cell line-derived neurotrophic factor gene transfer in the 1-methyl-4-phenyl-1,2,3,6-tetrahydropyridine model of Parkinson's disease. *J. Neurosci.*, **20**, 9126–34.

Ellis, H. M. and Horvitz, H. R. (1986). Genetic control of programmed cell death in the nematode *C. elegans*. *Cell*, **44**, 817–29.

Endrizzi, M. G., Hadinoto, V., Growney, J. D., Miller, W., and Dietrich, W. F. (2000). Genomic sequence analysis of the mouse Naip gene array. *Genome Res.*, **10**, 1095–102.

Enyedy, I. J., Ling, Y., Nacro, K. *et al.* (2001). Discovery of small-molecule inhibitors of Bcl-2 through structure-based computer screening. *J. Med. Chem.*, **44**, 4313–24.

Erlanson, D. A., Lam, J. W., Wiesmann, C. *et al.* (2003). In situ assembly of enzyme inhibitors using extended tethering. *Nat. Biotechnol.*, **21**, 308–14.

Fairbrother, W. J., Gordon, N. C., Humke, E. W. *et al.* (2001). The PYRIN domain: a member of the death domain-fold superfamily. *Protein Sci.*, **10**, 1911–18.

Ferreira, C. G., Epping, M., Kruyt, F. A., and Giaccone, G. (2002). Apoptosis: target of cancer therapy. *Clin. Cancer Res.*, **8**, 2024–34.

Ferri, K. F. and Kroemer, G. (2001). Organelle-specific initiation of cell death pathways. *Nat. Cell Biol.*, **3**, E255–63.

Fesik, S. W. (2000). Insights into programmed cell death through structural biology. *Cell*, **103**, 273–82.

Fischer, U., Janicke, R. U., and Schulze-Osthoff, K. (2003). Many cuts to ruin: a comprehensive update of caspase substrates. *Cell Death Differ.*, **10**, 76–100.

Fisher, P. B., Gopalkrishnan, R. V., Chada, S. *et al.* (2003). mda-7/IL-24, a novel cancer selective apoptosis inducing cytokine gene: from the laboratory into the clinic. *Cancer Biol. Ther.*, **2**, S23–37.

Fong, W. G., Liston, P., Rajcan-Separovic, E., St Jean, M., Craig, C., and Korneluk, R. G. (2000). Expression and genetic analysis of XIAP-associated factor 1 (XAF1) in cancer cell lines. *Genomics*, **70**, 113–22.

Forcet, C., Ye, X., Granger, L., Corset, V., Shin, H., Bredesen, D. E., and Mehlen, P. (2001). The dependence receptor DCC (deleted in colorectal cancer) defines an alternative mechanism for caspase activation. *Proc. Natl. Acad. Sci. USA*, **98**, 3416–21.

Foster, B. A., Coffey, H. A., Morin, M. J., and Rastinejad, F. (1999). Pharmacological rescue of mutant p53 conformation and function. *Science*, **286**, 2507–10.

Frankfurt, O. S. and Krishan, A. (2001a). Enzyme-linked immunosorbent assay (ELISA) for the specific detection of apoptotic cells and its application to rapid drug screening. *J. Immunol. Methods*, **253**, 133–44.

(2001b). Identification of apoptotic cells by formamide-induced DNA denaturation in condensed chromatin. *J. Histochem. Cytochem.*, **49**, 369–78.

French, L. E. and Tschopp, J. (2003). Protein-based therapeutic approaches targeting death receptors. *Cell Death Differ.*, **10**, 117–23.

Friedlander, R. M. (2003). Apoptosis and caspases in neurodegenerative diseases. *N. Engl. J. Med.*, **348**, 1365–75.

Fussenegger, M., Bailey, J. E., and Varner, J. (2000). A mathematical model of caspase function in apoptosis. *Nat. Biotechnol.*, **18**, 768–74.

Gabay, C. (2003). IL-1 trap. Regeneron/Novartis. *Curr. Opin. Investig. Drugs*, **4**, 593–7.

Garrity, M. M., Burgart, L. J., Riehle, D. L., Hill, E. M., Sebo, T. J., and Witzig, T. (2003). Identifying and quantifying apoptosis: navigating technical pitfalls. *Mod. Pathol.*, **16**, 389–94.

Geiger, T., Husken, D., Weiler, J. *et al.* (2000). Consequences of the inhibition of Hdm2 expression in human osteosarcoma cells using antisense oligonucleotides. *Anticancer Drug Des.*, **15**, 423–30.

Gil-Gomez, G. and Brady, H. J. (1998). Transgenic mice in apoptosis research. *Apoptosis*, **3**, 215–28.

Glaser, M., Collingridge, D. R., Aboagye, E. O. *et al.* (2003a). Iodine-124 labelled annexin-V as a potential radiotracer to study apoptosis using positron emission tomography. *Appl. Radiat. Isot.*, **58**, 55–62.

Glaser, M., Luthra, S. K., and Brady, F. (2003b). Applications of positron-emitting halogens in PET oncology. *Int. J. Oncol.*, **22**, 253–67.

Glover, C. J., Hite, K., DeLosh, R. *et al.* (2003). A high-throughput screen for identification of molecular mimics of Smac/DIABLO utilizing a fluorescence polarization assay. *Anal. Biochem.*, **320**, 157–69.

Goldstein, J. C., Kluck, R. M., and Green, D. R. (2000a). A single cell analysis of apoptosis. Ordering the apoptotic phenotype. *Ann. NY Acad. Sci.*, **926**, 132–41.

Goldstein, J. C., Waterhouse, N. J., Juin, P., Evan, G. I., and Green, D. R. (2000b). The coordinate release of cytochrome c during apoptosis is rapid, complete and kinetically invariant. *Nat. Cell Biol.*, **2**, 156–62.

Grabarek, J., Ardelt, B., Kunicki, J., and Darzynkiewicz, Z. (2002). Detection of in situ activation of transglutaminase during apoptosis: correlation with the cell cycle phase by multiparameter flow and laser scanning cytometry. *Cytometry*, **49**, 83–9.

Green, D. R. and Evan, G. I. (2002). A matter of life and death. *Cancer Cell*, **1**, 19–30.

Grodzicky, T. and Elkon, K. B. (2002). Apoptosis: a case where too much or too little can lead to autoimmunity. *Mt. Sinai. J. Med.*, **69**, 208–19.

Gromeier, M., Lachmann, S., Rosenfeld, M. R., Gutin, P. H., and Wimmer, E. (2000). Intergeneric poliovirus recombinants for the treatment of malignant glioma. *Proc. Natl. Acad. Sci. USA*, **97**, 6803–8.

Gross, A., McDonnell, J. M., and Korsmeyer, S. J. (1999). BCL-2 family members and the mitochondria in apoptosis. *Genes. Dev.*, **13**, 1899–911.

Growney, J. D. and Dietrich, W. F. (2000). High-resolution genetic and physical map of the Lgn1 interval in C57BL/6J implicates Naip2 or Naip5 in *Legionella pneumophila* pathogenesis. *Genome. Res.*, **10**, 1158–71.

Gu, J., Kagawa, S., Takakura, M. *et al.* (2000). Tumor-specific transgene expression from the human telomerase reverse transcriptase promoter enables targeting of the therapeutic effects of the Bax gene to cancers. *Cancer Res.*, **60**, 5359–64.

Gujral, J. S., Farhood, A., and Jaeschke, H. (2003). Oncotic necrosis and caspase-dependent apoptosis during galactosamine-induced liver injury in rats. *Toxicol. Appl. Pharmacol.*, **190**, 37–46.

Gumucio, D. L., Diaz, A., Schaner, P. *et al.* (2002). Fire and ICE: the role of pyrin domain-containing proteins in inflammation and apoptosis. *Clin. Exp. Rheumatol.*, **20**, S45–53.

Haefner, B. (2002). NF-kappaB: arresting a major culprit in cancer. *Drug Discov. Today*, **7**, 653–63.

Hakumaki, J. M. and Brindle, K. M. (2003). Techniques: visualizing apoptosis using nuclear magnetic resonance. *Trends Pharmacol. Sci.*, **24**, 146–9.

Han, B. H., Xu, D., Choi, J. *et al.* (2002). Selective, reversible caspase-3 inhibitor is neuroprotective and reveals distinct pathways of cell death after neonatal hypoxic-ischemic brain injury. *J. Biol. Chem.*, **277**, 30128–36.

Hanahan, D. and Weinberg, R. A. (2000). The hallmarks of cancer. *Cell*, **100**, 57–70.

Harlin, H., Reffey, S. B., Duckett, C. S., Lindsten, T., and Thompson, C. B. (2001). Characterization of XIAP-deficient mice. *Mol. Cell Biol.*, **21**, 3604–8.

Harton, J. A., Linhoff, M. W., Zhang, J., and Ting, J. P. (2002). Cutting edge: CATERPILLER – a large family of mammalian genes containing CARD, pyrin, nucleotide-binding, and leucine-rich repeat domains. *J. Immunol.*, **169**, 4088–93.

Hawkins, L. K., Lemoine, N. R., and Kirn, D. (2002). Oncolytic biotherapy: a novel therapeutic platform. *Lancet. Oncol.*, **3**, 17–26.

Hayward, R. L., Macpherson, J. S., Cummings, J., Monia, B. P., Smyth, J. F., and Jodrell, D. I. (2003). Antisense Bcl-xl down-regulation switches the response to topoisomerase I inhibition from senescence to apoptosis in colorectal cancer cells, enhancing global cytotoxicity. *Clin. Cancer Res.*, **9**, 2856–65.

Hegde, R., Srinivasula, S. M., Datta, P. *et al.* (2003). The polypeptide chain-releasing factor GSPT1/eRF3 is proteolytically processed into an IAP-binding protein. *J. Biol. Chem.*, **15**, 15.

Hengartner, M. O., Eelis, R. E., and Horvitz, H. R. (1992). *Caenorhabditis elegans* gene ced-9 protects cells from programmed cell death. *Nature*, **356**, 494–9.

Hinz, M., Loser, P., Mathas, S., Krappmann, D., Dorken, B., and Scheidereit, C. (2001). Constitutive NF-kappaB maintains high expression of a characteristic gene network, including CD40, CD86, and a set of antiapoptotic genes in Hodgkin/Reed–Sternberg cells. *Blood*, **97**, 2798–807.

Hirase, N., Yufu, Y., Abe, Y. *et al.* (2000). Primary macroglobulinemia with t(11;18)(q21;q21). *Cancer Genet. Cytogenet.*, **117**, 113–17.

Hoeberichts, F. A., and Woltering, E. J. (2003). Multiple mediators of plant programmed cell death: interplay of conserved cell death mechanisms and plant-specific regulators. *Bioessays*, **25**, 47–57.

Hofmann, K., Bucher, P., and Tschopp, J. (1997). The CARD domain: a new apoptotic signalling motif. *Trends Biochem. Sci.*, **22**, 155–6.

Holcik, M., Thompson, C. S., Yaraghi, Z., Lefebvre, C. A., MacKenzie, A. E., and Korneluk, R. G. (2000). The hippocampal neurons of neuronal apoptosis inhibitory protein 1 (NAIP1)-deleted mice display increased vulnerability to kainic acid-induced injury. *Proc. Natl. Acad. Sci. USA*, **97**, 2286–90.

Hornsby, P. J. and Didenko, V. V. (2002). In situ DNA ligation as a method for labeling apoptotic cells in tissue sections. An overview. *Methods Mol. Biol.*, **203**, 133–41.

Hotchkiss, R. S., Chang, K. C., Swanson, P. E. *et al.* (2000). Caspase inhibitors improve survival in sepsis: a critical role of the lymphocyte. *Nat. Immunol.*, **1**, 496–501.

Hoves, S., Krause, S. W., Scholmerich, J., and Fleck, M. (2003). The JAM-assay: optimized conditions to determine death-receptor-mediated apoptosis. *Methods*, **31**, 127–34.

Hu, Y., Cherton-Horvat, G., Dragowska, V. *et al.* (2003). Antisense oligonucleotides targeting XIAP induce apoptosis and enhance chemotherapeutic activity against human lung cancer cells in vitro and in vivo. *Clin. Cancer Res.*, **9**, 2826–36.

Huang, P. and Oliff, A. (2001). Signaling pathways in apoptosis as potential targets for cancer therapy. *Trends Cell Biol.*, **11**, 343–8.

Huang, Z. (2002). The chemical biology of apoptosis. Exploring protein–protein interactions and the life and death of cells with small molecules. *Chem. Biol.*, **9**, 1059–72.

Huettenbrenner, S., Maier, S., Leisser, C. *et al.* (2003). The evolution of cell death programs as prerequisites of multicellularity. *Mutat. Res.*, **543**, 235–49.

Igney, F. H. and Krammer, P. H. (2002). Death and anti-death: tumour resistance to apoptosis. *Nat. Rev. Cancer*, **2**, 277–88.

Imoto, I., Tsuda, H., Hirasawa, A. *et al.* (2002). Expression of cIAP1, a target for 11q22 amplification, correlates with resistance of cervical cancers to radiotherapy. *Cancer Res.*, **62**, 4860–6.

Imoto, I., Yang, Z. Q., Pimkhaokham, A. *et al.* (2001). Identification of cIAP1 as a candidate target gene within an amplicon at 11q22 in esophageal squamous cell carcinomas. *Cancer Res.*, **61**, 6629–34.

Inohara, N. and Nunez, G. (2003). NODs: intracellular proteins involved in inflammation and apoptosis. *Nat. Rev. Immunol.*, **3**, 371–82.

Isabel, E., Black, W. C., Bayly, C. I. *et al.* (2003). Nicotinyl aspartyl ketones as inhibitors of caspase-3. *Bioorg. Med. Chem. Lett.*, **13**, 2137–40.

Ishizaki, Y., Jacobson, M. D., and Raff, M. C. (1998). A role for caspases in lens fiber differentiation. *J. Cell Biol.*, **140**, 153–8.

Jakubczak, J. L., Ryan, P., Gorziglia, M. *et al.* (2003). An oncolytic adenovirus selective for retinoblastoma tumor suppressor protein pathway-defective tumors: dependence on E1A, the E2F-1 promoter, and viral replication for selectivity and efficacy. *Cancer Res.*, **63**, 1490–9.

Jansen, B. and Zangemeister-Wittke, U. (2002). Antisense therapy for cancer – the time of truth. *Lancet. Oncol.*, **3**, 672–83.

Jansen, B., Wacheck, V., Heere-Ress, E. *et al.* (2000). Chemosensitisation of malignant melanoma by BCL2 antisense therapy. *Lancet.*, **356**, 1728–33.

Jiang, X., Kim, H. E., Shu, H. *et al.* (2003). Distinctive roles of PHAP proteins and prothymosin-alpha in a death regulatory pathway. *Science*, **299**, 223–6.

Johnson, L., Shen, A., Boyle, L. *et al.* (2002). Selectively replicating adenoviruses targeting deregulated E2F activity are potent, systemic antitumor agents. *Cancer Cell*, **1**, 325–37.

Johnstone, R. W., Ruefli, A. A., and Lowe, S. W. (2002). Apoptosis: a link between cancer genetics and chemotherapy. *Cell*, **108**, 153–64.

Joyce, D. E., Gelbert, L., Ciaccia, A., DeHoff, B., and Grinnell, B. W. (2001). Gene expression profile of antithrombotic protein c defines new mechanisms modulating inflammation and apoptosis. *J. Biol. Chem.*, **276**, 11199–203.

Joza, N., Kroemer, G., and Penninger, J. M. (2002). Genetic analysis of the mammalian cell death machinery. *Trends Genet.*, **18**, 142–9.

Joza, N., Susin, S. A., Daugas, E. *et al.* (2001). Essential role of the mitochondrial apoptosis-inducing factor in programmed cell death. *Nature*, **410**, 549–54.

Karin, M., Cao, Y., Greten, F. R., and Li, Z. W. (2002). NF-kappaB in cancer: from innocent bystander to major culprit. *Nat. Rev. Cancer*, **2**, 301–10.

Kennedy, N. J., Kataoka, T., Tschopp, J., and Budd, R. C. (1999). Caspase activation is required for T cell proliferation. *J. Exp. Med.*, **190**, 1891–6.

Kerr, J. F. (2002). History of the events leading to the formulation of the apoptosis concept. *Toxicology*, **181–2**, 471–4.

Kerr, J. F., Wyllie, A. H., and Currie, A. R. (1972). Apoptosis: a basic biological phenomenon with wide-ranging implications in tissue kinetics. *Br. J. Cancer*, **26**, 239–57.

Kiechle, F. L., and Zhang, X. (2002). Apoptosis: biochemical aspects and clinical implications. *Clin. Chim. Acta.*, **326**, 27–45.

Kishore, N., Sommers, C., Mathialagan, S. *et al.* (2003). A selective IKK-2 inhibitor blocks NF-kappa B-dependent gene expression in interleukin-1 beta-stimulated synovial fibroblasts. *J. Biol. Chem.*, **278**, 32861–71.

Kohler, C., Orrenius, S., and Zhivotovsky, B. (2002). Evaluation of caspase activity in apoptotic cells. *J. Immunol. Methods*, **265**, 97–110.

Komarov, P. G., Komarova, E. A., Kondratov, R. V. *et al.* (1999). A chemical inhibitor of p53 that protects mice from the side effects of cancer therapy. *Science*, **285**, 1733–7.

Koonin, E. V. and Aravind, L. (2000). The NACHT family – a new group of predicted NTPases implicated in apoptosis and MHC transcription activation. *Trends Biochem. Sci.*, **25**, 223–4.

(2002). Origin and evolution of eukaryotic apoptosis: the bacterial connection. *Cell Death Differ.*, **9**, 394–404.

Koulov, A. V., Stucker, K. A., Lakshmi, C., Robinson, J. P, and Smith, B. D. (2003). Detection of apoptotic cells using a synthetic fluorescent sensor for membrane surfaces that contain phosphatidylserine. *Cell Death Differ.*, **10**, 1357–9.

Kovalenko, A., Chable-Bessia, C., Cantarella, G., Israel, A., Wallach, D., and Courtois, G. (2003). The tumour suppressor CYLD negatively regulates NF-kappaB signalling by deubiquitination. *Nature*, **424**, 801–5.

Kugler, S., Straten, G., Kreppel, F., Isenmann, S., Liston, P., and Bahr, M. (2000). The X-linked inhibitor of apoptosis (XIAP) prevents cell death in axotomized CNS neurons in vivo. *Cell Death Differ.*, **7**, 815–24.

Kylarova, D., Prochazkova, J., Mad'arova, J., Bartos, J., and Lichnovsky, V. (2002). Comparison of the TUNEL, lamin B and annexin V methods for the detection of apoptosis by flow cytometry. *Acta. Histochem.*, **104**, 367–70.

LaCasse, E. (in press). Apoptosis control based on down-regulating the inhibitor-of-apoptosis (IAP) proteins: XIAP antisense and other approaches. In *Cell Engineering*, volume 4, ed. M. Al-Rubeai and M. Fussenegger. Dordrecht: Kluwer.

LaCasse, E. C., Baird, S., Korneluk, R. G., and MacKenzie, A. E. (1998). The inhibitors of apoptosis (IAPs) and their emerging role in cancer. *Oncogene*, **17**, 3247–59.

Lamkanfi, M., Declercq, W., Kalai, M., Saelens, X., and Vandenabeele, P. (2002). Alice in caspase land. A phylogenetic analysis of caspases from worm to man. *Cell Death Differ.*, **9**, 358–61.

Lawen, A. (2003). Apoptosis – an introduction. *Bioessays*, **25**, 888–96.

Lecoeur, H., de Oliveira-Pinto, L. M., and Gougeon, M. L. (2002). Multiparametric flow cytometric analysis of biochemical and functional events associated with apoptosis and oncosis using the 7-aminoactinomycin D assay. *J. Immunol. Methods*, **265**, 81–96.

Li, F. (2003). Survivin study: what is the next wave? *J. Cell Physiol.*, **197**, 8–29.

Liang, M. C., Bardhan, S., Li, C., Pace, E. A., Porco, Jr. J. A., and Gilmore, T. D. (2003). Jesterone dimer, a synthetic derivative of the fungal metabolite jesterone, blocks activation of transcription factor nuclear factor kappaB by inhibiting the inhibitor of kappaB kinase. *Mol. Pharmacol.*, **64**, 123–31.

Lieberman, J. (2003). The ABCs of granule-mediated cytotoxicity: new weapons in the arsenal. *Nat. Rev. Immunol.*, **3**, 361–70.

Liming, P., Bradley, C. J., and Liu, J. J. (1999). The correlativity analysis of six methods of detecting apoptosis. *Chin. Med. Sci. J.*, **14**, 145–51.

Linton, S. D., Karanewsky, D. S., Ternansky, R. J. *et al.* (2002). Acyl dipeptides as reversible caspase inhibitors. Part 1: initial lead optimization. *Bioorg. Med. Chem. Lett.*, **12**, 2969–71.

Liston, P., Fong, W. G., Kelly, N. L. *et al.* (2001). Identification of XAF1 as an antagonist of XIAP anti-caspase activity. *Nat. Cell Biol.*, **3**, 128–33.

Liston, P., Roy, N., Tamai, K. *et al.* (1996). Suppression of apoptosis in mammalian cells by NAIP and a related family of IAP genes. *Nature*, **379**, 349–53.

Liu, H., Ye, H., Ruskone-Fourmestraux, A. *et al.* (2002). T(11;18) is a marker for all stage gastric MALT lymphomas that will not respond to *H. pylori* eradication. *Gastroenterology*, **122**, 1286–94.

Liu, T., Rojas, A., Ye, Y., and Godzik, A. (2003). Homology modeling provides insights into the binding mode of the PAAD/DAPIN/pyrin domain, a fourth member of the CARD/DD/DED domain family. *Protein Sci.*, **12**, 1872–81.

Lockshin, R. A. and Zakeri, Z. (2001). Programmed cell death and apoptosis: origins of the theory. *Nat. Rev. Mol. Cell Biol.*, **2**, 545–50.

(2002). Caspase-independent cell deaths. *Curr. Opin. Cell Biol.*, **14**, 727–33.

Loo, D. T. (2002). TUNEL assay. An overview of techniques. *Methods Mol. Biol.*, **203**, 21–30.

Los, M., Burek, C. J., Stroh, C., Benedyk, K., Hug, H., and Mackiewicz, A. (2003). Anticancer drugs of tomorrow: apoptotic pathways as targets for drug design. *Drug Discov. Today*, **8**, 67–77.

Mahlamaki, E. H., Barlund, M., Tanner, M. *et al.* (2002). Frequent amplification of 8q24-, 11q-, 17q-, and 20q-specific genes in pancreatic cancer. *Genes Chromosomes Cancer*, **35**, 353–8.

Makin, G. and Dive, C. (2003). Recent advances in understanding apoptosis: new therapeutic opportunities in cancer chemotherapy. *Trends Mol. Med.*, **9**, 251–5.

Makower, D., Rozenblit, A., Kaufman, H. *et al.* (2003). Phase II clinical trial of intralesional administration of the oncolytic adenovirus ONYX-015 in patients with hepatobiliary tumors with correlative p53 studies. *Clin. Cancer Res.*, **9**, 693–702.

Malkin, D., Li, F. P., Strong, L. C. *et al.* (1990). Germ line p53 mutations in a familial syndrome of breast cancer, sarcomas, and other neoplasms. *Science*, **250**, 1233–8.

Marsden, V. S., and Strasser, A. (2003). Control of apoptosis in the immune system: Bcl-2, BH3-only proteins and more. *Annu. Rev. Immunol.*, **21**, 71–105.

Martinon, F., Burns, K., and Tschopp, J. (2002). The inflammasome: a molecular platform triggering activation of inflammatory caspases and processing of proIL-beta. *Mol. Cell*, **10**, 417–26.

Mathis, D., Vence, L., and Benoist, C. (2001). Beta-cell death during progression to diabetes. *Nature*, **414**, 792–8.

McDermott, M. F. (2002). Genetic clues to understanding periodic fevers, and possible therapies. *Trends Mol. Med.*, **8**, 550–4.

McIntyre, K. W., Shuster, D. J., Gillooly, K. M. *et al.* (2003). A highly selective inhibitor of I kappa B kinase, BMS-345541, blocks both joint inflammation and destruction in collagen-induced arthritis in mice. *Arthritis Rheum*, **48**, 2652–9.

McKinnon, S. J., Lehman, D. M., Tahzib, N. G. *et al.* (2002). Baculoviral IAP repeat-containing-4 protects optic nerve axons in a rat glaucoma model. *Mol. Ther.*, **5**, 780–7.

Meier, P., Finch, A., and Evan, G. (2000). Apoptosis in development. *Nature*, **407**, 796–801.

Melino, G. (2002). The meaning of death. *Cell Death Differ.*, **9**, 347–8.

Micoud, F., Mandrand, B., and Malcus-Vocanson, C. (2001). Comparison of several techniques for the detection of apoptotic astrocytes in vitro. *Cell Prolif.*, **34**, 99–113.

Miller, L. K. (1999). An exegesis of IAPs: salvation and surprises from BIR motifs. *Trends Cell Biol.*, **9**, 323–8.

Mirkes, P. E. (2002). 2001 Warkany lecture: to die or not to die, the role of apoptosis in normal and abnormal mammalian development. *Teratology*, **65**, 228–39.

Mori, Y., Selaru, F. M., Sato, F. *et al.* (2003). The impact of microsatellite instability on the molecular phenotype of colorectal tumors. *Cancer Res.*, **63**, 4577–82.

Mulder, N. J., Apweiler, R., Attwood, T. K. *et al.* (2003). The InterPro Database, 2003, brings increased coverage and new features. *Nucleic Acids Res.*, **31**, 315–18.

Mullauer, L., Gruber, P., Sebinger, D., Buch, J., Wohlfart, S., and Chott, A. (2001). Mutations in apoptosis genes: a pathogenetic factor for human disease. *Mutat. Res.*, **488**, 211–31.

Murga-Penas, E., Hinz, K., Roser, K. *et al.* (2003). Translocations t(11;18)(q21;q21) and t(14;18)(q32;q21) are the main chromosomal abnormalities involving MLT/MALT1 in MALT lymphomas. *Leukemia*, **21**, 21.

Murphy, F. J., Hayes, I., and Cotter, T. G. (2003). Targeting inflammatory diseases via apoptotic mechanisms. *Curr. Opin. Pharmacol.*, **3**, 412–9.

Nakagawa, T., Zhu, H., Morishima, N. *et al.* (2000). Caspase-12 mediates endoplasmic-reticulum-specific apoptosis and cytotoxicity by amyloid-beta. *Nature*, **403**, 98–103.

Nathan, C. (2002). Points of control in inflammation. *Nature*, **420**, 846–52.

Natori, S., Higuchi, H., Contreras, P., and Gores, G. J. (2003). The caspase inhibitor IDN-6556 prevents caspase activation and apoptosis in sinusoidal endothelial cells during liver preservation injury. *Liver Transpl.*, **9**, 278–84.

Newmeyer, D. D. and Ferguson-Miller, S. (2003). Mitochondria: releasing power for life and unleashing the machineries of death. *Cell*, **112**, 481–90.

Newton, K. and Strasser, A. (2003). Caspases signal not only apoptosis but also antigen-induced activation in cells of the immune system. *Genes Dev.*, **17**, 819–25.

Nguyen, J. T. and Wells, J. A. (2003). Direct activation of the apoptosis machinery as a mechanism to target cancer cells. *Proc. Natl. Acad. Sci. USA*, **100**, 7533–8.

Nicholson, D. W. (1999). Caspase structure, proteolytic substrates, and function during apoptotic cell death. *Cell Death Differ.*, **6**, 1028–42.

Nygren, P. and Larsson, R. (2003). Overview of the clinical efficacy of investigational anticancer drugs. *J. Intern. Med.*, **253**, 46–75.

Oberhammer, F., Wilson, J. W., Dive, C. *et al.* (1993). Apoptotic death in epithelial cells: cleavage of DNA to 300 and/or 50 kb fragments prior to or in the absence of internucleosomal fragmentation. *EMBO J.*, **12**, 3679–84.

Okamoto, Y., Anan, H., Nakai, E. *et al.* (1999). Peptide based interleukin-1 beta converting enzyme (ICE) inhibitors: synthesis, structure activity relationships and crystallographic study of the ICE-inhibitor complex. *Chem. Pharm. Bull. (Tokyo)*, **47**, 11–21.

Olie, R. A., Hall, J., Natt, F., Stahel, R. A, and Zangemeister-Wittke, U. (2002). Analysis of ribosyl-modified, mixed backbone analogs of a bcl-2/bcl-xL antisense oligonucleotide. *Biochim. Biophys. Acta.*, **1576**, 101–9.

Olijslagers, S., Dege, A. Y., Dinsart, C. *et al.* (2001). Potentiation of a recombinant oncolytic parvovirus by expression of apoptin. *Cancer Gene. Ther.*, **8**, 958–65.

Olson, N. E., Graves, J. D., Shu, G. L., Ryan, E. J., and Clark, E. A. (2003). Caspase activity is required for stimulated B lymphocytes to enter the cell cycle. *J. Immunol.*, **170**, 6065–72.

Orlowski, R. Z. and Baldwin, Jr. A. S. (2002). NF-kappaB as a therapeutic target in cancer. *Trends Mol. Med.*, **8**, 385–9.

Otsuki, Y., Li, Z., and Shibata, M. A. (2003). Apoptotic detection methods – from morphology to gene. *Prog. Histochem. Cytochem.*, **38**, 275–339.

Ott, M., Robertson, J. D., Gogvadze, V., Zhivotovsky, B., and Orrenius, S. (2002). Cytochrome c release from mitochondria proceeds by a two-step process. *Proc. Natl. Acad. Sci. USA*, **99**, 1259–63.

Pawlowski, K., Pio, F., Chu, Z., Reed, J. C, and Godzik, A. (2001). PAAD – a new protein domain associated with apoptosis, cancer and autoimmune diseases. *Trends Biochem. Sci.*, **26**, 85–7.

Pecora, A. L., Rizvi, N., Cohen, G. I. *et al.* (2002). Phase I trial of intravenous administration of PV701, an oncolytic virus, in patients with advanced solid cancers. *J. Clin. Oncol.*, **20**, 2251–66.

Peng, Y., Li, C., Chen, L., Sebti, S., and Chen, J. (2003). Rescue of mutant p53 transcription function by ellipticine. *Oncogene*, **22**, 4478–87.

Perfettini, J. L., and Kroemer, G. (2003). Caspase activation is not death. *Nat. Immunol.*, **4**, 308–10.

Perkins, D. (2002). Targeting apoptosis in neurological disease using the herpes simplex virus. *J. Cell Mol. Med.*, **6**, 341–56.

Perrelet, D., Ferri, A., Liston, P., Muzzin, P., Korneluk, R. G., and Kato, A. C. (2002). IAPs are essential for GDNF-mediated neuroprotective effects in injured motor neurons in vivo. *Nat. Cell Biol.*, **4**, 175–9.

Perrelet, D., Ferri, A., MacKenzie, A. E. *et al.* (2000). IAP family proteins delay motoneuron cell death in vivo. *Eur. J. Neurosci.*, **12**, 2059–67.

Peter, M. E. and Krammer, P. H. (2003). The CD95(APO-1/Fas) DISC and beyond. *Cell Death Differ.*, **10**, 26–35.

Petrin, D., Baker, A., Coupland, S. G. *et al.* (2003). Structural and functional protection of photoreceptors from MNU-induced retinal degeneration by the X-linked inhibitor of apoptosis. *Invest. Ophthalmol. Vis. Sci.*, **44**, 2757–63.

Philchenkov, A. A. (2003). Caspases as regulators of apoptosis and other cell functions. *Biochemistry (Mosc.)*, **68**, 365–76.

Pirollo, K. F., Rait, A., Sleer, L. S., and Chang, E. H. (2003). Antisense therapeutics: from theory to clinical practice. *Pharmacol. Ther.*, **99**, 55–77.

Pozarowski, P., Huang, X., Halicka, D. H., Lee, B., Johnson, G., and Darzynkiewicz, Z. (2003). Interactions of fluorochrome-labeled caspase inhibitors with apoptotic cells: a caution in data interpretation. *Cytometry*, **55A**, 50–60.

Print, C. G. and Lakoski-Loveland, K. (2000). Germ cell suicide: new insights into apoptosis during spermatogenesis. *Bioessays*, **22**, 423–30.

Proskuryakov, S. Y., Konoplyannikov, A. G., and Gabai, V. L. (2003). Necrosis: a specific form of programmed cell death? *Exp. Cell Res.*, **283**, 1–16.

Ramaswamy, S., Ross, K. N., Lander, E. S., and Golub, T. R. (2003). A molecular signature of metastasis in primary solid tumors. *Nat. Genet.*, **33**, 49–54.

Ramaswamy, S., Tamayo, P., Rifkin, R. *et al.* (2001). Multiclass cancer diagnosis using tumor gene expression signatures. *Proc. Natl. Acad. Sci. USA*, **98**, 15149–54.

Rampino, N., Yamamoto, H., Ionov, Y. *et al.* (1997). Somatic frameshift mutations in the BAX gene in colon cancers of the microsatellite mutator phenotype. *Science*, **275**, 967–9.

Ranger, A. M., Malynn, B. A., and Korsmeyer, S. J. (2001). Mouse models of cell death. *Nat. Genet.*, **28**, 113–18.

Rao, R. V., Castro-Obregon, S., Frankowski, H. *et al.* (2002). Coupling endoplasmic reticulum stress to the cell death program. An Apaf-1-independent intrinsic pathway. *J. Biol. Chem.*, **277**, 21836–42.

Rasmussen, H., Rasmussen, C., Lempicki, M. *et al.* (2002). TNFerade Biologic: preclinical toxicology of a novel adenovector with a radiation-inducible promoter, carrying the human tumor necrosis factor alpha gene. *Cancer Gene. Ther.*, **9**, 951–7.

Rathmell, J. C., and Thompson, C. B. (2002). Pathways of apoptosis in lymphocyte development, homeostasis, and disease. *Cell*, **109**, S97–107.

Read, S. H., Baliga, B. C., Ekert, P. G., Vaux, D. L., and Kumar, S. (2002). A novel Apaf-1-independent putative caspase-2 activation complex. *J. Cell Biol.*, **159**, 739–45.

Reed, J. C. (2002). Apoptosis-based therapies. *Nat. Rev. Drug Discov.*, **1**, 111–21.

(2003). Apoptosis-targeted therapies for cancer. *Cancer Cell*, **3**, 17–22.

Reed, J. C., Doctor, K., Rojas, A. *et al.* (2003). Comparative analysis of apoptosis and inflammation genes of mice and humans. *Genome Res.*, **13**, 1376–88.

Rich, T., Watson, C. J., and Wyllie, A. (1999). Apoptosis: the germs of death. *Nat. Cell Biol.*, **1**, E69–71.

Rieux-Laucat, F., Le Deist, F., and Fischer, A. (2003). Autoimmune lymphoproliferative syndromes: genetic defects of apoptosis pathways. *Cell Death Differ.*, **10**, 124–33.

Robert, A., Miron, M. J., Champagne, C., Gingras, M. C., Branton, P. E., and Lavoie, J. N. (2002). Distinct cell death pathways triggered by the adenovirus early region 4 ORF 4 protein. *J. Cell Biol.*, **158**, 519–28.

Robertson, G. S., Crocker, S. J., Nicholson, D. W., and Schulz, J. B. (2000). Neuroprotection by the inhibition of apoptosis. *Brain Pathol.*, **10**, 283–92.

Robertson, J. D., Fadeel, B., Zhivotovsky, B., and Orrenius, S. (2002). 'Centennial' Nobel Conference on apoptosis and human disease. *Cell Death Differ.*, **9**, 468–75.

Roshak, A. K., Callahan, J. F., and Blake, S. M. (2002). Small-molecule inhibitors of NF-kappaB for the treatment of inflammatory joint disease. *Curr. Opin. Pharmacol.*, **2**, 316–21.

Roux, P. P., Dorval, G., Boudreau, M. *et al.* (2002). K252a and CEP1347 are neuroprotective compounds that inhibit mixed-lineage kinase-3 and induce activation of Akt and ERK. *J. Biol. Chem.*, **277**, 49473–80.

Roy, N., Mahadevan, M. S., McLean, M. *et al.* (1995). The gene for neuronal apoptosis inhibitory protein is partially deleted in individuals with spinal muscular atrophy. *Cell*, **80**, 167–78.

Sakahira, H., Enari, M., Ohsawa, Y., Uchiyama, Y., and Nagata, S. (1999). Apoptotic nuclear morphological change without DNA fragmentation. *Curr. Biol.*, **9**, 543–6.

Salomon, A. R., Voehringer, D. W., Herzenberg, L. A., and Khosla, C. (2000). Understanding and exploiting the mechanistic basis for selectivity of polyketide inhibitors of F(0)F(1)-ATPase. *Proc. Natl. Acad. Sci. USA*, **97**, 14766–71.

Salvesen, G. S., and Duckett, C. S. (2002). IAP proteins: blocking the road to death's door. *Nat. Rev. Mol. Cell Biol.*, **3**, 401–10.

Sanchez-Izquierdo, D., Buchonnet, G., Siebert, R. *et al.* (2003). MALT1 is deregulated by both chromosomal translocation and amplification in B-cell non-Hodgkin lymphoma. *Blood*, **101**, 4539–46.

Sapolsky, R. M. (2003). Neuroprotective gene therapy against acute neurological insults. *Nat. Rev. Neurosci.*, **4**, 61–9.

Saraste, A. (1999). Morphologic criteria and detection of apoptosis. *Herz*, **24**, 189–95.

Sasaki, H., Sheng, Y., Kotsuji, F., and Tsang, B. K. (2000). Down-regulation of X-linked inhibitor of apoptosis protein induces apoptosis in chemoresistant human ovarian cancer cells. *Cancer Res.*, **60**, 5659–66.

Scapin, G., Patel, S. B., Lisnock, J., Becker, J. W., and LoGrasso, P. V. (2003). The structure of JNK3 in complex with small molecule inhibitors: structural basis for potency and selectivity. *Chem. Biol.*, **10**, 705–12.

Schultz, D. R. and Harrington, W. J. Jr. (2003). Apoptosis: programmed cell death at a molecular level. *Semin. Arthritis Rheum.*, **32**, 345–69.

Scorrano, L. and Korsmeyer, S. J. (2003). Mechanisms of cytochrome c release by proapoptotic BCL-2 family members. *Biochem. Biophys. Res. Commun.*, **304**, 437–44.

Scott, C. W., Sobotka-Briner, C., Wilkins, D. E. *et al.* (2003). Novel small molecule inhibitors of caspase-3 block cellular and biochemical features of apoptosis. *J. Pharmacol. Exp. Ther.*, **304**, 433–40.

Semra, Y. K., Seidi, O. A., and Sharief, M. K. (2002). Disease activity in multiple sclerosis correlates with T lymphocyte expression of the inhibitor of apoptosis proteins. *J. Neuroimmunol.*, **122**, 159–66.

Shabbits, J. A., Hu, Y., and Mayer, L. D. (2003). Tumor chemosensitization strategies based on apoptosis manipulations. *Mol. Cancer Ther.*, **2**, 805–13.

Shangary, S. and Johnson, D. E. (2003). Recent advances in the development of anticancer agents targeting cell death inhibitors in the Bcl-2 protein family. *Leukemia*, **17**, 1470–81.

Sharief, M. K. and Semra, Y. K. (2001a). Heightened expression of survivin in activated T lymphocytes from patients with multiple sclerosis. *J. Neuroimmunol.*, **119**, 358–64.

(2001b). Upregulation of the inhibitor of apoptosis proteins in activated T lymphocytes from patients with multiple sclerosis. *J. Neuroimmunol.*, **119**, 350–7.

(2002). Down-regulation of survivin expression in T lymphocytes after interferon beta-1a treatment in patients with multiple sclerosis. *Arch. Neurol.*, **59**, 1115–21.

Sharief, M. K., Noori, M. A., Douglas, M. R., and Semra, Y. K. (2002a). Upregulated survivin expression in activated T lymphocytes correlates with disease activity in multiple sclerosis. *Eur. J. Neurol.*, **9**, 503–10.

Sharief, M. K., Noori, M. A., and Zoukos, Y. (2002b). Reduced expression of the inhibitor of apoptosis proteins in T cells from patients with multiple sclerosis following interferon-beta therapy. *J. Neuroimmunol.*, **129**, 224–31.

Shi, Y. (2001). A structural view of mitochondria-mediated apoptosis. *Nat. Struct. Biol.*, **8**, 394–401.

(2002). Mechanisms of caspase activation and inhibition during apoptosis. *Mol. Cell*, **9**, 459–70.

Silke, J. and Vaux, D. L. (2001). Two kinds of BIR-containing protein – inhibitors of apoptosis, or required for mitosis. *J. Cell Sci.*, **114**, 1821–7.

Sloviter, R. S. (2002). Apoptosis: a guide for the perplexed. *Trends Pharmacol. Sci.*, **23**, 19–24.

Smahi, A., Courtois, G., Rabia, S. H. *et al.* (2002). The NF-kappaB signalling pathway in human diseases: from incontinentia pigmenti to ectodermal dysplasias and immune-deficiency syndromes. *Hum. Mol. Genet.*, **11**, 2371–5.

Sperandio, S., de Belle, I., and Bredesen, D. E. (2000). An alternative, nonapoptotic form of programmed cell death. *Proc. Natl. Acad. Sci. USA*, **97**, 14376–81.

Srivastava, S., Zou, Z. Q., Pirollo, K., Blattner, W., and Chang, E. H. (1990). Germ-line transmission of a mutated p53 gene in a cancer-prone family with Li–Fraumeni syndrome. *Nature*, **348**, 747–9.

Stadelmann, C. and Lassmann, H. (2000). Detection of apoptosis in tissue sections. *Cell Tissue Res.*, **301**, 19–31.

Staub, E., Dahl, E., and Rosenthal, A. (2001). The DAPIN family: a novel domain links apoptotic and interferon response proteins. *Trends Biochem. Sci.*, **26**, 83–5.

Steed, P. M., Tansey, M. G., Zalevsky, J. *et al.* (2003). Inactivation of TNF signaling by rationally designed dominant-negative TNF variants. *Science*, **301**, 1895–8.

Stennicke, H. R., Ryan, C. A., and Salvesen, G. S. (2002). Reprieval from execution: the molecular basis of caspase inhibition. *Trends Biochem. Sci.*, **27**, 94–101.

Stojdl, D. F., Lichty, B., Knowles, S. *et al.* (2000). Exploiting tumor-specific defects in the interferon pathway with a previously unknown oncolytic virus. *Nat. Med.*, **6**, 821–5.

Straten, G., Schmeer, C., Kretz, A. *et al.* (2002). Potential synergistic protection of retinal ganglion cells from axotomy-induced apoptosis by adenoviral administration of glial cell line-derived neurotrophic factor and X-chromosome-linked inhibitor of apoptosis. *Neurobiol. Dis.*, **11**, 123–33.

Streubel, B., Lamprecht, A., Dierlamm, J. *et al.* (2003). T(14;18)(q32;q21) involving IGH and MALT1 is a frequent chromosomal aberration in MALT lymphoma. *Blood*, **101**, 2335–9.

Suzuki, Y., Imai, Y., Nakayama, H., Takahashi, K., Takio, K., and Takahashi, R. (2001). A serine protease, HtrA2, is released from the mitochondria and interacts with XIAP, inducing cell death. *Mol. Cell*, **8**, 613–21.

Swinney, D. C., Xu, Y. Z., Scarafia, L. E. *et al.* (2002). A small molecule ubiquitination inhibitor blocks NF-kappa B-dependent cytokine expression in cells and rats. *J. Biol. Chem.*, **277**, 23573–81.

Tamm, I., Kornblau, S. M., Segall, H. *et al.* (2000). Expression and prognostic significance of IAP-family genes in human cancers and myeloid leukemias. *Clin. Cancer Res.*, **6**, 1796–803.

Tamm, I., Schriever, F., and Dorken, B. (2001). Apoptosis: implications of basic research for clinical oncology. *Lancet. Oncol.*, **2**, 33–42.

Tan, Y. J., Teng, E., and Ting, A. E. (2003). A small inhibitor of the interaction between Bax and Bcl-X(L) can synergize with methylprednisolone to induce apoptosis in Bcl-X(L)-overexpressing breast-cancer cells. *J. Cancer Res. Clin. Oncol.*, **129**, 437–48.

Tawa, P., Tam, J., Cassady, R., Nicholson, D. W, and Xanthoudakis, S. (2001). Quantitative analysis of fluorescent caspase substrate cleavage in intact cells and identification of novel inhibitors of apoptosis. *Cell Death Differ.*, **8**, 30–7.

Thome, M. and Tschopp, J. (2001). Regulation of lymphocyte proliferation and death by FLIP. *Nat. Rev. Immunol.*, **1**, 50–8.

Thompson, C. B. (1995). Apoptosis in the pathogenesis and treatment of disease. *Science*, **267**, 1456–62.

Thompson, T. G., DiDonato, C. J., Simard, L. R. *et al.* (1995). A novel cDNA detects homozygous microdeletions in greater than 50% of type I spinal muscular atrophy patients. *Nat. Genet.*, **9**, 56–62.

Thornberry, N. A. and Lazebnik, Y. (1998). Caspases: enemies within. *Science*, **281**, 1312–16.

Tibbetts, M. D., Zheng, L., and Lenardo, M. J. (2003). The death effector domain protein family: regulators of cellular homeostasis. *Nat. Immunol.*, **4**, 404–9.

Tracey, L., Villuendas, R., Dotor, A. M. *et al.* (2003). Mycosis fungoides shows concurrent deregulation of multiple genes involved in the TNF signaling pathway: an expression profile study. *Blood*, **102**, 1042–50.

Trapp, T., Korhonen, L., Besselmann, M., Martinez, R., Mercer, E. A., and Lindholm, D. (2003). Transgenic mice overexpressing XIAP in neurons show better outcome after transient cerebral ischemia. *Mol. Cell Neurosci.*, **23**, 302–13.

Trompouki, E., Hatzivassiliou, E., Tsichritzis, T., Farmer, H., Ashworth, A., and Mosialos, G. (2003). CYLD is a deubiquitinating enzyme that negatively regulates NF-kappaB activation by TNFR family members. *Nature*, **424**, 793–6.

Troy, C. M., Rabacchi, S. A., Hohl, J. B., Angelastro, J. M., Greene, L. A., and Shelanski, M. L. (2001). Death in the balance: alternative participation of the caspase-2 and -9 pathways in neuronal death induced by nerve growth factor deprivation. *J. Neurosci.*, **21**, 5007–16.

Tschopp, J., Martinon, F., and Burns, K. (2003). NALPs: a novel protein family involved in inflammation. *Nat. Rev. Mol. Cell Biol.*, **4**, 95–104.

Tsujimoto, Y. (2003). Cell death regulation by the Bcl-2 protein family in the mitochondria. *J. Cell Physiol.*, **195**, 158–67.

Tsujimoto, Y. and Croce, C. M. (1986). Analysis of the structure, transcripts, and protein products of bcl-2, the gene involved in human follicular lymphoma. *Proc. Natl. Acad. Sci. USA*, **83**, 5214–18.

Tzung, S. P., Kim, K. M., Basanez, G. *et al.* (2001). Antimycin A mimics a cell-death-inducing Bcl-2 homology domain 3. *Nat. Cell Biol.*, **3**, 183–91.

van Bokhoven, H. and McKeon, F. (2002). Mutations in the p53 homolog p63: allele-specific developmental syndromes in humans. *Trends Mol. Med.*, **8**, 133–9.

van der Eb, M. M., Pietersen, A. M., Speetjens, F. M. *et al.* (2002). Gene therapy with apoptin induces regression of xenografted human hepatomas. *Cancer Gene. Ther.*, **9**, 53–61.

Vanderluit, J. L., McPhail, L. T., Fernandes, K. J., Kobayashi, N. R., and Tetzlaff, W. (2003). In vivo application of mitochondrial pore inhibitors blocks the induction of apoptosis in axotomized neonatal facial motoneurons. *Cell Death Differ.*, **10**, 969–76.

van de Wiele, C., Lahorte, C., Vermeersch, H. *et al.* (2003). Quantitative tumor apoptosis imaging using technetium-99m-HYNIC annexin V single photon emission computed tomography. *J. Clin. Oncol.*, **21**, 3483–7.

Vaux, D. L. (2002). Apoptosis timeline. *Cell Death Differ.*, **9**, 349–54.

Vaux, D. L., Cory, S., and Adams, J. M. (1988). Bcl-2 gene promotes haemopoietic cell survival and cooperates with c-myc to immortalize pre-B cells. *Nature*, **335**, 440–2.

Venter, J. C., Adams, M. D., Myers, E. W. *et al.* (2001). The sequence of the human genome. *Science*, **291**, 1304–51.

Verhagen, A. M., Coulson, E. J., and Vaux, D. L. (2001). Inhibitor of apoptosis proteins and their relatives: IAPs and other BIRPs. *Genome. Biol.*, **2**, REVIEWS3009. 1–10.

Verhagen, A. M., Ekert, P. G., Pakusch, M. *et al.* (2000). Identification of DIABLO, a mammalian protein that promotes apoptosis by binding to and antagonizing IAP proteins. *Cell*, **102**, 43–53.

Vezina, J., Grossmuller, F., and Muller, K. (2001). Influence of a transiently transfected gene on apoptosis, measurements guided by cotransfected GFP. *J. Immunol. Methods*, **252**, 163–9.

Vila, M. and Przedborski, S. (2003). Targeting programmed cell death in neurodegenerative diseases. *Nat. Rev. Neurosci.*, **4**, 365–75.

Wajant, H., Pfizenmaier, K., and Scheurich, P. (2003). Tumor necrosis factor signaling. *Cell Death Differ.*, **10**, 45–65.

Waldmeier, P. C. (2003). Prospects for antiapoptotic drug therapy of neurodegenerative diseases. *Prog. Neuropsychopharmacol. Biol. Psychiatry.*, **27**, 303–21.

Waldmeier, P. C., Feldtrauer, J. J., Qian, T., and Lemasters, J. J. (2002). Inhibition of the mitochondrial permeability transition by the nonimmunosuppressive cyclosporin derivative NIM811. *Mol. Pharmacol.*, **62**, 22–9.

Walker, P. R., Carson, C., Leblanc, J., and Sikorska, M. (2002). Labeling DNA damage with terminal transferase. Applicability, specificity, and limitations. *Methods Mol. Biol.*, **203**, 3–19.

Wang, J. L., Liu, D., Zhang, Z. J. *et al.* (2000). Structure-based discovery of an organic compound that binds Bcl-2 protein and induces apoptosis of tumor cells. *Proc. Natl. Acad. Sci. USA*, **97**, 7124–9.

Wang, X. (2001). The expanding role of mitochondria in apoptosis. *Genes Dev.*, **15**, 2922–33.

Waterhouse, N. J. and Trapani, J. A. (2003). A new quantitative assay for cytochrome c release in apoptotic cells. *Cell Death Differ.*, **10**, 853–5.

Watts, J. A. and Kline, J. A. (2003). Bench to bedside: the role of mitochondrial medicine in the pathogenesis and treatment of cellular injury. *Acad. Emerg. Med.*, **10**, 985–97.

Weil, M., Jacobson, M. D., and Raff, M. C. (1998). Are caspases involved in the death of cells with a transcriptionally inactive nucleus? Sperm and chicken erythrocytes. *J. Cell Sci.*, **111**, 2707–15.

Wencker, D., Chandra, M., Nguyen, K. *et al.* (2003). A mechanistic role for cardiac myocyte apoptosis in heart failure. *J. Clin. Invest.*, **111**, 1497–504.

Willis, T. G., Jadayel, D. M., Du, M. Q. *et al.* (1999). Bcl10 is involved in t(1;14)(p22;q32) of MALT B cell lymphoma and mutated in multiple tumor types. *Cell*, **96**, 35–45.

Wright, E. K., Goodart, S. A., Growney, J. D. *et al.* (2003). Naip5 affects host susceptibility to the intracellular pathogen *Legionella pneumophila*. *Curr. Biol.*, **13**, 27–36.

Wu, T. Y., Wagner, K. W., Bursulaya, B., Schultz, P. G., and Deveraux, Q. L. (2003). Development and characterization of nonpeptidic small molecule inhibitors of the XIAP/caspase-3 interaction. *Chem. Biol.*, **10**, 759–67.

Xu, D. G., Crocker, S. J., Doucet, J. P. *et al.* (1997). Elevation of neuronal expression of NAIP reduces ischemic damage in the rat hippocampus. *Nat. Med.*, **3**, 997–1004.

Xu, M., Okada, T., Sakai, H. *et al.* (2002). Functional human NAIP promoter transcription regulatory elements for the NAIP and PsiNAIP genes. *Biochim. Biophys. Acta.*, **1574**, 35–50.

Yang, W., Guastella, J., Huang, J. C. *et al.* (2003). MX1013, a dipeptide caspase inhibitor with potent in vivo antiapoptotic activity. *Br. J. Pharmacol.*, **140**, 402–12.

Yaraghi, Z., Korneluk, R. G., and MacKenzie, A. (1998). Cloning and characterization of the multiple murine homologues of NAIP (neuronal apoptosis inhibitory protein). *Genomics*, **51**, 107–13.

Yasuhara, S., Zhu, Y., Matsui, T. *et al.* (2003). Comparison of comet assay, electron microscopy, and flow cytometry for detection of apoptosis. *J. Histochem. Cytochem.*, **51**, 873–85.

Yuan, J. and Yankner, B. A. (2000). Apoptosis in the nervous system. *Nature*, **407**, 802–9.

Zakeri, Z. and Lockshin, R. A. (2002). Cell death during development. *J. Immunol. Methods*, **265**, 3–20.

Zeiss, C. J. (2003). The apoptosis–necrosis continuum: insights from genetically altered mice. *Vet. Pathol.*, **40**, 481–95.

Zeuner, A., Eramo, A., Peschle, C., and De Maria, R. (1999). Caspase activation without death. *Cell Death Differ.*, **6**, 1075–80.

Zhang, J. Y. (2002). Apoptosis-based anticancer drugs. *Nat. Rev. Drug Discov.*, **1**, 101–2.

Zhang, Q., Siebert, R., Yan, M. *et al.* (1999). Inactivating mutations and overexpression of BCL10, a caspase recruitment domain-containing gene, in MALT lymphoma with t(1;14)(p22;q32). *Nat. Genet.*, **22**, 63–8.

Zhao, M., Beauregard, D. A., Loizou, L., Davletov, B., and Brindle, K. M. (2001). Non-invasive detection of apoptosis using magnetic resonance imaging and a targeted contrast agent. *Nat. Med.*, **7**, 1241–4.

Zheng, T. S., Hunot, S., Kuida, K. *et al.* (2000). Deficiency in caspase-9 or caspase-3 induces compensatory caspase activation. *Nat. Med.*, **6**, 1241–7.

Zhivotovsky, B. and Orrenius, S. (2003). Defects in the apoptotic machinery of cancer cells: role in drug resistance. *Semin. Cancer Biol.*, **13**, 125–34.

Zhu, S., Stavrovskaya, I. G., Drozda, M. *et al.* (2002). Minocycline inhibits cytochrome c release and delays progression of amyotrophic lateral sclerosis in mice. *Nature*, **417**, 74–8.

Zornig, M., Hueber, A., Baum, W., and Evan, G. (2001). Apoptosis regulators and their role in tumorigenesis. *Biochim. Biophys. Acta.*, **1551**, F1–37.

Developmental apoptosis in health and disease

Hyung Don Ryoo and Hermann Steller

Howard Hughes Medical Institute
The Rockefeller University
New York, USA

2.1 Introduction

Animal development starts from a single fertilized cell which subsequently multiplies and grows into an elaborate adult body. Amid such massive growth, negative regulation of growth plays an important role in orchestrating the development of intricate body structures. Programmed cell death is an integral part of such negative regulation, important in sculpting body parts, eliminating unnecessary cells, controlling organ size and cell number, proper wiring of the nervous system, and development of the immune system.

Historically, developmental biology has played a pivotal role in forming the basic concepts of programmed cell death. Soon after cells were discovered more than a century and half ago, Vogt observed that cell death occurs in a predictable pattern during amphibian metamorphosis, such as the disappearance of tadpole tails through cell death (Vogt, 1842). In 1951, Glücksmann had proposed a radical hypothesis that cell death might be an active part of development (Glücksmann, 1951). He suggested three distinct functions of apoptosis: (1) deleting unneeded structures (phylogenetic cell death); (2) controlling cell number (histogenetic cell death); and (3) sculpting structures (morphogenetic cell death). Later, the term "programmed cell death" was coined to describe a predicted pattern of cell death associated with insect metamorphosis (Lockshin and Williams, 1964). The evidence that new protein synthesis is required to trigger programmed cell death came in an experiment during which it was shown that treatment with cycloheximide, a protein synthesis inhibitor, blocked tadpole tail cell death during frog metamorphosis (Tata, 1966).

Kerr, Wyllie, and Currie coined the term "apoptosis" to describe normal cell death that occurs with distinctive and reproducible morphologic changes (Kerr *et al.*,

Apoptosis in Health and Disease: Clinical and Therapeutic Aspects, ed. Martin Holcik, Alex E. MacKenzie,
Robert G. Korneluk, and Eric C. LaCasse. Published by Cambridge University Press. © Cambridge University Press 2004.

1972). They distinguished such normal cell death from accidental cell death that occurs at the center of acute lesions such as trauma and ischemia. The latter case, a process termed "necrosis," is associated with cell swelling and rupture that eventually lead to an inflammatory response. By contrast, normal cell death is associated with cell shrinkage and condensation, and the organelles and plasma membrane retain their integrity until they are cleanly engulfed by phagocytes. This prevents any undesirable leakage of cellular contents and inflammatory response.

Apoptosis research attracted a wider range of scientists, however, only after the detailed documentation by Horvitz and colleagues that healthy cells die during *Caenorhabditis elegans* development with the typical morphologic changes attributed to apoptosis (Sulston and Horvitz, 1977; Horvitz *et al.*, 1982), and the subsequent identification of genes that are dedicated to triggering apoptosis during development (Ellis and Horvitz, 1986; Hengartner *et al.*, 1992; Yuan *et al.*, 1993). Their discovery that *C. elegans* apoptotic genes have human homologs with similar function opened the way for molecular studies of mammalian apoptosis. With the advent of the post-genome era, new apoptotic genes have been discovered based on sequence homology, many of which still remain to be characterized. It is now clear that the central apoptotic genes are well conserved through evolution, with the complexity of apoptotic pathways increasing from *C. elegans*, to *Drosophila*, to vertebrates (Aravind *et al.*, 2001). Each experimental organism has contributed overlapping and distinct information to our understanding of apoptosis. In this chapter, the conserved apoptotic genes in several model organisms and various examples of apoptosis that they regulate during development will be reviewed.

2.2 Apoptosis in *C. elegans*

C. elegans is a popular genetic model system with unique features of development. It has contributed significantly to the understanding of how cell fate decisions are made during development. In the 1970s, Sulston and Horvitz used Nomarski optics to visualize live cells of *C. elegans* and trace all cell lineages during *C. elegans* development (Sulston and Horvitz, 1977). By doing so, they found that, of the 1090 cells that are generated during the development of the adult hermaphrodite soma, 131 cells undergo apoptosis. The identities of the 131 cells were nearly invariant from animal to animal, indicating that these cells die as part of a predetermined developmental program. Such a notion led to a search for genes that are dedicated to regulating programmed cell death. Through systematic mutant screens, Horvitz and colleagues elucidated a pathway of genes that is devoted to triggering apoptosis (reviewed in Horvitz *et al.*, 1994). Surprisingly, blocking cell death in *C. elegans* does

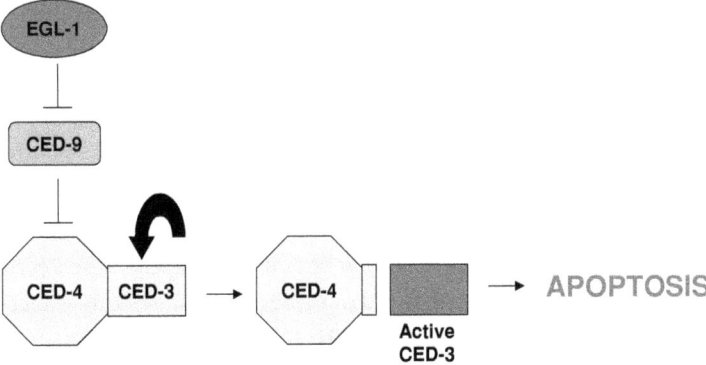

Figure 2.1   Apoptotic pathway in *C. elegans*. EGL-1 is a BH3- only domain protein. EGL-1 binds to CED-9, a homolog of human BCL-2. CED-9 binds CED-4, a homolog of human APAF-1. When CED-9 is inhibited by EGL-1, CED-4 is free to activate CED-3, the *C. elegans* caspase. CED-3 undergoes auto-cleavage for full enzymatic activation.

not interfere with overall normal development and leads to viable and fertile adults. This greatly facilitates the isolation and characterization of cell death defective mutants, but it also indicates a more limited and specialized role of apoptosis for the development of this organism.

Here, the genes regulating caspase activation, engulfment of dead corpses, and some examples of apoptosis during animal development will be reviewed.

2.2.1   The apoptotic pathway in *C. elegans*

The current understanding of apoptosis in *C. elegans* is delineated in Figure 2.1. In the soma, various developmental cues induce the expression of *egl-1*, a BH3-only domain gene. EGL-1 binds to and inhibits CED-9, a crucial anti-apoptotic protein. CED-9, which resides on the mitochondrial outer membrane, inhibits apoptosis by sequestering CED-4 to the mitochondria. When CED-9 is inhibited by EGL-1, CED-4 is free to bind and activate CED-3. CED-3 is a member of a distinct protease family with an essential cysteine in its active site. This family of proteases, termed caspases, cut after aspartate residues in substrates. The *C. elegans* genome encodes four putative caspases (Aravind *et al.*, 2001), but CED-3 is the only *C. elegans* caspase shown to control apoptosis. In vitro studies suggest that CED-3, like all other known caspases, is initially synthesized as a zymogen. Upon binding to CED-4, CED-3 is activated by auto-cleavage. Activation of CED-3 by auto-proteolysis is the triggering point of apoptosis, as it leads to irreversible cleavage of CED-3 and its downstream substrates. This leads to the distinct morphologic changes associated with apoptosis. Blocking CED-3 with inhibitors prevents apoptosis in most cells. In hermaphrodites, loss-of-function *ced-3*, *ced-4*, or *egl-1* lead to the survival of all

131 cells originally fated to die. One exception is the death of the "linker cell" which occurs specifically in males and is partially independent of *ced-3*.

2.2.2 Engulfment of corpses in *C. elegans*

Apoptotic cells are efficiently recognized by phagocytes and engulfed, therefore preventing any leakage of cellular content and potential inflammation response. Apoptotic cells in developing tissues are almost always found inside other cells, suggesting that phagocytosis occurs at the onset of apoptosis. Recent studies in *C. elegans* have significantly enhanced our understanding of this aspect of apoptosis regulation (reviewed in Hengartner, 2001; Conradt, 2002).

C. elegans cell corpses can be visualized with Nomarski microscopy and distinguished from living cells. Cell corpses are engulfed by neighboring cells at the onset of apoptosis and disappear quickly in wild-type worms, usually within an hour. Unlike *Drosophila* or mammals, *C. elegans* does not have dedicated phagocytes that engulf dying cells. Instead, cell corpses are engulfed by neighboring cells. The dying cells are, therefore, likely to expose an "eat me" signal, which is activated by *ced-3*. The nature of this "eat me" signal remains unknown, but candidates include membrane lipids that are asymmetrically distributed in living cells and exposed to the outer surface only in apoptotic cells.

C. elegans genetics has revealed cell death mutants in which cell corpses persist. Analysis of these mutants has thus far revealed two partially redundant pathways, defined by *ced-1*, *ced-6*, and *ced-7* on the one hand and *ced-2*, *ced-5*, *ced-10*, and *ced-12* on the other. *ced-1*, *ced-6*, and *ced-7* encode a scavenger receptor-like protein, a phospho-tyrosine binding (PTB)-domain containing adaptor protein, and an ATP binding cassette (ABC) transporter, respectively. These genes are required in the engulfing cells to recognize an as yet unidentified corpse signal. *ced-7* also functions in apoptotic cells, where it is possibly involved in the generation of such a recognition signal. *ced-2*, *ced-5*, *ced-10*, and *ced-12*, which define the second pathway, encode homologs of mammalian CrkII, Dock 180, Rac, and ELMO, respectively. These genes are involved in mobilizing the actin cytoskeleton of the phagocytic cells to engulf the cell corpses. Mutants defective of cell corpse engulfment do not cause lethality, at least under laboratory conditions. Nevertheless, *C. elegans* has maintained throughout evolution an efficient mechanism to ensure the removal of cell corpses.

Interestingly, recent evidence indicates that phagocytosis is not merely a means to clean up dead corpses but can also affect the decision to die itself. For example, the male linker cell death can be prevented by inactivating one of the two engulfment pathways or by ablating neighboring cells (Hedgecock *et al.*, 1983). Similarly, blocking engulfment prevents a few of the apoptotic cell death events that normally occur after embryogenesis (Reddien *et al.*, 2001). Moreover, recent work has revealed a positive feedback loop between the engulfment machinery in neighboring cells and the cell death machinery in apoptotic cells; mutations which block engulfment strongly enhance the partial loss-of-function mutations of pro-apoptotic genes (Hoeppner *et al.*, 2001; Reddien *et al.*, 2001). This demonstrates that phagocytosis does not occur only after cell death, but, in some cells, influences the decision to activate *ced-3*. The molecular mechanism behind this is still unclear.

2.2.3   Role of apoptosis in the *C. elegans* soma

A well-characterized example of apoptosis in *C. elegans* includes the death of sexually dimorphic neurons such as the two hermaphrodite-specific neurons (HSNs). Animals differentiating as males instead of hermaphrodites are programmed to eliminate these HSNs through apoptosis. Conversely, males have four cephalic companion neurons (CEMs) that are eliminated in hermaphrodites through apoptosis. Whether the survival of these sexually dimorphic neurons causes behavioral defects is unclear.

As mentioned earlier, unlike in more complex organisms, apoptosis in *C. elegans* is not essential for the organism's viability. Apoptosis-defective mutants, such as *ced-3*, *ced-4*, or gain-of-function *ced-9*, survive to adulthood with extra cells mostly in their nervous system (Ellis and Horutz, 1986; Hengartner *et al.*, 1992). Despite the absence of developmental apoptosis, adults exhibit largely normal behavioral responses; these include coordinated locomotion, moving in the opposite direction after touch, and egg laying in response to serotonin (Ellis and Horutz, 1986). This indicates that the primary role of apoptosis is to eliminate unnecessary cells.

2.2.4   Apoptosis in the *C. elegans* germline

During *C. elegans* somatic development, cell lineage and cell number is predetermined at the onset of embryonic development. This is not the case with *C. elegans* germline development, where more than one-half of potential oocytes are estimated to die by apoptosis and the number of surviving germ cells is not fixed.

In *C. elegans* hermaphrodites, germline development begins with syncitial cells undergoing mitosis. The syncitial nuclei gradually migrate toward the uterus and enter meiosis. During this meiotic phase, apoptosis is triggered in many developing nuclei. The surviving nuclei undergo cellularization and enter the uterus. Similar to apoptosis in the soma, germ cell apoptosis requires *ced-3* and *ced-4*. However, unlike apoptosis in the soma, germ cell apoptosis is not dependent on *egl-1* (Gumienny *et al.*, 1999).

The purpose of such massive apoptosis in the germline is unclear, but it has been suggested that those cells undergoing germline apoptosis are nurse cells (Gumienny *et al.*, 1999). In mammals and *Drosophila*, nurse cells provide proteins and RNA to the cytoplasm of oocytes, and these materials are used to fuel early embryonic development. Although *C. elegans* nurse cells cannot be morphologically distinguished from oocytes, it is likely that nurse cells provide RNA and protein to the common cytoplasm of syncitial cells. And once their role has been accomplished, they are eliminated by apoptosis.

The germline is also used to study DNA damage-induced apoptosis in *C. elegans*. Whereas the development of the soma is fast and less likely to accumulate extensive damages, germline DNA damage is passed on to future generations and the germline needs to be vigorously protected from accumulating mutations and damages. In the case of DNA damage, apoptotic machinery operates to eliminate these cells in the *C. elegans* germline (Gartner *et al.*, 2000). Here, radiation induces damage response genes such as the p53 homolog *cep-1* and, subsequently, apoptosis (Derry *et al.*, 2001; Schumacher *et al.*, 2001).

2.2.5 Conclusion

The importance of *C. elegans* apoptosis research was fully realized when the core components of the apoptotic program were shown to be conserved in all multicellular animals. Many aspects of apoptosis are still poorly understood, and *C. elegans* genetics continues to play an important role in addressing a wide range of questions, such as corpse engulfment and its communication with dying cells, the downstream or parallel components of *ced-3* that lead to morphologic changes associated with apoptosis, caspase-independent programmed cell death, and specification of death as a developmental fate.

2.3   Apoptosis in *Drosophila*

Drosophila is another popular organism with facile genetics that is widely used to study animal development. Its developmental process is far more complex compared with that of *C. elegans*, with cell lineage and number subject to change upon the change of surroundings. In addition, if the role of apoptosis during *C. elegans* development is mainly restricted to eliminating unnecessary cells, apoptosis in *Drosophila* plays a broader role similar to that in mammals. This includes eliminating cells for morphologic changes, control of cell number and body size, and eliminating damaged cells or those with confused developmental fate. Due to its broader role, apoptosis is essential for the viability of embryos and developmental progression, and, to coordinate growth, patterning, and apoptosis of complex tissues, secreted growth factors and morphogens play important roles in *Drosophila* apoptosis. Reflecting an increased complexity in biology, *Drosophila* apoptotic genes show an increased molecular complexity, compared with that of *C. elegans* (reviewed in Aravind *et al.*, 2001).

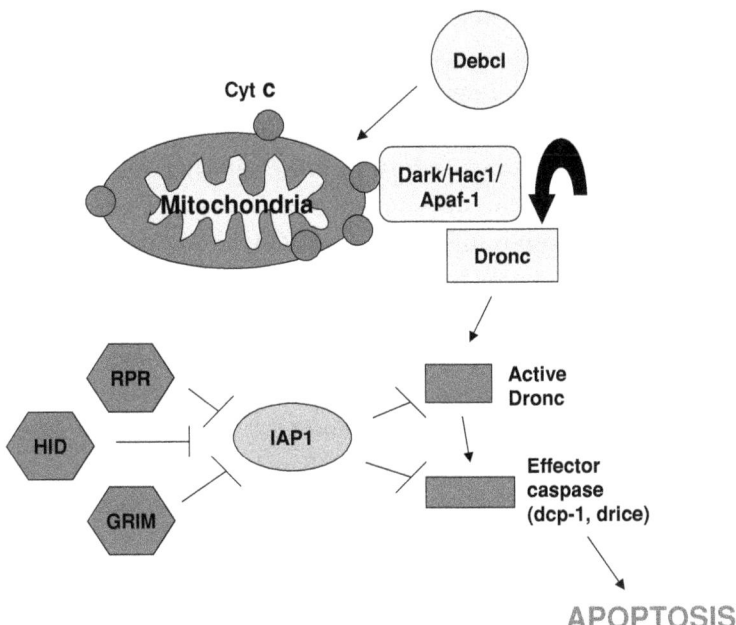

Figure 2.2   Apoptotic pathway in *Drosophila* (gas and brake model). This diagram is based on genetic and biochemical evidence from *Drosophila* as well as extrapolating from the biochemical evidence from humans and *C. elegans*. Similar to *C. elegans*, BCL-2 family member proteins regulate apoptosis. One pro-apoptotic member, debcl, is believed to promote cytochrome c (shaded ball) exposure from the mitochondria. Cytochrome c is believed to bind to *Drosophila* Dark/hac-1/Apaf-1, a homolog of CED-4 and human APAF-1. Dark/hac-1/Apaf-1 is believed to activate Dronc, similar to the human upstream caspase, caspase-9. Dronc activates effector caspases such as DCP-1 and DRICE. Parallel to this mitochondrial pathway (gas), *Drosophila* IAP1 inhibits caspases (brake). In dying cells, IAP1 is blocked by REAPER, HID, and GRIM, which bind to IAP1 and promote their degradation. *reaper, hid,* and *grim* expression is induced in cells doomed to die.

2.3.1 Apoptotic pathways in *Drosophila*

The core cell death mechanism in *Drosophila* is delineated in Figure 2.2. Central to apoptosis in *Drosophila* are caspases, the executioners of apoptosis homologous to *ced-3* of *C. elegans*. The *Drosophila* genome encodes seven putative caspases, including the upstream caspase *dronc* and the effector caspases *drice* and *dcp1* (Kumar and Doumanis, 2000). Extrapolating from studies in the human and *C. elegans*, effector caspases are regulated by upstream caspases. Upstream caspases are activated upon binding to an activator, termed *Dark/hac-1/apaf-1*, a protein which is similar to *ced-4* in *C. elegans* (Zhou *et al.*, 1999; reviewed in White, 2000). *Dark/hac-1/apaf-1* itself is likely to be regulated by mitochondrial proteins similar to *ced-9*. In fact, a pro-apoptotic member of the *ced-9* family gene named *debcl/dborg/drob* has been reported in *Drosophila* (Brachmann *et al.*, 2000; Colussi *et al.*, 2000; Igaki *et al.*, 2000). Although an anti-apoptotic member of *ced-9* family is yet to be identified, its human homolog Bcl-2 can strongly suppress cell death in *Drosophila* (Gaumer *et al.*, 2000), indicating that a functional homolog of Bcl-2 may exist.

In mammals, cytochrome c release from the mitochondria is a central event during apoptosis, whereas this does not occur in *C. elegans*. The role of cytochrome c in *Drosophila* apoptosis is still controversial. On the one hand, although the release of cytochrome c has not yet been demonstrated in fly cells, cytochrome c immunoreactivity is enhanced in cells undergoing apoptosis. In addition, *Dark/hac-1/apaf-1* has a cytochrome c-binding domain conserved with human Apaf-1 (Varkey *et al.*, 1999; White, 2000). On the other hand, decreasing cytochrome c levels by RNA interference (RNAi) reportedly does not affect stress-induced apoptosis in SL2 cell cultures (Zimmermann *et al.*, 2002). Based on this, it has been suggested that cytochrome c is not involved in apoptosis. However, it remains possible that small amounts of cytochrome c remain upon RNAi treatment and are sufficient to trigger apoptosis. Furthermore, SL2 tissue culture cells may not be representative of many non-immortalized *Drosophila* cells in situ. Despite these unresolved issues, the "intrinsic cell death pathway" appears to be conserved between *C. elegans*, *Drosophila*, and humans.

Drosophila also has apoptotic regulators that are not found in *C. elegans*. At the center of a second pathway is *Drosophila* inhibitor of apoptosis protein 1 (DIAP1) (Hay *et al.*, 1995). IAPs are a family of baculovirus IAP repeat (BIR) domain-containing proteins found in various animals and viruses. Some members, including DIAP1, inhibit apoptosis by binding to caspases and promoting their ubiquitination (Wilson *et al.*, 2002). In cells that are doomed to die, DIAP1 has to be inhibited for full activation of caspases. This is done by three pro-apoptotic genes: *reaper*, *hid*, and *grim* (White *et al.*, 1994; Grether *et al.*, 1995; Chen *et al.*, 1996). These three genes, collectively termed reaper hid grim (RHG) genes, are expressed in cells that are doomed to die and inhibit DIAP1 by binding to their BIR domains. The binding of RHG proteins to DIAP1 has at least two consequences. First, caspases are "liberated" from DIAP1 and hence no longer inhibited (Wang *et al.*, 1999; Goyal *et al.*, 2000) and, second, binding of Reaper and Grim stimulates DIAP1 auto-ubiquitination and destruction, thereby removing the "brake on death" and ensuring the irreversibility of the apoptotic program (Ryoo *et al.*, 2002; Yoo *et al.*, 2002). The decision to live or die in *Drosophila* almost always involves RHG genes and IAP1; no apoptosis is observed in embryos deficient of RHG genes (White *et al.*, 1994). Conversely, all cells undergo apoptosis in embryos deficient of IAP1 genes (Wang *et al.*, 1999; Goyal *et al.*, 2000).

A third pathway that triggers apoptosis is the tumor necrosis factor receptor (TNFR) pathway. Whereas TNFR-induced apoptosis is well characterized in humans, the *Drosophila* pathway is only beginning to be understood. In mammals, TNFR activation often triggers apoptosis independent of mitochondria, and thereby is defined as an independent cell death pathway termed "the extrinsic pathway." Human TNFRs contain multiple repeats of an extracellular cysteine-rich domain and an intracellular death domain. The *Drosophila* genome encodes only one TNFR family member. It has a single copy of the cysteine-rich domain and is not considered a TNFR ortholog (Aravind *et al.*, 2001). A possible ligand for this receptor is EIGER, the only TNF family protein encoded in the *Drosophila* genome (Moreno *et al.*, 2002b). When over-expressed, *eiger* activates the Jun N-terminal kinase (JNK), which in turn leads to the transcriptional activation of *hid*. Therefore, in contrast to the findings in humans, the TNFR signaling pathway does not appear to constitute an independent pathway in *Drosophila*. Other *Drosophila* homologs of the extrinsic pathway are encoded in the genome, such as FADD and caspase-8 (*dredd*). However, they have been characterized mostly as mediators of the *Drosophila* innate immune response (Leulier *et al.*, 2002) and their role in apoptosis remains unclear.

Taken together, the *Drosophila* cell death program is a combination of two antagonistic and parallel pathways that regulate caspase activation. On one side is *Dark1/hac-1/apaf-1*, which acts to activate caspases. On the other side is the negative regulation of caspases by DIAP1. The strong phenotype of RHG genes and *diap1* also indicates that the caspase activation pathway by *Dark/hac-1/apaf-1* might be constitutively active in *Drosophila* embryos. The role of apoptosis in various aspects of *Drosophila* development will be reviewed below.

2.3.2 Engulfment of corpses in *Drosophila*

After apoptosis is triggered, apoptotic cells in *Drosophila* are engulfed by professional macrophages, hemocytes that become phagocytic in the presence of dying cells (Abrams *et al.*, 1993; Tepass *et al.*, 1994). In mutants that lack hemocytes and phagocytes, the pattern of apoptotic cells remains largely normal. Therefore, engulfment is not essential for induction of apoptosis in most cells. However, dying cells trigger cells to differentiate into hemocytes. In embryos lacking *reaper*, *hid*, and *grim* and therefore exhibiting no apoptotic cells, hemocytes fail to differentiate. Conversely, induction of ectopic apoptosis results in an increased number of phagocytes. Therefore, the differentiation of phagocytes in *Drosophila* is dependent on signals from dying cells. To date, only one gene in *Drosophila* has been characterized as being involved in phagocytic activity. *croquemort* (*crq*) encodes a *Drosophila* scavenger receptor related to human CD36 receptor, believed to detect an "eat me" signal from apoptotic cells. Embryos deficient in CRQ have phagocytes that cannot engulf apoptotic corpses (Franc *et al.*, 1999). CRQ expression appears to be induced by the presence of apoptotic cells, as mutants lacking any apoptosis have very low CRQ expression. Similarly, the human CD36 scavenger receptor expression is induced by its ligand, low-density lipoprotein (LDL). The true ligand of CRQ remains unknown.

2.3.3 Apoptosis in sculpting morphology

Perhaps reflecting a more complex pathway of genes, apoptosis in *Drosophila* plays a broader role during the animal's development, compared with that of *C. elegans*. Similar to *C. elegans*, many cells in the *Drosophila* embryonic nervous system die during development. But unlike *C. elegans*, such cell death is essential for sculpting morphologic structures and developmental progression. An important example is head involution, a process in which the dorsal fold migrates anteriorly to cause dramatic morphologic changes in the head (Abbott and Lengyel, 1991). One of the mutations that block this process is *head involution defective* (*hid*), an RHG gene that induces apoptosis. *hid* is expressed, among other places, in the embryonic head cells to eliminate cells and facilitate such morphologic changes (Grether *et al.*, 1995).

Another example is the Hox gene, *Deformed* (*Dfd*). Hox genes are evolutionarily conserved transcription factors that determine segment identity in *Drosophila*. *Dfd* controls the mandibular and maxillary segment identity in the embryonic head. As its name implies, *Dfd* mutants exhibit morphologic defects in head structures such as the mouth hook and fail to form a cleft that divides the mandibular and maxillary segments. The formation of such segmental boundaries is now thought to be a result of apoptosis induced by *rpr* expression (Lohmann *et al.*, 2002). *Dfd* directly induces the expression of *rpr*, thereby killing off the cells, and sculpting the segment boundary. Similarly, segment boundaries in the embryonic abdomen are sculpted by another Hox gene, *Abd-B*, which also induces *rpr* expression to kill off cells.

2.3.4 Eliminating unnecessary cells

Eliminating unnecessary cells during development is another major purpose of apoptosis defined by Glücksmann. Such an example is found during metamorphosis, and was one of the first examples of programmed cell death reported (Vogt, 1842). In a wide variety of animals that undergo metamorphosis, larval tissues that are no longer necessary are eliminated by apoptosis while the adult body parts emerge. Examples include the loss of tadpole tails in adult frogs, or the death of larval tissues in insect pupae, which makes room for adult tissues to grow. Although cell death during metamorphosis was the first example of cell death to be described, its underlying mechanism is only beginning to be understood. During its late larval stage, *Drosophila* secretes the hormone ecdysone to signal a global change that is to accompany metamorphosis (reviewed in Baehrecke, 2000). The larval cells respond to ecdysone by activating a set of early transcription factors, including E74A and E93, which in turn coordinate the induction of *rpr* and *hid* transcription. The observation here is akin to *C. elegans* cell death, in which apoptosis is induced in cells that are no longer necessary and induced by transcription factors that contribute to their cell fate. But unlike *C. elegans*, metamorphic cell death is signaled by secreted hormones that act globally rather than specifying cell death independently in each individual cell.

2.3.5 Matching cell numbers by apoptosis

Drosophila cell number is heavily regulated by secreted growth factors and signaling molecules, something that is commonly seen in mammals but not in *C. elegans*. Rather than having a fixed number of cells by a precise developmental program, *Drosophila* and more complex animals often regulate cell numbers by limiting the amount of survival factors in a given tissue or group of cells. A primary example is the role of epidermal growth factor (EGF) signaling in controlling embryonic glial cell numbers (Zhou *et al.*, 1995; Bergmann *et al.*, 1998, 2002). Embryonic mid-line glia development is initiated by the generation of 12 glial cells per segment. Over the next few hours, nine cells on average fail to make proper contacts with the neighboring neurons and undergo apoptosis. Glial cells have a default fate to die because they express a combination of *reaper* and *hid*. All cells that express *reaper* (six on average) die invariantly, but some cells that express *hid* survive. In surviving cells, HID is phosphorylated by mitogen-activated protein (MAP) kinase and its pro-apoptotic activity is neutralized (Bergmann *et al.*, 1998). These cells have MAP kinase activated because the adjacent neurons are secreting *spitz*, a *Drosophila* EGF molecule. As a result, only those glial cells that make contacts with a nearby neuron can overcome apoptosis and survive. Therefore, instead of determining individual cell fate, secretion of survival signal controls the proper cell number and cell–cell interaction at the same time.

2.3.6 Coordinating growth, patterning, and apoptosis

A more complex problem in development and apoptosis is how cell death is regulated in coordination with complex pattern formation. In *Drosophila*, cell lineage, size control, and patterning are regulated at the level of compartments. For example, developing wing tissue can be subdivided into an anterior (A) compartment and a posterior (P) compartment (Garcia-Bellido *et al.*, 1973). Regarding compartment growth, there have been puzzling observations that altering cellular growth rate does not alter compartment size or patterning. This is because, when cells are genetically manipulated to grow faster, there is compensating cell death so that the compartment size is normal (e.g. Johnston *et al.*, 1999). Moreover, increased cell death is carried out in such an orderly way that patterning is not disrupted and the resulting compartment appears to be indistinguishable from wild-type. Related to this is a phenomenon called cell competition, used to describe the observation that slow growing cells are eliminated by apoptosis, which is induced by the faster growing neighbor cells. Again, although many cells disappear, the resulting wing is indistinguishable from wild-type in terms of size and patterning. Cell competition occurs only between cells in the same compartment (Simpson and Morata, 1981). This indicates that growth, patterning, and apoptosis are all coupled within a compartment, the underlying mechanism of which remains poorly understood.

Providing a possible link between pattern formation and control of cell death, a patterning molecule was recently shown to have a dual function as a survival factor (Moreno *et al.*, 2002a). In the *Drosophila* wing, *decapentaplegic* (*dpp*), a transforming growth factor (TGF)-β family signaling molecule, is secreted from an organizer along the A/P compartment boundary. *dpp* acts as a morphogen and induces different target genes depending on its distance from the A/P boundary. Upon ectopic expression in a clone of cells, it is able to induce an additional wing with all the correct patterning (Zecca *et al.*, 1995; Nellen *et al.*, 1996).

In clones of cells that are being eliminated by cell competition, *dpp* signaling is reduced prior to apoptosis (Moreno *et al.*, 2002b). Artificial attenuation of *dpp* signaling also induces apoptosis. This work provides two important cases of developmental apoptosis. First, as *dpp* has a dual role as a morphogen and a survival factor, it provides a framework for understanding how patterning and apoptosis can be coordinated. Second, it provides an example of apoptosis when a cell is receiving confusing cell fate signals. In the case of a developing wing, it has been subsequently observed that the cells destined to die due to low *dpp* signaling can be rescued by over-expression of a cell adhesion protein that is normally expressed in cells receiving very low *dpp* signal (Milan *et al.*, 2002). Based on this, one can hypothesize that cells know their position in a developing tissue by at least two independent means – in this case, the amount of *dpp* signaling and the identity of the cell adhesion molecule. It follows that upon receiving conflicting positional information, the cell would destroy itself to prevent developmental defects. Therefore, in addition to coordinating growth, patterning, and size control, apoptosis provides means to eliminate dangerous cells that receive conflicting developmental signals.

2.3.7 Conclusion

The biology of apoptosis in *Drosophila* is more complex than *C. elegans* and has made identification of certain apoptotic genes more difficult. For example, the identification of the anti-apoptotic Bcl-2 homologs has so far been elusive, although there is evidence indicating their presence. Another area that is incompletely understood is the engulfment of dead corpses, with the identification of only one gene, *crq*, so far.

However, *Drosophila* has proven to be a powerful system to study apoptosis in the context of developmental biology. It has many features of apoptosis similar to mammals, including the role of secreted factors. *Drosophila* research is also providing some clues as to how cells coordinate growth, patterning, and apoptotis, and how developmental confusion triggers apoptosis. Some of these experiments are only practical in *Drosophila* and provide great promise for future progress.

2.4   Apoptosis during mammalian development

In the past decade, a large number of laboratories have discovered and characterized apoptotic genes in mammals. It is now established that failure to trigger apoptosis contributes significantly to cancer progression and auto-immune disease (see Chapters 3 and 6). Mammalian apoptosis studies have been greatly aided by human mutations found in cancer, leukemia, and other diseases attributed to abnormal apoptosis, and new candidate genes can be analyzed quickly in cultured cells. The availability of the human genome sequence in the recent years has dramatically increased the number of candidate genes that remain to be characterized, and targeted gene knockout techniques in mice have greatly aided studying their role during mammalian development.

There are a few genes implicated in caspase-independent cell death in mammals. Despite impressive progress, many fundamental questions remain unanswered regarding their role during development. For example, we do not understand the extent to which caspase-independent cell death plays a role during apoptosis. Also, unlike *C. elegans* and *Drosophila*, in which certain mutant lines block virtually all apoptosis during development, no comparable genetic conditions have been observed in mammals. Rather, targeted disruption of various apoptotic genes in mice gives a spectrum of phenotypes. Therefore, analyses of various mutant animals provide only a partial picture of how apoptotic pathways are regulated during mammalian embryo development.

While the role of individual apoptotic genes in development is less clear, the pattern of apoptosis during development has been thoroughly studied. Apoptotic assays reveal prominent apoptosis in the developing nervous system and the immune system, and during morphogenesis of various body structures. Similar to *Drosophila*, the apoptosis that occurs during development can be subdivided into four categories: (1) sculpting of body parts for morphogenesis; (2) elimination of cells which are no longer necessary; (3) elimination of dangerous cells; and

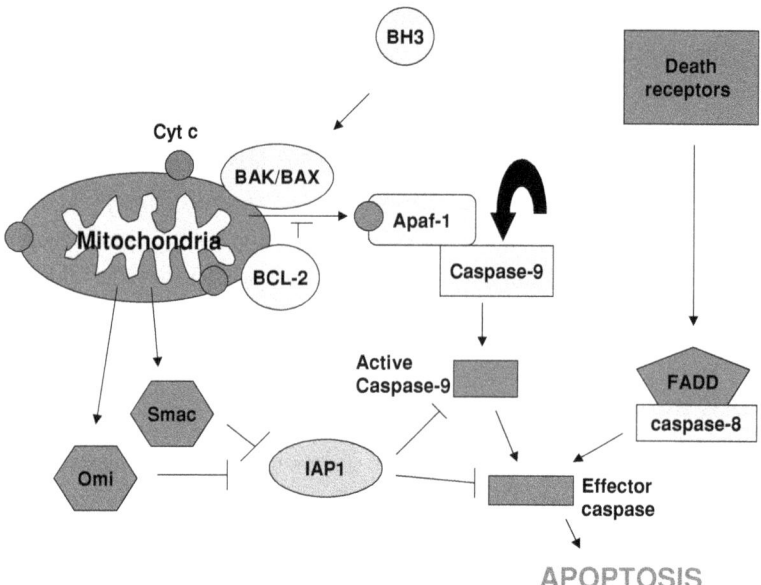

APOPTOSIS

Figure 2.3   *Apoptosis pathway in mammals.* Similar to *C. elegans*, BH3 domain proteins induce the release of mitochondrial cytochrome c by antagonizing anti-apoptotic BCL-2 family proteins, homologous to *C. elegans* CED-9. Pro-apoptotic members of the BCL-2 family, BAK and BAX, are necessary for this cytochrome c release. The released cytochrome c forms an apoptosome complex along with Apaf-1 (CED-4 homolog) and caspase-9, which leads to caspase-9 auto-cleavage and activation. Active caspase-9 cleaves effector caspases to trigger apoptosis. These caspases are inhibited by IAPs through physical interaction. In dying cells, IAPs are inactivated by the release of IAP antagonists, Smac and Omi, from the mitochondria. A mitochondria-independent apoptotic pathway is regulated by death receptor-containing proteins such as TNFR. They recruit the adaptor protein FADD and caspase-8 (which is subsequently auto-cleaved). Activated caspase-8 cleaves effector caspases to trigger apoptosis.

(4) control of body size or proper cell number. First, the genetic components of the apoptotic pathway will be reviewed and then a few selected examples of their role in development will be considered.

2.4.1   Apoptotic pathways in mammals

The embryonic development of mammals is far more complicated than that of invertebrates, such as *C. elegans* and *Drosophila*, but many basic aspects are conserved. The current understanding of the mammalian apoptotic pathway is depicted in Figure 2.3. Similar to *C. elegans*, the mitochondrial proteins play a primary role in the regulation of apoptosis (reviewed in Wang, 2001). Various pro- and anti-apoptotic Bcl-2 family of proteins similar to *C. elegans ced-9* regulate mitochondrial integrity. In cells that are doomed to die, the pro-apoptotic members are either transcriptionally activated or their sub-cellular localizations are altered, leading to the release of cytochrome c and other apoptotic molecules from the mitochondria. Essential in this mitochondrial release are two pro-apoptotic Bcl-2 family proteins Bax and Bak, without which no mitochondrial release of cytochrome c occurs (Wei *et al.*, 2001). Cytochrome c in the cytoplasm triggers the formation of apoptosome, a complex comprising of cytochrome c, Apaf-1, and caspase-9 (Acehan *et al.*, 2002). Whereas the pro-caspase-9 is enzymatically inactive, the binding of Apaf-1 allosterically stimulates caspase-9 activity. Once activated, caspase-9 undergoes auto-cleavage, and is converted to a constitutively active form. This, in turn, goes on to cleave effector caspases, such as caspase-3, thereby activating them. Additionally, there are multiple IAPs that inhibit caspase activity in living cells and their activity is blocked by functional analogs of *reaper* such as Smac or Omi/HtrA2 (Goyal, 2001; Suzuki *et al.*, 2001). Smac and Omi/HtrA2 are also mitochondrial proteins; therefore, the IAP inhibition pathway may relate to the mitochondrial pathway. Another pro-apoptotic protein residing in mitochondria is apoptosis-related protein in the TGF-β signaling pathway (ARTS) (Larisch *et al.*, 2000). ARTS promotes caspase activation and translocates from mitochondria to the nucleus during apoptosis, but its mechanism of action has not been determined. Although the mitochondrial pathway is extensively regulated by intercellular signaling, this pathway is often referred to as the "intrinsic pathway."

A pathway independent of the "intrinsic pathway" is the "death receptor pathway" or the "extrinsic pathway" (reviewed in Ashkenazi and Dixit, 1999). Here, secreted ligands such as Fas ligand or TRAIL activate death receptors directly by inducing trimerization of the receptors. The trimerized receptors recruit the adaptor protein FADD, which in turn recruits caspase-8. Caspase-8 is activated by auto-cleavage and goes on to activate effector caspases or other substrates. Although activated caspase-8 stimulates mitochondrial protein release by a Bcl-2 family protein, Bid, experimentally blocking mitochondrial protein release does not block apoptosis in some cell types. Caspase-8 is able to activate directly effector caspases, such as caspase-3, -6, and -7.

Finally, although caspase-independent cell death appears to have only a minor role during *C. elegans* or *Drosophila* development, its role in mammals may be broader. A prime example is the "apoptosis-inducing factor" or AIF (Susin *et al.*, 1999), a mitochondrial protein that is released along with cytochrome c upon death stimuli. It translocates to the nucleus and triggers chromatin condensation and phosphatidylserine exposure on the cell surface. Another example is endonuclease G, a Dnase that resides in the mitochondria of living cells, but translocates to the nucleus in dying cells to fragment the nucleosomal DNA during apoptosis (Li *et al.*, 2001). Since these caspase-independent cell death proteins are, at the same time, mitochondrial proteins, this pathway is termed mitochondrial dysgenesis-induced cell death.

2.4.2 Engulfment of corpses in mammals

In mammals, cellular content leaking from dying cells can cause inflammation. Upon microbial infection, macrophages secrete molecules that are protective, but which also potentially may cause an injurious inflammatory response. However, macrophages are also able to recognize the "eat me" signals from apoptotic cells and can rapidly eliminate them through phagocytosis, thereby preventing inflammation.

The nature of the "eat me" signals is somewhat better characterized in humans, and is complemented by the information from other genetic organisms. One of the best characterized is phosphatidylserine (PtdSer), a phospholipid that is normally restricted to the inner-membrane leaflet of living cells. Membrane asymmetry is lost early on in apoptotic cells, and they exhibit PtdSer in the outer-membrane leaflet. ABC1 transporter promotes Ca^{2+}-induced PtdSer exposure to the outer membrane, and might be required in apoptotic cells (Hamon *et al.*, 2000). Interestingly, ABC1 is a homolog of the *C. elegans* protein CED-7, which is required to generate an "eat me" signal in apoptotic cells of *C. elegans*. Mammalian ABC1 and *C. elegans* CED-7 are required also in the phagocytes to detect the "eat me" signal. Reflecting this, PtdSer has been shown to appear at the outer-membrane leaflet, not only in the apoptotic cells but also in the phagocytic cells (Marguet *et al.*, 1999).

In parallel to the PtdSer, other "eat me" signals are implicated in the engulfment of corpses, including surface sugars, bridging molecules such as C1q and other components of the complement system, oxidized LDLs, and ICAM-3.

Macrophages are believed to detect PtdSer through the PtdSer receptor (PSR) (Fadok *et al.*, 2000), MFG-E8 (Scott *et al.*, 2001), and some scavenger receptors such as CD36 and SR-BI, which directly bind to PtdSer. CD36, a homolog of the *Drosophila* CRQ, interacts with a PTB-containing adaptor protein similar to *C. elegans* CED-6. MGF-E8 is a secreted glycoprotein with an RGD (Arg–Gly–Asp) motif that can be recognized by some members of the integrin family, including those found on the surface of macrophages. Integrins expressed in macrophages such as $\alpha_\nu\beta_5$, in turn, can direct cytoskeletal changes that would envelop apoptotic cells by recruiting the CrkII/DOCK180/Rac1 complex, consisting of the homologs of the *C. elegans* engulfment proteins CED-2, CED-5, and CED-10 (reviewed in Savill and Fadok, 2000). Taken together, the basic machinery of apoptotic corpse engulfment is conserved between species.

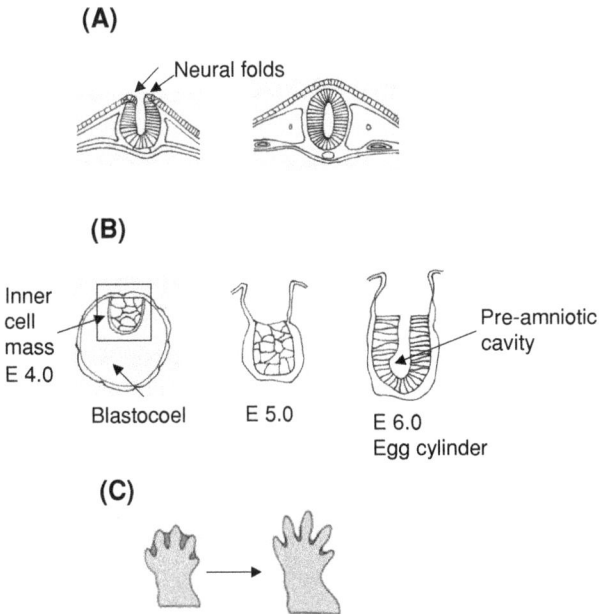

Figure 2.4   *Apoptosis in mammalian body morphogenesis*. Three examples of apoptosis during mammalian body sculpting are described. (A) Neural tube formation. Initially, neural plate invaginates to form a tube-like structure. The neural fold subsequently pinches off and fuses the other end to form the neural tube. Many apoptotic cells are detected in the neural fold during the pinching off process, and can be blocked by caspase inhibitors. (B) Pre-amniotic cavity formation. During early stages of mammalian development, inner cell mass (ICM) is formed to become the future embryo and yolk sac. ICM is soon hollowed by programmed cell death, which gives rise to a lumen and a two-layered egg cylinder. (C) Finger interdigit cell death. Fingers are formed by the carving out of interdigit cells through apoptosis. Interdigit cells are induced to die by TGF-β family proteins expressed in the interdigits, which trigger cytochrome c release and apoptosis. In ducks, which maintain interdigit webs, BMP inhibitors such as gremlin are thought to block TGF-β signaling pathway and inhibit apoptosis.

2.4.3 Body sculpting and morphogenesis

Cell death as a normal aspect of mammalian development was first postulated more than 50 years ago. Glücksmann identified several examples of developing organs in which cell death is correlated with morphogenesis (Glücksmann, 1951). For example, formation of tube structures in various organs can be subdivided into two different mechanisms: first, the rolling up of an established epithelial sheet, which occurs during neural tube and primitive gut development, and, second, programmed cell death of inner cells of a solid structure, which is termed canalization or cavitation. As we shall see, apoptosis is implicated to play a role for both mechanisms during mammalian development (Figure 2.4).

The primary example of the first mechanism is the neural tube formation. Neural tube development starts with the invagination of the neural plate, which generates "neural folds" at its folded edges. In a pinching off process, neural tube separates from the overlying epidermis and forms the tubular structure (Figure 2.4A). Assays for cell death markers demonstrate many apoptotic cells at the neural folds during the pinching off process. Administering anti-caspase drugs in neural tube blocks tube formation (Weil *et al.*, 1997).

An example for the second mechanism can be found during early mouse embryogenesis (Coucouvanis and Martin, 1995). At the 32-cell stage of mammalian embryogenesis, inner cell mass is generated, which is an aggregate of cells that is to become the embryo and its associated yolk sac, allantois, and amnion (Figure 2.4B). Initially, inner cell mass is a solid structure, but bone morphogenic proteins such as BMP-2 or BMP-4 are expressed and the pro-amniotic cavity begins to appear by programmed cell death, resulting in the formation of a hollow, two-layered egg cylinder (Figure 2.4B). Such programmed cell death shows morphologic features of apoptosis. Blocking BMP signaling with dominant–negative BMP receptors is able to block apoptosis and cavity formation (Coucouvanis and Martin, 1999).

Targeted gene knock-out studies to this date, however, have identified only one apoptotic gene essential for this pre-amniotic cavity formation – AIF, which is thought to mediate a caspase-independent cell death (Joza *et al.*, 2001). Targeted disruption of major apoptosis genes including Apaf-1, Bax and Bak double mutants, FADD, and various caspases all bypass this early stage without defects. Because it is thought that Bax and Bak are required to release mitochondrial proteins like AIF, the normal pre-amniotic cavity formation in Bak–Bax double mutant mice remains unexplained, revealing our insufficient understanding of caspase-independent cell death and pre-amniotic cavity formation. Other tube structures, such as the mouse vagina, are known to require apoptosis, as Bax and Bak double mutant mice show vaginas that are not hollowed by programmed cell death (Lindsten *et al.*, 2000).

Another example of body sculpting and apoptosis is finger interdigit cell death. Vertebrate fingers are formed during embryogenesis, not by an outgrowth of finger digits, but by the death of interdigit cells (Figure 2.4C). These interdigit cells express TGF-β member proteins including BMP molecules that are thought to induce the apoptotic program. Inhibition of this signaling pathway by expressing dominant–negative BMP receptor blocks interdigital cell death (Zou and Niswander, 1996). In ducks, which have surviving interdigital webs, it is thought that BMP antagonists, such as gremlin, are expressed in the interdigits and guarantee their survival (Merino *et al.*, 1999). However, which of the various BMP family proteins are responsible for induction of apoptosis in interdigital cells remains unclear. BMP-2, BMP-4, and BMP-7 are all expressed in the interdigital cells and, therefore, are potential candidates. Recently, TGF-β-2, TGF-β-3 double knock-out mice were reported to exhibit interdigit webs indicating the role of TGB-β in apoptosis induction (Dunker *et al.*, 2002). Mice deficient of key apoptotic proteins, such as Bak, Bax double knockouts, or Apaf-1 knock-outs, exhibit surviving interdigits (Cecconi *et al.*, 1998; Yoshida *et al.*, 1998; Lindsten *et al.*, 2000). Therefore, we can infer that TGF-β signaling eventually activates the mitochondrial release of cytochrome c.

2.4.4 Apoptosis in controlling cell numbers

As we have seen in *Drosophila*, limited amounts of survival factors can effectively control proper cell number and matching between neurons and glia. This mode of regulation, also termed the trophic theory, was first observed with neuronal cultures that required nerve growth factor (NGF) for their survival (reviewed in Levi-Montalcini, 1987). If developing neurons are given exogenous NGF, the normal death of sympathetic neurons is prevented. Conversely, if they are treated with anti-NGF antibody or their receptor functions are blocked, all of these neurons undergo apoptosis. Many trophic factors similar to NGF have been identified to date, which include brain-derived neurotrophic factor (BDNF), ciliary neurotrophic factor (CNTF), glial-derived neurotrophic factor (GDNF), neurotrophins-3 (NT-3) and -4/5 (NT-4/-5), and cardiotrophin-1 (CT-1) (reviewed in Yuan and Yankner, 2000). These trophic factors are expressed locally, and only neurons or glial cells that make proper connections will bind trophic factors to neurotrophin receptors. Neurotrophin receptors are receptor tyrosine kinases, TrkA, TrkB, and TrkC. Ligand binding induces their dimerization and auto-phosphorylation. Downstream of this signaling pathway lie the PI3K and Ras/MAPK pathways. Both PI3K and Ras/MAPK are implicated in general inhibition of cell death in other mammalian tissues and other species. An anti-apoptotic role of MAPK in *Drosophila* has already been reviewed in this chapter. PI3K and MAPK appear to inhibit apoptosis by blocking the mitochondrial pathway. This is evident in Bax −/− animals where sympathetic neurons survive in the absence of NGF-independent survival. Taken together, neurons have the apoptotic machinery in cells ready to operate at all times and, in the absence of such survival factors, death is the default pathway.

In the nervous system, this mechanism is very effectively used for proper wiring. The vast majority of neurons generated during early development undergo apoptosis later. Neurons that fail to make proper connections with their target tissues die, since the target tissue is often the source of survival signals. Therefore, it is common to get a neuronal hyperplasia phenotype in mice deficient in pro-apoptotic molecules.

2.4.5   Apoptosis in eliminating dangerous cells during development

The developing immune system uses extensive apoptosis to cull out self-reactive immune cells. During development, immune cells undergo recombination in their immunoglobin genes and T-cell receptor (TCR) genes to generate a diverse repertoire that would bind to most, if not all, potential antigens. What is initially generated to defend the host now becomes a potential threat to the organism as it can attack host cells. Therefore, an important step of the immune system development is to eliminate self-reactive T-cells or B-cells (reviewed in Rathmell and Thompson, 2002). In the case of TCRs, they are designed to recognize foreign antigens that are bound to major histocompatibility complex (MHC) molecules. In the absence of TCR/MHC interaction, such lymphocytes fail to differentiate. By contrast, if TCR/MHC interaction is too strong, such as may occur in auto-reactive species, these cells are deemed dangerous and are eliminated by apoptosis. The mechanism underlying such negative selection is still poorly understood, but several genes have been implicated; among them is the p53-related gene p73, E2F1, nuclear steroid hormone receptor Nur77, Mef2, and Bcl-2-related BH3 domain protein Bim. Lymphocytes deficient in, or those over-expressing dominant–negative forms of, these genes fail to undergo apoptosis after TCR/MHC ligation. Among these central cell death regulators, Bim is transcriptionally activated after TCR/MHC ligation, and deletion of auto-reactive species is defective in Bim $-/-$ mice (Bouillet *et al.*, 2002).

2.4.6   Apoptosis-like events during development

Apoptosis-like events are involved in the terminal differentiation of certain cell types. These include skin keratinocytes, lens epithelial cells, and mammalian red blood cells, which lose their nuclei and other sub-cellular organelles in the process of terminal differentiation. During this process, their chromatin undergoes the same symptomatic changes as those observed during apoptosis, including chromatin condensation and DNA cleavage into oligonucleosomes (e.g. Counis *et al.*, 1998). Strikingly, lens and red blood cells continue to live and metabolize, whereas differentiated keratinocytes die and form a layer of corpses on the surface of the skin. Caspases are also implicated in the negative regulation of red blood cell differentiation by cleaving GATA-1 (De Maria *et al.*, 1999). Whereas the withdrawal of erythropoietin blocks red blood cell differentiation, this program can be restored by caspase inhibitors.

Similarly, skeletal muscle differentiation exhibits some features reminiscent of apoptosis (Fernando *et al.*, 2002). These include actin fiber disassembly and re-organization, and the involvement of myosin light chain kinase, which is also required for membrane blebbing in apoptotic cells. Caspase-3-null mice exhibit dramatic reduction of myotube and myofiber formation. Mammalian sterile twenty-like kinase is a target of caspase-3 in these cells, and restoration of this kinase rescues myoblast differentiation.

2.4.7 Conclusions

A decade of intense gene discovery and the advent of the post-genomic era have revealed that the basic apoptotic machinery is largely conserved between *C. elegans*, *Drosophila*, and humans, with the complexity dramatically increasing in this order. Each experimental system has distinct advantages and has made unique contributions, but each organism also has its own inherent limitations. In the future, it is likely that cross-species studies that integrate approaches in different organisms will help overcome such limitations, thereby providing complementary means to address problems difficult to study in any given organism. Years of research have brought us to a point where the initial hypothesis by Glücksmann on the role of cell death in development has been largely confirmed.

However, there are many more basic questions that remain to be answered. For example, some cells are very resistant to apoptosis and, in such cells, caspases do not trigger apoptosis but, rather, differentiation. What makes these cells differ in their apoptotic response is poorly understood. Also, there is controversy regarding the role of caspase-independent cell death during development. Therefore, despite the great advances in isolating apoptotic genes in the past decade, much remains to be discovered regarding the role and regulation of apoptotic genes during animal development.

ACKNOWLEDGEMENTS

We thank James Darnell, Shai Shaham, Holger Kissel, Caesar Mendez, and Ann Tang for carefully reading the manuscript and Tae Hee Kim for discussion and comments regarding mammalian development. H. D. R. is a fellow of the Leukemia-Lymphoma Society. H. S. is an investigator at the Howard Hughes Medical Institute.

REFERENCES

Abbott, M. K. and Lengyel, J. A. (1991). Embryonic head involution and rotation of male terminalia require the *Drosophila* locus *head involution defective*. *Genetics*, **129**, 783–9.

Abrams, J. M., White, K., Fessler, L. I., and Steller, H. (1993). Programmed cell death during *Drosophila* embryogenesis. *Development*, **117**, 29–43.

Acehan, D., Jiang, X., Morgan, D. G., Heuser, J. E., Wang, X., and Akey, C. W. (2002). Three-dimensional structure of the apoptosome: implications for assembly, procaspase-9 binding and activation. *Mol. Cell*, **9**, 423–32.

Aravind, L., Dixit, V. M., and Koonin, E. V. (2001). Apoptotic molecular machinery: vastly increased complexity in vertebrates revealed by genome comparisons. *Science*, **291**, 1279–84.

Ashkenazi, A. and Dixit, V. M. (1999). Apoptosis control by death and decoy receptors. *Curr. Opin. Cell Biol.*, **11**, 255–60.

Baehrecke, E. H. (2000). Steroid regulation of programmed cell death during *Drosophila* development. *Cell Death Differ.*, **7**, 1057–62.

Bergmann, A., Agapite, J., McCall, K., and Steller, H. (1998). The *Drosophila* gene *hid* is a direct molecular target of Ras-dependent survival signaling. *Cell*, **95**, 331–41.

Bergmann, A., McCall, K., and Steller, H. (2002). Regulation of cell number by MAPK-dependent control of apoptosis: a mechanism for trophic survival signaling. *Dev. Cell*, **2**, 159–70.

Bouillet, P., Purton, J. F., Godfrey, D. I. *et al.* (2002). BH3-only Bcl-2 family member Bim is required for apoptosis of autoreactive thymocytes. *Nature*, **415**, 922–6.

Brachmann, C. B., Jassim, O. W., Wachsmuth, B. D., and Cagan, R. L. (2000). The *Drosophila* bcl-2 family member dBorg-1 functions in the apoptotic response to UV-irradiation. *Curr. Biol.*, **10**, 547–50.

Cecconi, F., Alvarez-Bolado, G., Meyer, B. I., Roth, K. A., and Gruss, P. (1998). Apaf1 (ced-4 homolog) regulates programmed cell death in mammalian development. *Cell*, **94**, 727–37.

Chen, P., Nordstrom, W., Gish, B., and Abrams, J. M. (1996). Grim, a novel cell death gene in *Drosophila*. *Genes Dev.*, **10**, 1773–82.

Colussi, P. A., Quinn, L. M., Huang, D. C. *et al.* (2000). Debcl, a proapoptotic Bcl-2 homologue, is a component of the *Drosophila melanogaster* cell death machinery. *J. Cell Biol.*, **148**, 625–7.

Conradt, B. (2002). With a little help from your friends: cells don't die alone. *Nat. Cell. Biol.*, **4**, E139–43.

Coucouvanis, E. and Martin, G. R. (1995). Signals for death and survival: a two-step mechanism for cavitation in the vertebrate embryo. *Cell*, **83**, 279–87.

(1999). BMP signaling plays a role in visceral endoderm differentiation and cavitation in the early mouse embryo. *Development*, **126**, 535–46.

Counis, M. F., Chaudun, E., Arruti, C. *et al.* (1998). Analysis of nuclear degradation during lens cell differentiation. *Cell Death Differ.*, **5**, 251–61.

De Maria, R., Zeuner, A., Eramo, A. *et al.* (1999). Negative regulation of erythropoiesis by caspase-mediated cleavage of GATA-1. *Nature*, **401**, 489–93.

Derry, W. B., Putzke, A. P., and Rothman, J. H. (2001). *Caenorhabditis elegans* p53: role in apoptosis, meiosis, and stress resistance. *Science*, **294**, 591–5.

Dunker, N., Schmitt, K., and Krieglstein, K. (2002). TGF-beta is required for programmed cell death in interdigital webs of the developing mouse limb. *Mech. Dev.*, **113**, 111–20.

Ellis, H. M. and Horvitz, H. R. (1986). Genetic control of programmed cell death in the nematode *C. elegans*. *Cell*, **44**, 817–29.

Fadok, V. A., Bratton, D. L., Rose, D. M., Pearson, A., Ezekewitz, R. A. B., and Henson, P. M. (2000). A receptor for phosphatidylserine-specific clearance of apoptotic cells. *Nature*, **405**, 85–90.

Fernando, P., Kelly, J. F., Balazsi, K., Slack, R. S., and Megeney, L. A. (2002). Caspase-3 activity is required for skeletal muscle differentiation. *Proc. Natl. Acad. Sci. USA*, **99**, 11025–30.

Franc, N. C., Heitzler, P., Ezekowitz, R. A., and White, K. (1999). Requirement for *croquemort* in phagocytosis of apoptotic cells in *Drosophila. Science*, **284**, 1991–4.

Garcia-Bellido, A., Ripoll, P., and Morata, G. (1973). Developmental compartmentalisation of the wing disk of *Drosophila. Nat. New Biol.*, **245**, 251–3.

Gartner, A., Milstein, S., Ahmed, S., Hodgkin, J., and Hengartner, M. O. (2000). A conserved checkpoint pathway mediates DNA damage-induced apoptosis and cell cycle arrest in *C. elegans. Mol. Cell*, **5**, 435–43.

Gaumer, S., Guenal, I., Brun, S., Theodore, L., and Mignotte, B. (2000). Bcl-2 and bax mammalian regulators of apoptosis are functional in *Drosophila. Cell Death Differ.*, **7**, 804–14.

Glücksmann, A. (1951). Cell deaths in normal vertebrate ontogeny. *Biol. Rev.*, **26**, 5986.

Goyal, L. (2001). Cell death inhibition: keeping caspases in check. *Cell*, **104**, 805–8.

Goyal, L., McCall, K., Agapite, J., Hartwieg, E., and Steller, H. (2000). Induction of apoptosis by *Drosophila reaper, hid* and *grim* through inhibition of IAP function. *EMBO J.*, **19**, 589–97.

Grether, M. E., Abrams, J. M., Agapite, J., White, K., and Steller, H. (1995). The *head involution defective* gene of *Drosophila melanogaster* functions in programmed cell death. *Genes Dev.*, **9**, 1694–708.

Gumienny, T. L., Lambie, E., Hartwieg, E., Horvitz, H. R., and Hengartner, M. O. (1999). Genetic control of programmed cell death in the *Caenorhabditis elegans* hermaphrodite germline. *Development*, **126**, 1011–22.

Hamon, Y., Broccardo, C., Chambenoit, O. *et al.* (2000). ABC1 promotes engulfment of apoptotic cells and transbilayer redistribution of phosphatidylserine. *Nat. Cell Biol.*, **2**, 399–406.

Hay, B. A., Wassarman, D. A., and Rubin, G. M. (1995). *Drosophila* homologs of baculovirus inhibitor of apoptosis proteins function to block cell death. *Cell*, **83**, 1253–62.

Hedgecock, E. M., Sulston, J. E., and Thomson, J. N. (1983). Mutations affecting programmed cell death in the nematode *Caenorhabditis elegans. Science*, **220**, 1277–9.

Hengartner, M. O. (2001). Apoptosis: corralling the corpses. *Cell*, **104**, 325–8.

Hengartner, M. O., Ellis, R. E., and Horvitz, H. R. (1992). *Caenorhabditis elegans* gene *ced-9* protects cells from programmed cell death. *Nature*, **356**, 494–9.

Hoeppner, D. J., Hengartner, M. O., and Schnabel, R. (2001). Engulfment genes cooperate with *ced-3* to promote cell death in *Caenorhabditis elegans. Nature*, **412**, 202–6.

Horvitz, H. R., Ellis, H., and Sternberg, P. (1982). Programmed cell death in nematode development. *Neurosci. Comment*, **1**, 56–65.

Horvitz, H. R., Shaham, S., and Hengartner, M. O. (1994). The genetics of programmed cell death in the nematode *Caenorhabditis elegans. Cold Spring Harb. Quant. Biol.*, **59**, 377–85.

Igaki, T., Kanuka, H., Inohara, N. *et al.* (2000). Drob-1, a *Drosophila* member of the Bcl-2/CED-9 family that promotes cell death. *Proc. Natl. Acad. Sci. USA*, **97**, 662–7.

Johnston, L. A., Prober, D. A., Edgar, B. A., Eisenman, R. N., and Gallant, P. (1999). *Drosophila* myc regulates cellular growth during development. *Cell*, **98**, 779–90.

Joza, N., Susin, S. A., Daugas, E. *et al.* (2001). Essential role of the mitochondrial apoptosis-inducing factor in programmed cell death. *Nature*, **410**, 549–54.

Kerr, J. F. R., Wyllie, A. H., and Currie, A. R. (1972). Apoptosis: a basic biological phenomenon with wide-ranging implication in tissue kinetics. *Br. J. Cancer*, **26**, 239–57.

Kumar, S. and Doumanis, J. (2000). The fly caspases. *Cell Death Differ.*, **7**, 1039–44.

Larisch, S., Yi, Y., Lotan, R. *et al.* (2000). A novel mitochondrial septin-like protein, ARTS, mediates apoptosis dependent on its P-loop motif. *Nat. Cell. Biol.*, **2**, 915–21.

Leulier, F., Vidal, S., Saigo, K., Ueda, R., and Lemaitre, B. (2002). Inducible expression of double-stranded RNA reveals a role for dFADD in the regulation of the antibacterial response in *Drosophila* adults. *Curr. Biol.*, **12**, 996–1000.

Levi-Montalcini, R. (1987). The nerve growth factor 35 years later. *Science*, **237**, 1154–62.

Li, L. Y., Luo, X., and Wang, X. (2001). Endonuclease G is an apoptotic DNase when released from mitochondria. *Nature*, **412**, 95–9.

Lindsten, T., Ross, A. J., King, A. *et al.* (2000). The combined functions of proapoptotic Bcl-2 family members bak and bax are essential for normal development of multiple tissues. *Mol. Cell*, **6**, 1389–99.

Lockshin, R. A. and Williams, C. M. (1964). Programmed cell death. II. Endocrine potentiation of the breakdown of the intersegmental muscles of silkmoths. *J. Insect Physiol.*, **10**, 643.

Lohmann, I., McGinnis, N., Bodmer, M., and McGinnis, W. (2002). The *Drosophila* Hox gene *Deformed* sculpts head morphology via direct regulation of the apoptosis activator reaper. *Cell*, **110**, 457–66.

Marguet, D., Luciani, M. F., Moynault, A., Williamson, P., and Chimini, G. (1999). Engulfment of apoptotic cells involves the redistribution of membrane phosphatidylserine on phagocyte and prey. *Nat. Cell Biol.*, **1**, 454–6.

Merino, R., Rodriguez-Leon, J., Macias, D., Ganan, Y., Economides, A. N., and Hurle, J. M. (1999). The BMP antagonist Gremlin regulates outgrowth, chondrogenesis and programmed cell death in the developing limb. *Development*, **126**, 5515–22.

Milan, M., Perez, L., and Cohen, S. M. (2002). Short-range cell interactions and cell survival in the *Drosophila* wing. *Dev. Cell*, **2**, 797–805.

Moreno, E., Basler, K., and Morata, G. (2002a). Cells compete for *decapentaplegic* survival factor to prevent apoptosis in *Drosophila* wing development. *Nature*, **416**, 755–9.

Moreno, E., Yan, M., and Basler, K. (2002b). Evolution of TNF signaling mechanisms. JNK-dependent apoptosis triggered by Eiger, the *Drosophila* homolog of the TNF Superfamily. *Curr. Biol.*, **12**, 1263–8.

Nellen, D., Burke, R., Struhl, G., and Basler, K. (1996). Direct and long-range action of a DPP morphogen gradient. *Cell*, **85**, 357–68.

Rathmell, J. C. and Thompson, C. B. (2002). Pathways of apoptosis in lymphocyte development, homeostasis, and disease. *Cell*, **109**, S97–107.

Reddien, P. W., Cameron, S., and Horvitz, H. R. (2001). Phagocytosis promotes programmed cell death in *C. elegans*. *Nature*, **412**, 198–202.

Ryoo, H. D., Bergmann, A., Gonen, H., Ciechanover, A., and Steller, H. (2002). Regulation of *Drosophila* IAP1 degradation and apoptosis by *reaper* and *ubcD1*. *Nat. Cell Biol.*, **4**, 432–8.

Savill, J. and Fadok, V. (2000). Corpse clearance defines the meaning of cell death. *Nature*, **407**, 784–8.

Schumacher, B., Hofmann, K., Boulton, S., and Gartner, A. (2001). The *C. elegans* homolog of the p53 tumor suppressor is required for DNA damage induced apoptosis. *Curr. Biol.*, **11**, 1722–7.

Scott, R. S., McMahon, E. J., Pop, S. M. *et al.* (2001). Phagocytosis and clearance of apoptotic cells is mediated by MER. *Nature*, **411**, 207–11.

Simpson, P. and Morata, G. (1981). Differential mitotic rates and patterns of growth in compartments in the *Drosophila* wing. *Dev. Biol.*, **85**: 299–308.

Sulston, J. E. and Horvitz, H. R. (1977). Post-embryonic cell lineages of the nematode, *Caenorhabditis elegans*. *Dev. Biol.*, **56**, 110–56.

Susin, S. A., Lorenzo, H. K., Zamzami, N. *et al.* (1999). Molecular characterization of mitochondrial apoptosis-inducing factor. *Nature*, **397**, 441–6.

Suzuki, Y., Imai, Y., Nakayama, H., Takahashi, K., Takio, K., and Takahashi, R. (2001). A serine protease, HtrA2, is released from the mitochondria and interacts with XIAP, inducing cell death. *Mol. Cell*, **8**, 613–21.

Tata, J. R. (1966). Requirement for RNA and protein synthesis for induced regression of the tadpole tail in organ culture. *Dev. Biol.*, **13**, 77–94.

Tepass, U., Fessler, L. I., Aziz, A., and Hartenstein, V. (1994). Embryonic origin of hemocytes and their relationship to cell death in *Drosophila*. *Development*, **120**, 1829–37.

Varkey, J., Chen, P., Jemmerson, R., and Abrams, J. M. (1999). Altered cytochrome c display precedes apoptotic cell death in *Drosophila*. *J. Cell Biol.*, **144**, 701–10.

Vogt, C. (1842). Untersuchungen über die Entwicklungsgeschichte der Geburtshelferkroete (Alytes obstetricans).

Wang, S. L., Hawkins, C. J., Yoo, S. J., Muller, H. A., and Hay, B. A. (1999). The *Drosophila* caspase inhibitor DIAP1 is essential for cell survival and is negatively regulated by HID. *Cell*, **98**, 453–63.

Wang, X. (2001). The expanding role of mitochondria in apoptosis. *Genes Dev.*, **15**, 2922–33.

Wei, C., Zong, W.-X., Cheng, E. H.-Y. *et al.* (2001). Proapoptotic BAX and BAK: a requisite gateway to mitochondrial dysfunction and death. *Science*, **292**, 727–30.

Weil, M., Jacobson, M. D., and Raff, M. C. (1997). Is programmed cell death required for neural tube closure? *Curr. Biol.*, **7**, 281–4.

White, K. (2000). Cell death: *Drosophila* Apaf-1 – no longer in the (d)Ark. *Curr. Biol.*, **10**, R167–9.

White, K., Grether, M. E., Abrams, J. M., Young, L., Farrell, K., and Steller, H. (1994). Genetic control of programmed cell death in *Drosophila*. *Science*, **264**, 677–83.

Wilson, R., Ditzel, M., Zachariou, A. *et al.* (2002). The DIAP1 RING finger mediates ubiquitination of Dronc and is indispensable for regulating apoptosis. *Nat. Cell Biol.*, **4**, 445–50.

Yoshida, H., Kong, Y. Y., Yoshida, R. *et al.* (1998). Apaf1 is required for mitochondrial pathways of apoptosis and brain development. *Cell*, **94**, 739–50.

Yoo, S. J., Huh, J. R., Muro, I. *et al.* (2002). Hid, Rpr and Grim negatively regulate DIAP1 levels through distinct mechanisms. *Nat. Cell Biol.*, **4**: 416–24.

Yuan, J. and Yankner, B. A. (2000). Apoptosis in the nervous system. *Nature*, **407**, 802–9.

Yuan, J., Shaham, S., Ledoux, S., Ellis, H. M., and Horvitz, H. R. (1993). The *C. elegans* cell death gene *ced-3* encodes a protein similar to mammalian interleukin-1-beta-converting enzyme. *Cell*, **75**, 641–52.

Zecca, M., Basler, K., and Struhl, G. (1995). Sequential organizing activities of *engrailed*, *hedgehog* and *decapentaplegic* in the *Drosophila* wing. *Development*, **121**, 2265–78.

Zhou, L., Hashimi, H., Schwartz, L. M., and Nambu, J. R. (1995). Programmed cell death in the *Drosophila* central nervous system midline. *Curr. Biol.*, **5**, 784–90.

Zhou, L., Song, Z., Tittel, J., and Steller, H. (1999). HAC-1, a *Drosophila* homolog of APAF-1 and CED-4 functions in developmental and radiation-induced apoptosis. *Mol. Cell*, **4**, 745–55.

Zimmermann, K. C., Ricci, J. E., Droin, N. M., and Green, D. R. (2002). The role of ARK in stress-induced apoptosis in *Drosophila* cells. *J. Cell Biol.*, **156**, 1077–87.

Zou, H. and Niswander, L. (1996). Requirement for BMP signaling in interdigital apoptosis and scale formation. *Science*, **272**, 738–41.

Apoptosis and cancer

Erinn L. Soucie[1], Gerard Evan[2], and Linda Z. Penn[1]

[1]Department of Medical Biophysics, University of Toronto, Ontario Cancer Institute, Princess Margaret Hospital, Toronto, Canada
[2]Cancer Research Institute, University of California San Francisco, San Francisco, USA

3.1 Introduction

The body of knowledge concerning apoptosis and the molecular mechanisms involved in this cellular process continues to expand at an unprecedented rate. This knowledge has impacted not only on our understanding of cancer development, but also on diagnosis and treatment approaches for malignant diseases. This chapter summarizes some of the ways in which tumor cells have successfully corrupted cell survival and death machinery in order to evade the penalty of apoptosis. This is followed by a discussion of novel therapeutics that have begun to emerge and which are designed to address directly these mechanistic changes that can confound current treatment regimens. The advent of current technologies such as laser capture micro-dissection of tumor tissues and microarray expression profiling have also allowed for accurate diagnosis of cancer genotypes. With these and other technologies, the potential now exists to develop new treatment regimens that can directly target individual tumor susceptibilities and, further, to predict the likelihood of response of a particular tumor to specific treatments. Together, advances in the field of apoptosis research have resulted in a revolution in the design of novel anti-cancer therapeutics for cancer treatment. In future, diagnosis of the nature of the anti-apoptotic lesions within a given tumor cell could be directly translated into a rational treatment strategy to eradicate cancers.

3.2 Defining cancer in apoptotic terms

The word "cancer" has evolved to become an umbrella term which encompasses the many incarnations of the disease. At its root, cancer is the result of genetic lesions that give rise to a dissident phenotype in an affected cell. Having overcome normal growth and death regulatory controls, a cancerous cell can continue to

Apoptosis in Health and Disease: Clinical and Therapeutic Aspects, ed. Martin Holcik, Alex E. MacKenzie, Robert G. Korneluk, and Eric C. LaCasse. Published by Cambridge University Press. © Cambridge University Press 2004.

evolve, selecting for additional mutations to support its continued survival amid persistent attempts by the host organism to combat its deleterious effects and to reinstate social order to the affected tissue.

How do tumor cells become transformed in light of the strict codes that govern cellular behavior? Mutations or lesions in the genome can be inherited or acquired as a result of environmental hazards such as radiation or toxic chemicals. Such assaults can impact an organism through ambient exposure over time, or may be received in acute doses. These agonists are generally genotoxic and can incur random DNA lesions throughout the genome. Such genetic mutations are not necessarily detrimental as the cell is armed with defenses to protect it from just such insults and can repair most DNA damage quite readily. In cases where the cell's natural defenses are temporarily overcome, such as in the case of high doses of acute carcinogen exposure, DNA damage can accumulate and the chance of non-benign lesions increases. Moreover, if by chance mutations arise in genes regulating key cellular functions, the cell can become transformed and, if not halted, take on a disastrous malignant trajectory.

The exact nature of the series or combination of genetic lesions that needs to occur for a transformed cell to become fully malignant is debatable. Whether or not it is the mutation of specific key regulatory genes or the summation of a particular series of mutagenic events that causes a given cell to surpass a tolerable threshold of dysfunction may depend on the individual context of these mutations in the cell (Hanahan and Weinberg, 2000; Evan and Vousden, 2001). The bottom line is that, one way or another, the cell reaches a point where it can no longer compensate for these mutations and is overcome by the tumor phenotype. Importantly, for a tumor cell to survive, it must acquire the fundamental abilities that have come to define the cancer – namely, the ability to proliferate ad infinitum, to recruit vasculature when local provisions become inadequate, to metastasize, to evade immune surveillance, and, perhaps most importantly, to inhibit or suppress the cell's innate apoptotic response normally triggered when any or all of the above occur (Hanahan and Weinberg, 2000).

Tumor cells differ from their healthy counterparts not only in their unlimited proliferative capacity, but also in their susceptibility to cell death. The intrinsic mechanisms at work to promote rapid expansion and survival of the malignant cells inevitably put such cells on the edge of the apoptotic threshold (Evan and Vousden, 2001). Indeed, some models suggest that a constant baseline of survival inputs is a corequisite for cell division, and that, in the absence of such survival signaling, cells default along an apoptotic pathway concomitantly engaged as part of the cell's cycle checkpoints (Ishizaki *et al.*, 1995). As such, although typically rapidly growing tumor cells have evolved to stay the apoptotic process, they are more sensitive to additional stresses such as a decrease in survival inputs or exposure to toxic insults

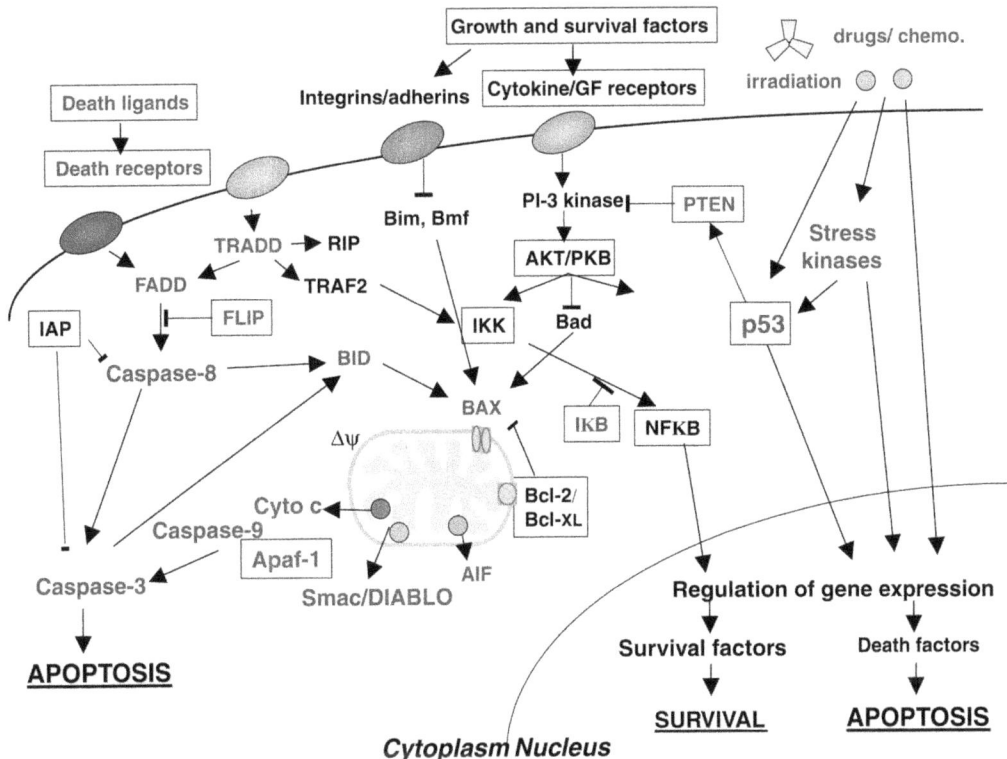

Figure 3.1   *A schematic diagram highlighting key regulators of apoptosis and survival signaling networks.* Pro-apoptotic molecules are shown in outlined text and anti-apoptotic molecules are shown in plain text.

than most normal cells. This Achilles heel provides a means for intervention that can distinguish between the tumor cell and the normal cell, and thus potentially for specific eradication of cancer upon treatment to target these defined pathways in cells.

The genius that is nature provides a preponderance of apoptotic signaling pathways within a single cell as a means of ensuring activation of a compensatory apoptotic response when one pathway fails (Figure 3.1). Intrinsically, cancer cells have the ability, or rather the genetic instability, to select for mutations in genes that negate one or many crucial aspects of the apoptotic program. Indeed, the induction of apoptosis can involve both the lack of survival signals as well as the liberation of pro-apoptotic effectors. As such, a cancer cell has essentially two means of bypassing its deathly fate to become fully transformed: by increasing survival signals or by selecting for deletions or mutations in those genes essential for the death program to proceed, thereby effectively short-circuiting the cell's death machinery. Both tactics are rampant among cancers and contribute to the individual genotype and subsequent etiology of each disease.

Normally, the viability of a cell requires ubiquitous or specific survival signals depending on a cell type or context. Similarly, the individual layers of regulatory molecules superimposed on the apoptotic machinery can be global or, again, specific. Consequently, a malignant cell will often only evolve to select for mutations that hinder the particular aspect of apoptotic signaling relevant to that cell, and the imposing stress. Importantly, this can leave other aspects of the apoptotic machinery intact, although probably dormant. Therefore, although the tumor may have temporarily succeeded in evading the apoptotic response, the challenge to its continued survival remains as it faces renewed attack through these other avenues of death still available to the host cell. This highlights the importance of detecting and defining the nature of the anti-apoptotic lesions in cancers in order to direct a point of attack for treatment of that particular cancer type. The goal, then, has become to identify and to dissect further these apoptotic pathways as a means to identify novel therapeutic targets aimed at either bypassing or reactivating the apoptotic programs in these affected cells.

3.3   Molecular mechanisms of apoptotic deregulation in cancers

The number of known mutations and combinations of mutations that have evolved in cancer cells to evade programmed cell death reflects the multiplicity and complexity of these signaling cascades. The number of genetic lesions required to enhance survival, or to inhibit apoptosis, of cancer cells can vary depending on the nature and importance of the given mutation. The consequences of any one mutation can be amplified as a result of that molecule's impact or involvement in multiple facets of the cell's normal function as well as through the additional roles that mutation may have acquired in the context of the tumor's biology (Pelengaris *et al.*, 2000; Elenbaas *et al.*, 2001). Moreover, the stage at which a malignant cell must evolve or die in response to apoptotic stress may also vary due to the availability of survival factors in the cell's environment. The accumulation of lesions up to this point can, in turn, shape the form of the resistance necessary for that tumor to carry on. Overall, the enhancement of survival signals to outweigh the death signals or corruption of the death signal itself are the two main strategies for apoptotic evasion that have emerged as common themes in cancers (Figure 3.2).

3.3.1   Enhanced survival

In addition to growth and proliferative inputs, almost all cells are dependent on the input of survival signals to suppress the apoptotic tendencies of actively dividing cells. Such survival signals are usually transmitted through surface receptors on the cell, which respond to the various factors in the tissue environment. The means by which some tumor cells avoid succumbing to deathly pathways upon deregulated

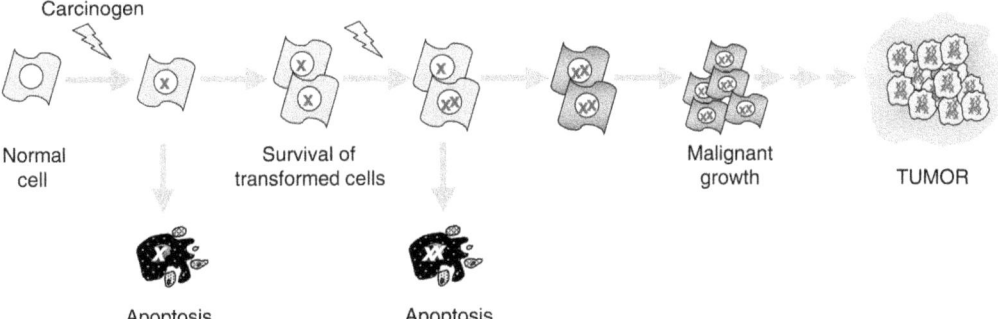

Figure 3.2   *One model for the stepwise progression of cancer is the genesis of a single cell which has accumulated a series of genetic mutations to allow for its continued survival and expansion despite its inherent suicidal tendencies.* A transformed cell can accumulate additional mutations in the absence of external carcinogens should it acquire those lesions which promote genetic instability, promoting further selection for those mutations which enable the cancer cell to overcome natural barriers to its expansion and to evade immune detection.

proliferation is to compensate by sustaining or upregulating survival signals. In tumors, this includes deregulated expression of the survival factors insulin-like growth factors (IGF)-1 and -II, activation of chemokine receptors, and alterations in the PI3K/Akt pathway which normally transmits these survival inputs within the cell (Collins *et al.*, 1994; West *et al.*, 2002). Oncogenes such as Ras or Bcr-Abl can increase Akt activity and an amplification of the PI3K catalytic subunit has also been detected in some forms of cancer (Khwaja *et al.*, 1997; Skorski *et al.*, 1997). An increase in Akt expression itself is also common to many tumor types. Moreover, PTEN, the suppressor of Akt function, is frequently mutated or deleted in many advanced tumors (Hill and Hemmings, 2002).

In addition, tumor cells which express oncogenic tyrosine kinases are usually resistant to DNA damage-induced apoptosis and this has been associated with increases in genomic instability and resistance to genotoxic therapies. Members of this family of oncogenes include Fgfr, Kit, ErbB2, Abl, PdgfrB, Scf, and Jak2. At a molecular level, tumors harboring lesions in this family of signaling molecules can stem the apoptotic cascade initiated following therapy by three mechanisms: through enhanced repair of DNA damage, through delayed cell cycle checkpoints, which allows for extended periods for DNA repair, and through upregulation of anti-apoptotic factors such as Bcl-2 and Bcl-XL (Skorski, 2002).

An important regulator of cell survival downstream of these signal transduction pathways is the transcription factor NF-κB. The activation of NF-κB is associated with the regulation of genes that encode chemokines, cytokines, adhesion factors, proteases, oncogenes, and tumor suppressors, as well as inhibitors of apoptosis (Richmond, 2002). NF-κB normally resides in the cell's cytoplasm where it is retained through binding to the inhibitors of NF-κB, namely the IκB proteins. Signaling through a variety of extracellular stimuli can result in the degradation of IκB proteins, whereupon NF-κB subunits are released and migrate to the nucleus in order to effect the regulation of specific target genes. The consequence of activated NF-κB in tumor cells is, therefore, the constitutive expression of these target genes to allow for the continued growth and escape from apoptosis of these cells. In humans, the list of tumors with elevated NF-κB is extensive and includes melanoma, mammary carcinoma, leukemia, lymphoma, non-small-cell lung carcinoma, and pancreatic cancer, to name a few (Darnell, 2002; Gilmore *et al.*, 2002; Karin *et al.*, 2002).

3.3.2 Blocking apoptosis

In addition to increased survival signaling, the increased expression of molecules that can directly interfere with the apoptotic pathway is also common to many tumor types. As alluded to above, the result of many oncogenic lesions is the deregulation of downstream target genes including members of the Bcl-2 family, inhibitor of apoptosis proteins (IAPs), as well as a plethora of other genes involved in cell growth, proliferation, and apoptosis.

The aberrant expression or activity of IAP proteins has been observed in many different human diseases, among them cancer. For example, survivin and the melanoma IAP (MLIAP or livin) which are expressed during human development but not in normal adult tissue are found at high levels in certain types of malignant cells (Adida *et al.*, 1998; Kasof and Gomes, 2001). Of the IAP proteins implicated thus far in cancers, survivin has attracted considerable attention due to its altered expression profile in the majority of cancers (Altieri, 2003).

Probably one of the more famous oncogenes, and one of the earliest examples of a gene which directly contributes to a tumor's pro-survival, or anti-apoptotic phenotype, is Bcl-2. Since its discovery as an inhibitor of apoptosis, Bcl-2 has become the founding member of a family of homologous proteins involved in the regulation of the apoptotic process (Figure 3.1). Other anti-apoptotic family members, each considered oncogenes in their own rights, include Bcl-xL, Bcl-w, A1, and Mcl-1 (Gross *et al.*, 1999). Although they are thought to function primarily at the mitochondria, the presence of Bcl-2 proteins at other cellular membranes such as the endoplasmic reticulum suggests a broader role for these factors in the cell (Rudner *et al.*, 2002). Indeed, the mitochondria represent a nodal point for many death signals that converge to induce both permeabilization of mitochondrial membranes, thereby disrupting the electrochemical gradient essential for ATP synthetase activity in the oxidative phosphorylation pathways, and the release of mitochondrial proteins (Olson and Kornbluth, 2001; Ravagnan *et al.*, 2002). Once released, these mitochondrial proteins are involved in the amplification and propagation of the apoptotic signaling cascade. Although the biochemical action of Bcl-2 proteins is still controversial, it is believed that these molecules function to keep mitochondrial membranes intact and to prevent the release of pro-apoptotic factors from the mitochondria downstream of apoptotic signaling in the cell. The overall abundance of these proteins within a cell appears to dictate the ability of that cell to survive in response to a greater number of apoptotic insults.

It is now becoming apparent that the anti-apoptotic action of these molecules occurs in part through their ability to interact with pro-apoptotic members of the same family. Indeed, the Bcl-2 members constitute a large part of the machinery responsible for transmitting the apoptotic signal emanating from the extracellular milieu (Puthalakath and Strasser, 2002). Interestingly, the presence of either Bax or Bak seems to be essential for apoptosis in many cell types (Cheng *et al.*, 2001). A third member of this family of apoptotic regulators is formed of the BH3-only molecules, which, in contrast to Bax and Bak family members, share only one region of homology to Bcl-2 and are otherwise unique. These include Bim, Bid, Bad, Bik, Noxo, Puma, and Bmf. Importantly, it is through this BH3 domain that these proteins are able to interact with the pro-survival Bcl-2 proteins and this interaction is key to their ability to effect the apoptotic signal (Chittenden, 2002). Mounting evidence has implied that individual BH3-only proteins might transduce specific death signals in response to specific developmental cues or intracellular damage in different cell types (Puthalakath and Strasser, 2002).

As central regulators of the apoptotic response, the Bcl-2 family of proteins is normally held in check by diverse mechanisms including transcriptional control and post-translational modification. Not surprisingly, it has followed that the deregulation of these proteins is almost universal during oncogenesis. Just as the pro-survival members of this family are putative oncogenes, anti-apoptotic family members can act as potent tumor suppressors. For example, Bax or Bak are mutated in some human gastric and colorectal cancers, as well as in leukemia and the absence of both Bak and Bax enhances transformation in model systems (Krajewska *et al.*, 1996; Rampino *et al.*, 1997; Meijerink *et al.*, 1998; Shibata *et al.*, 1999; Knudson *et al.*, 2001; Eischen *et al.*, 2002).

At the core of the apoptotic pathways lie the caspases. The members of this family of aspartate-directed cysteine proteases are synthesized as latent zymogens and undergo proteolytic activation during apoptosis. Once activated, the initiator caspases commence an irreversible and self-sustaining cascade of effector caspase activation, leading to the systematic disassembly of the cell. Although it is widely believed that, since most, if not all, apoptotic signals ultimately converge to activate members of the caspase family, caspase-independent cell death can occur in the context of some cancers (Lorenzo *et al.*, 1999; Elliott *et al.*, 2000). As such, although the pathologic consequences of caspase deactivation during cancer effectively is to suppress the apoptotic response downstream of many agonists, not all avenues of destruction are necessarily blocked in such cases.

Caspases are regulated at multiple levels including transcription, post-translational modification and cellular localization, and proteosome-mediated degradation, as well as through the enzymatic inhibition of their catalytic activity (Shi, 2002). As such, cancer cells have a variety of means at their disposal through which they can prevent the unleashing of this proteolytic pathway. For example, aggressive forms of neuroblastomas have been identified in which the caspase-8 gene is either deleted or hypermethylated (Hopkins-Donaldson et al., 2000; Teitz et al., 2000; Teitz et al., 2001). As such, caspase-8 is effectively silenced in these tumors, which is potentially at the root of the survival mechanism used by this cancer to evolve. More recently, it has been reported that both caspase-8 and caspase-10 become deregulated in a number of pediatric tumors and tumor cell lines (Harada et al., 2002). High levels of the caspase-8 inhibitor Flip have also been detected in human melanomas and murine B-cell lymphomas, and viral analogs of Flip (v-Flips) are encoded by some tumorigenic viruses (Hu et al., 1997; Irmler et al., 1997; Thome et al., 1997; Tepper and Seldin, 1999; Mueller and Scott, 2000). Apaf-1, an integral part of the caspase apoptosome complex involved in the activation of caspase-9, also shows a high rate of allelic loss in metastatic melanomas (Soengas et al., 2001). Moreover, it is often the case that the remaining allele becomes hypermethylated and thus transcriptionally inactivated. Melanomas that do not express Apaf-1 also fail to respond to chemotherapy, which is likely to be due to this breakdown in the apoptotic machinery.

A final strategy to bypass cell death is to downregulate or inactivate the death receptors. There are many examples of tumor types in which expression of the CD95 death receptor is reduced (Muschen et al., 2000). This may be the result of transcriptional regulation imposed downstream of oncogenic ras activation or through mutations affecting p53 activity. Furthermore, frequent deletions and mutations affecting the Trail receptors R1 and R2 occur in a wide variety of cancers (Ozoren et al., 2000; Fisher et al., 2001; Shin et al., 2001; Lee et al., 2001; Seitz et al., 2002). Together, such abnormalities can contribute to chemo-resistance and immune system evasion of these tumors.

3.4   Reactivating the death pathway to combat cancer

Currently, there exist two main classes of therapeutic agents that specifically target aspects of the apoptotic pathway in tumor cells to trigger cell death. One class of toxic agents exploits the oncogenic mutations used to drive growth and proliferation of a malignant cell to, in turn, drive the apoptotic process in the same cell. In effect, such therapies can target signaling cascades whose inhibition is non-lethal in normal cells, but trigger programmed death in tumor cells. More recently, a second class of non-genotoxic agonists has been developed which aim to modulate directly key apoptotic regulators in order to activate the death program.

3.4.1 Taking advantage of tumor-specific susceptibilities

Most traditional genotoxic therapies take advantage of the rapid proliferation innate to tumor cells to effect their toxicity. However, the molecular mechanism through which a tumor cell can acquire this proliferative capacity can also render it more susceptible to specific forms of apoptotic cell death. In fact, key oncogenic growth-promoting factors such as Myc and E2F concomitantly lower the apoptotic threshold when activated (Yamasaki, 1999; Pelengaris et al., 2000). It follows that, directly or indirectly, these mutations alter the apoptotic-signaling network of the tumor cell such that the tumor cell is rendered distinct from normal cells. For example, cells harboring an activated allele of the c-myc oncogene are sensitized to undergo apoptosis in response to a variety of cellular antagonists that in a normal cell are insufficient to elicit the full death response (Prendergast, 1999). Low doses of anti-proliferative or cytostatic agents normally induce the rapid downregulation of Myc expression and subsequent growth arrest. In the presence of activated Myc expression, however, these cells become unresponsive to these controls and instead are programmed to die. Interestingly, deregulation of the c-myc gene factors into the genesis of nearly all cancers, and so this popular mechanism used by tumor cells to drive proliferation may be readily exploited to contrive their demise (Boxer and Dang, 2001).

Cancer cells can also become sensitized to death triggers through the over-expression of factors that transmit these signals within the cell. An increase in the number of death receptors for tumor necrosis factor-related apoptosis-inducing ligand (TRAIL) and CD95 ligands on the surface of some tumor cells is common to many tumor types and has opened the door for novel protein therapeutics designed to trigger apoptosis through these death pathways. The difference in the ratio between death-inducing and decoy receptors characteristic of some tumor cells renders them more vulnerable to death ligands than a normal cell. Recombinant proteins and monoclonal antibodies that bind the TRAIL receptors, for instance, are in the works because of their tumor-specific killing action (Timmer et al., 2002; Smyth et al., 2003). Another factor, which may have an effect on the cellular susceptibility to these agents, is the level of expression of FLIP, a natural inhibitor of caspase-8 downstream of death receptor activation. The downregulating FLIP expression through antisense regulation may be a mechanism to sensitize cancer cells to this group of protein therapies (Perez and White, 2003).

Deregulated transcription, a symptom of most tumor cells, can also be taken advantage of to instigate the apoptotic program (Darnell, 2002). Deregulated gene expression can be the result of changes in transcription factor activity as well as of epigenetic modifications at the level of the genome. Such modifications include methylation and acetylation of histones. The stearic impact of the addition or removal of such moieties to the core components of the nucleosome results in an open and/or condensed DNA structure and, consequently, an active or repressed gene locus. Normally, histone modifications are highly regulated by the select recruitment of protein-modifying enzymes such as histone acetyltransferases (HATs), histone deacetylases (HDACs), or methyl-transferases by other transcription factors at the promoter site. Together, this combination of factors at specific promoters provides a platform for gene regulation.

A link between cancer and aberrant HDAC activity has now been shown in several instances. For example, gross chromosomal rearrangements involving the genes which encode transcription factors such as retinoic acid receptors (RARs) and ETO and TEL have been noted in several forms of leukemia. The fusion proteins resulting from such translocations possess an increased affinity for HDACs and hence can induce constitutive repression of target genes (Gelmetti *et al.*, 1998; Guidez *et al.*, 2000; Khan *et al.*, 2001). Indeed, misassociation with chromatin remodeling enzymes to effect inappropriate transcriptional regulation may be an important mechanism of action for many oncogenes implicated in the various aspects of cellular signaling. As such, it is not surprising that inhibitors of these modifying enzymes have a widespread impact on the various aspects of cell cycle regulation, differentiation, and apoptosis in both normal and tumor cells. Interestingly, analysis of the overall effects of HDAC inhibitors on gene transcription have shown that only about 10% of genes become deregulated upon exposure to such agonists (Mariadason *et al.*, 2000; Glaser *et al.*, 2003). Importantly, whereas normal cells have been shown to check the cell cycle upon exposure to HDAC inhibitors, tumor cells more often undergo apoptosis. Indeed, in vivo studies have shown that tumor cells are much more susceptible to the detrimental effects of HDAC inhibitors (Richon *et al.*, 1998; Butler *et al.*, 2000). This is likely to be the result of an acquired dependence by transformed cells for chromosome remodeling activity that limits their ability to absorb the effects of such inhibitors.

Protease inhibitors are emerging as another broad target inhibitor of protein regulation, but, in this case, at the level of degradation. Protease inhibitors serve as general managers in charge of turning over proteins that have been marked for degradation through ubiquitination. This type of regulation through protein turnover is common and applies to many of the key molecular regulators in the cell involved in aspects of cell proliferation, cell cycle checkpoints, adhesion, and angiogenesis, as well as apoptosis. Not surprisingly, inhibition of this machinery using proteosome inhibitors results in an upregulation in many of these proteins. Interestingly, the consequences of this upregulation are only detrimental to transformed cells. Several studies have now been able to show a tumor-specific apoptotic response upon exposure to protease inhibitors, at least in cell lines and animal models (Dou and Li, 1999). The molecular mechanism for this tumor specificity has been attributed to the stabilization of cell cycle inhibitors p21 and p27, the stabilization of inhibitors of NF-κB survival signaling, as well as the increase in both p53 and the pro-apoptotic molecule Bax in the affected cells (Dou and Li, 1999). However, due to the broad effects of such drugs on the cell's degradation machinery, the contribution of other molecules stabilized as a result of this treatment to the apoptotic phenotype cannot be ruled out. Nonetheless, it may be the case that cells are more dependent on proteases to regulate crucial survival signals in a transformed setting and so these cells are more susceptible to their inhibition.

Other tumor-selective therapies that target the basic operational machinery of the cell are inhibitors of Hsp90. Hsp90 is a member of a larger family of chaperone proteins and holds a unique role in the regulation of cell cycle progression and cell viability. Transformed cells often express high levels of this protein and this may be the reason why some tumor cells are more susceptible to these inhibitors. Moreover, recent evidence suggests a role for Hsp90 in the control of function of several important molecules along the death pathways including Akt, Apaf-1, and the death receptor-interacting protein RIP (Lewis *et al.*, 2000; Pandey *et al.*, 2000; Sato *et al.*, 2000). As such, inhibition of Hsp90 could have broader ramifications for cell survival, and thus Hsp90 has become a target of new anti-cancer therapeutics.

HMG-CoA, the rate-limiting enzyme of the mevalonate pathway, is another protein involved in regulating global aspects of cellular metabolism and has also become a prime target for tumor-specific therapies. Inhibitors of HMG-CoA have traditionally been used in patients to control hypercholesterolemia. However, in addition to controlling the cholesterol pathways, HMG-CoA reductase is also an essential regulator of cell signaling, protein synthesis, and control of the cell cycle, as well as affecting membrane integrity. Recently, it has been shown that members of the statin family of HMG-CoA inhibitors can induce apoptosis in several different types of cancers including acute myeloid leukemia, head and neck squamous cell carcinoma, and cervical carcinoma in a sensitive and specific manner (Wong *et al.*, 2002). The molecular mechanism of this specific killing action is thought to be the consequence of loss of isoprenylation of substrates downstream in the mevalonate pathway. Why these inhibitors appear to be non-lethal to the normal cells remains mysterious since presumably this pathway controlling the modification of proteins is important in most cell types. Most likely, the specificity with which these agents induce apoptosis results from the existence of feedback mechanisms along the mevalonate pathway which can compensate for loss of HMG-CoA reductase activity, mechanisms which may be lost or insufficient in cancer cells. Together, the evidence is mounting to support the potential success of such therapies in combination with current treatments in combating malignant disease.

3.4.2 Negating the tumor advantage

There has been a great interest in devising therapeutic strategies for directly modulating key molecules that make life-or-death decisions in cells. As more information becomes available with respect to the structure and mechanisms of action of apoptotic regulatory molecules, specific inhibitors can be rationally designed. So far, several approaches aimed at reversing the genetic lesions supposed to be blocking the apoptotic program in tumors have been met with some success. Also, small molecule inhibitors, which can engage the apoptotic machinery directly, have been created.

Inactivating mutations of key effectors of apoptosis can render cells insensitive to some agonists. The knowledge of such mutations can be used to direct therapies that are not dependent on such factors to elicit the apoptotic program. Alternatively, therapies to reinstate a critical regulator of both normal cellular proliferation and death have been suggested. For example, strategies for the re-introduction of an intact p53 tumor suppressor protein through gene transfer methods, including viral delivery systems, as well as synthetic peptides or monoclonal antibodies to rescue defective p53 transcriptional activity have been proposed (Foster *et al.*, 1999; Pruschy *et al.*, 2001). An upregulation of active p53 could also be achieved through the introduction of small peptides to block interactions between p53 and negative regulators such as Mdm-2 (Bottger *et al.*, 1997). However, these strategies are limited by the high levels of p53 protein in most tumor cells and the abundance of negative regulators and subversive lesions that must be overcome in order for reactivation of this protein to be effective – a problem common to such central modulators of key cellular functions.

Another major roadblock put up by tumor cells to escape apoptosis and to curtail traditional treatments is the upregulation of the anti-apoptotic protein, Bcl-2. Because it lies downstream of most apoptotic signaling cascades, the overexpression of this particular inhibitor of apoptosis is extremely potent at blocking a wide array of incoming cytotoxic agonists (Coultas and Strasser, 2003). As such, Bcl-2 antisense therapies have achieved a level of success in effectively re-sensitizing tumor cells to conventional therapies with few adverse side effects (Gutierrez-Puente *et al.*, 2002). Other antisense and inhibitor therapies are also being devised to inhibit the activity of IAPs to unleash the full potential of caspases in afflicted cells (Chen *et al.*, 2000). Together, these types of therapy have the potential to reinstate the corrupted apoptotic program of cancer cells such that they become sensitive to current treatment regimens including hypoxia, radiation, and chemotherapeutics.

Recently, binding and structural studies have provided insights for new approaches to bind and inhibit directly the function of anti-apoptotic Bcl-2 family members (Letai *et al.*, 2002). BH3 domain mimetics have been suggested as attractive agents as they would be non-toxic to normal cells since, in a manner similar to the antisense approaches mentioned above for Bcl-2 inhibition, a second apoptotic signal would be necessary to achieve the full apoptotic effect with these small molecule inhibitors.

As a key player in a universal survival pathway commonly upregulated in cancer, Akt serves as an attractive molecular target for the development of novel apoptotic modulators. So far, both antisense and dominant-negative approaches have been used in attempts to suppress the activity of this key protein kinase. Furthermore, Akt is activated downstream of several growth factor receptors, several of which are also targets for anti-cancer drugs currently under development (Datta *et al.*, 1999). However, the broader role of Akt in other aspects of cellular metabolism raises the issue of undesirable consequences for the inhibition of this molecule in normal cells. For example, ablation of Akt in mouse models resulted in insulin resistance and hyperglycemia, and the role for Akt in glycogen synthesis, protein translation, and cell size regulation would be likely to be deleterious if disrupted in a normal cell (Cho *et al.*, 2001).

In addition to Akt, other kinases have emerged as promising therapeutic targets to reinstate the apoptotic program in cancerous cells. The pharmacologic agent STI571 as well as several newer small molecule inhibitors act to inhibit Abl, an oncogenic tyrosine kinase associated with the Philadelphia chromosome in the blast cells of patients with chronic myelogenous leukemia (Nimmanapalli and Bhalla, 2002). ErB2/HER2/neu, a receptor tyrosine kinase over-expressed in a number of carcinomas, most prominently in breast, has also been targeted for the development of novel therapeutics (Drevs *et al.*, 2003). Antisense oligonucleotides, monoclonal antibodies, and small molecule inhibitors have all shown some degree of success, both alone and in combination with traditional drugs, in curbing the growth and transforming activity of this receptor in model systems. Indeed, some of these rationally designed drugs have now progressed into clinical trials.

Removing survival signals through modulation of NF-κB has emerged as another strategy to reactivate apoptotic signaling in cancer cells. Indirect inactivation of NF-κB has been proposed through upstream interference with either IKK or IkB protein regulation (Darnell, 2002). IKK is responsible for the phosphorylation and subsequent proteosome-mediated destruction of IkB proteins. The destruction of IkB proteins releases NF-κB from sequestration in the cytosol to then instigate survival signaling in the nucleus. Recently, a number of compounds including prostaglandins, certain non-steroidal anti-inflammatory drugs (NSAIDs), as well as other synthetic compounds have been reported to inhibit directly the catalytic activity of IKKs (Holmes-McNary and Baldwin, 2000; Rossi *et al.*, 2000; Sporn *et al.*, 2001). Proteosome inhibitors are also being explored for their potential to induce anti-tumor activity through their effects on preventing IkB degradation and, hence, NF-κB activity (Dou and Li, 1999). Once again, however, these agents are likely not to be tumor specific as these pathways are relevant to both transformed and non-transformed cells.

3.5   On the horizon

The importance of specific cancer phenotypes to the prediction of outcome upon treatment has focused researchers on the problem of more accurate disease diagnosis. In recent years, there has been an explosion of microarray technology that has advanced the ways in which individual tumors can be analyzed. These technologies have provided the means for high throughput analysis of changes at the level of the genome and the ways in which it is expressed at both the level of RNA and, more recently, at the protein level in both cells and tissues. Moreover, microarray technology has allowed, for the first time, simultaneous analysis of the expression of thousands of genes from a given sample population (Blohm and Guiseppi-Elie, 2001; Lakhani and Ashworth, 2001). These methods have now enabled researchers to profile the aberrant changes in the overall pattern of normal gene expression that may have contributed to the malignant state of a tumor cell. Building on two-way comparisons between normal and tumor cell profiles, efforts to correlate such profiles to a given disease and prognosis have allowed for better understanding and treatment of known cancer types as well as novel cancer subgroups (Figure 3.3).

Despite the level of error inherent to microarray processing and data interpretation, improvements in both the areas of experimental design and technology have substantially increased the amount of useable information gained by this approach. Moreover, new technologies including laser capture microdissection to obtain pure populations of transformed cells from within a heterogeneous tumor sample have lent to the greater accuracy of microarray outputs. By extension, recent advances in RNA amplification techniques have boosted the amount of information that can be derived from this type of analysis when tumor material is limited. Although there is an appeal for translation of this technology directly into the clinic for high throughput disease diagnosis and screening, so far these applications remain at the level of research. With the advent of these discovery tools, information surrounding the relevance and penetrance of specific molecular changes during the development of disease are beginning to emerge. However, there remain important issues surrounding both the verification and validation of these findings (Simon *et al.*, 2003). The enormity of these tasks remains a challenge to researchers, and partner technologies have begun to emerge alongside microarrays to support this

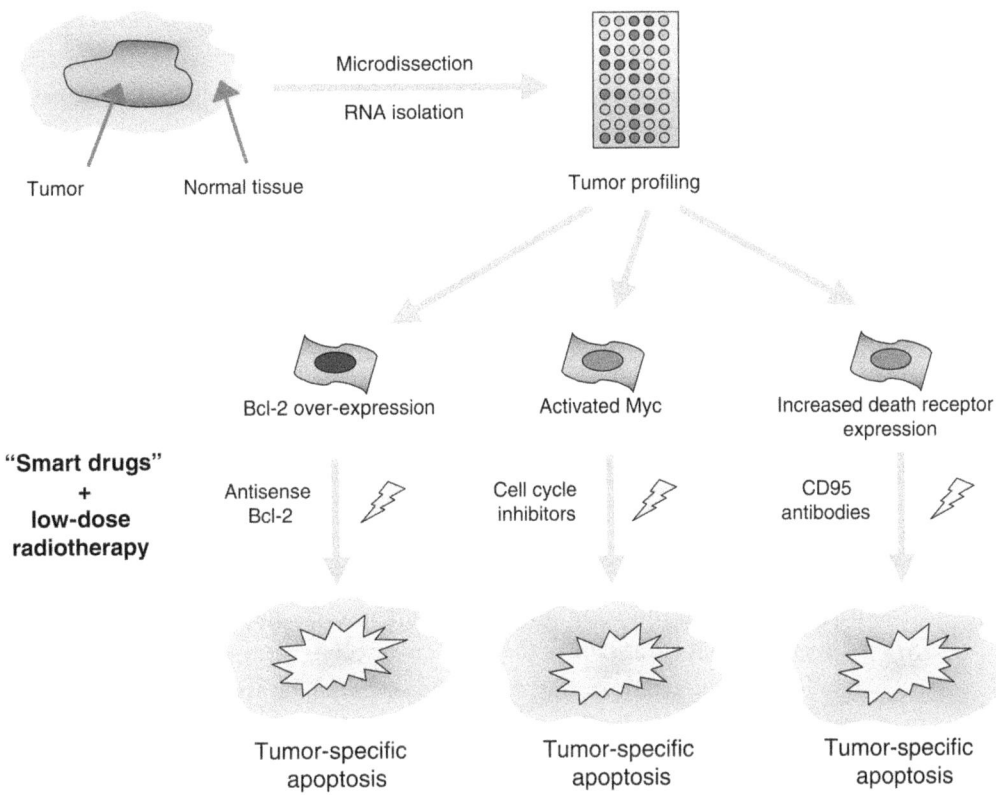

Figure 3.3   *New and evolving technologies allow for tumor sampling through laser capture microdissection for subsequent microarray analysis to diagnose tumor etiologies at a genetic level. Identification of the specific genetic lesions underlying the tumor phenotype can allow for the rational design of treatment regimens to exploit the particular sensitivities of the tumor cells specifically. These so-called "smart drugs," with predictable molecular targets within the tumor cell, can be administered in combination with lower doses of traditional therapeutics to reduce collateral damage to the surrounding normal tissues while still inducing a potent and specific apoptotic effect on the tumor.*

process. Therefore, in the foreseeable future, these methods will no doubt reach the level of ease and accuracy that will make them amenable to widespread usage in the clinics for cancer diagnosis and treatment.

3.6 Conclusions

Overall, the goal of cancer therapeutics is to kill selectively tumor cells that have been sensitized to undergo apoptosis beyond their normal counterparts. Historically, most cancer therapeutics have been genotoxic. Agonists such as radiation and chemotherapeutic drugs trigger the apoptotic response by virtue of attacking the integrity of the cellular genome. In most normal cells, the prescribed level of these insults is such that the cell's repair machinery is sufficient to overcome the incurred damage. Tumorigenic cells, however, where checkpoint controls to allow for such repair are defunct, cannot survive the damage induced by these agents and succumb to their effects. In cases where normal cells are also affected by such treatments, as in the case of actively dividing cells of the epithelia or blood, these cells can either be replenished between treatments or protected to some extent by supplementing treatment regimens with specific growth factors. However, genotoxic treatments carry the burden of incurring transforming mutations in otherwise healthy cells to prevent their repair subsequently, leading to secondary tumors as a result of treatment. In effect, it may be that, in the case of cancer, you cannot fight fire with fire.

In designing more current treatment technologies, researchers have begun to focus on developing new, rational, non-genotoxic agonists to arm further the battery of anti-cancer treatments. A greater understanding of cancer etiology as well as of the mechanisms of transformation has allowed for insight into novel therapeutic targets to trigger tumor-specific apoptotic cell death. These so-called "smart drugs" aim to exacerbate the cancerous state of the tumor to affect its demise. Moreover, this new generation of non-genotoxic agents has the potential for systemic administration as their effects would be limited to the tumor cells. In future, cancer therapy will be likely to reflect the multiplicity and diversity of the disease itself. Moreover, as tumor cells scramble to overcome apoptotic barriers to their survival and expansion, new treatments will adapt in parallel to overturn these obstacles towards the ultimate victory over cancer.

REFERENCES

Adida, C., Crotty, P. L., McGrath, J., Berrebi, D., Diebold, J., and Altieri, D. C. (1998). Developmentally regulated expression of the novel cancer anti-apoptosis gene survivin in human and mouse differentiation. *Am. J. Pathol.*, **152**, 43–9.

Altieri, D. C. (2003). Validating survivin as a cancer therapeutic target. *Nat. Rev. Cancer*, **3**, 46–54.

Blohm, D. H. and Guiseppi-Elie, A. (2001). New developments in microarray technology. *Curr. Opin. Biotechnol.*, **12**, 41–7.

Bottger, A., Bottger, V., Sparks, A., Liu, W. L., Howard, S. F., and Lane, D. P. (1997). Design of a synthetic Mdm2-binding mini protein that activates the p53 response in vivo. *Curr. Biol.*, **7**, 860–9.

Boxer, L. M. and Dang, C. V. (2001). Translocations involving c-myc and c-myc function. *Oncogene*, **20**, 5595–610.

Butler, L. M., Agus, D. B., Scher, H. I. *et al.* (2000). Suberoylanilide hydroxamic acid, an inhibitor of histone deacetylase, suppresses the growth of prostate cancer cells in vitro and in vivo. *Cancer Res.*, **60**, 5165–70.

Chen, J., Wu, W., Tahir, S. K. *et al.* (2000). Down-regulation of survivin by antisense oligonuc-leotides increases apoptosis, inhibits cytokinesis and anchorage-independent growth. *Neoplasia*, **2**, 235–41.

Cheng, E. H., Wei, M. C., Weiler, S. *et al.* (2001). BCL-2, BCL-X(L) sequester BH3 domain-only molecules preventing BAX- and BAK-mediated mitochondrial apoptosis. *Mol. Cell*, **8**, 705–11.

Chittenden, T. (2002). BH3 domains: intracellular death-ligands critical for initiating apoptosis. *Cancer Cell*, **2**, 165–6.

Cho, H., Mu, J., Kim, J. K. *et al.* (2001). Insulin resistance and a diabetes mellitus-like syndrome in mice lacking the protein kinase Akt2 (PKB beta). *Science*, **292**, 1728–31.

Collins, M. K., Perkins, G. R., Rodriguez-Tarduchy, G., Nieto, M. A., and Lopez-Rivas, A. (1994). Growth factors as survival factors: regulation of apoptosis. *Bioessays*, **16**, 133–8.

Coultas, L. and Strasser, A. (2003). The role of the Bcl-2 protein family in cancer. *Semin. Cancer Biol.*, **13**, 115–23.

Darnell, J. E., Jr. (2002). Transcription factors as targets for cancer therapy. *Nat. Rev. Cancer*, **2**, 740–9.

Datta, S. R., Brunet, A., and Greenberg, M. E. (1999). Cellular survival: a play in three Akts. *Genes Dev.*, **13**, 2905–27.

Dou, Q. P. and Li, B. (1999). Proteasome inhibitors as potential novel anticancer agents. *Drug Resist. Update*, **2**, 215–23.

Drevs, J., Medinger, M., Schmidt-Gersbach, C., Weber, R., and Unger, C. (2003). Receptor tyrosine kinases: the main targets for new anticancer therapy. *Curr. Drug Targets*, **4**, 113–21.

Eischen, C. M., Rehg, J. E., Korsmeyer, S. J., and Cleveland, J. L. (2002). Loss of Bax alters tumor spectrum and tumor numbers in ARF-deficient mice. *Cancer Res.*, **62**, 2184–91.

Elenbaas, B., Spirio, L., Koerner, F. *et al.* (2001). Human breast cancer cells generated by oncogenic transformation of primary mammary epithelial cells. *Genes Dev.*, **15**, 50–65.

Elliott, K., Ge, K., Du, W., and Prendergast, G. C. (2000). The c-Myc-interacting adaptor protein Bin1 activates a caspase-independent cell death program. *Oncogene*, **19**, 4669–84.

Evan, G. I. and Vousden, K. H. (2001). Proliferation, cell cycle and apoptosis in cancer. *Nature*, **411**, 342–8.

Fisher, M. J., Virmani, A. K., Wu, L. *et al.* (2001). Nucleotide substitution in the ectodomain of trail receptor DR4 is associated with lung cancer and head and neck cancer. *Clin. Cancer Res.*, **7**, 1688–97.

Foster, B. A., Coffey, H. A., Morin, M. J., and Rastinejad, F. (1999). Pharmacological rescue of mutant p53 conformation and function. *Science*, **286**, 2507–10.

Gelmetti, V., Zhang, J., Fanelli, M., Minucci, S., Pelicci, P. G., and Lazar, M. A. (1998). Aberrant recruitment of the nuclear receptor corepressor–histone deacetylase complex by the acute myeloid leukemia fusion partner ETO. *Mol. Cell Biol.*, **18**, 7185–91.

Gilmore, T., Gapuzan, M. E., Kalaitzidis, D., and Starczynowski, D. (2002). Rel/NF-kappa B/I kappa B signal transduction in the generation and treatment of human cancer. *Cancer Lett.*, **181**, 1–9.

Glaser, K. B., Staver, M. J., Waring, J. F., Stender, J., Ulrich, R. G., and Davidsen, S. K. (2003). Gene expression profiling of multiple histone deacetylase (HDAC) inhibitors: defining a common gene set produced by HDAC inhibition in T24 and MDA carcinoma cell lines. *Mol. Cancer Ther.*, **2**, 151–63.

Gross, A., McDonnell, J. M., and Korsmeyer, S. J. (1999). BCL-2 family members and the mitochondria in apoptosis. *Genes Dev.*, **13**, 1899–911.

Guidez, F., Petrie, K., Ford, A. M. *et al.* (2000). Recruitment of the nuclear receptor corepressor N-CoR by the TEL moiety of the childhood leukemia-associated TEL-AML1 oncoprotein. *Blood*, **96**, 2557–61.

Gutierrez-Puente, Y., Zapata-Benavides, P., Tari, A. M., and Lopez-Berestein, G. (2002). Bcl-2-related antisense therapy. *Semin. Oncol.*, **29**, 71–6.

Hanahan, D. and Weinberg, R. A. (2000). The hallmarks of cancer. *Cell*, **100**, 57–70.

Harada, K., Toyooka, S., Shivapurkar, N. *et al.* (2002). Deregulation of caspase 8 and 10 expression in pediatric tumors and cell lines. *Cancer Res.*, **62**, 5897–901.

Hill, M. M. and Hemmings, B. A. (2002). Inhibition of protein kinase B/Akt. Implications for cancer therapy. *Pharmacol. Ther.*, **93**, 243–51.

Holmes-McNary, M. and Baldwin, A. S., Jr. (2000). Chemopreventive properties of transresveratrol are associated with inhibition of activation of the IkappaB kinase. *Cancer Res.*, **60**, 3477–83.

Hopkins-Donaldson, S., Bodmer, J. L., Bourloud, K. B., Brognara, C. B., Tschopp, J., and Gross, N. (2000). Loss of caspase-8 expression in highly malignant human neuroblastoma cells correlates with resistance to tumor necrosis factor-related apoptosis-inducing ligand-induced apoptosis. *Cancer Res.*, **60**, 4315–19.

Hu, S., Vincenz, C., Buller, M., and Dixit, V. M. (1997). A novel family of viral death effector domain-containing molecules that inhibit both CD-95- and tumor necrosis factor receptor-1-induced apoptosis. *J. Biol. Chem.*, **272**, 9621–4.

Irmler, M., Thome, M., Hahne, M. *et al.* (1997). Inhibition of death receptor signals by cellular FLIP. *Nature*, **388**, 190–5.

Ishizaki, Y., Cheng, L., Mudge, A. W., and Raff, M. C. (1995). Programmed cell death by default in embryonic cells, fibroblasts, and cancer cells. *Mol. Biol. Cell*, **6**, 1443–58.

Karin, M., Cao, Y., Greten, F. R., and Li, Z. W. (2002). NF-kappaB in cancer: from innocent bystander to major culprit. *Nat. Rev. Cancer*, **2**, 301–10.

Kasof, G. M. and Gomes, B. C. (2001). Livin, a novel inhibitor of apoptosis protein family member. *J. Biol. Chem.*, **276**, 3238–46.

Khan, M. M., Nomura, T., Kim, H. *et al.* (2001). Role of PML and PML-RARalpha in Mad-mediated transcriptional repression. *Mol. Cell*, **7**, 1233–43.

Khwaja, A., Rodriguez-Viciana, P., Wennstrom, S., Warne, P. H., and Downward, J. (1997). Matrix adhesion and Ras transformation both activate a phosphoinositide 3-OH kinase and protein kinase B/Akt cellular survival pathway. *EMBO J.*, **16**, 2783–93.

Knudson, C. M., Johnson, G. M., Lin, Y., and Korsmeyer, S. J. (2001). Bax accelerates tumorigenesis in p53-deficient mice. *Cancer Res.*, **61**, 659–65.

Krajewska, M., Moss, S. F., Krajewski, S., Song, K., Holt, P. R., and Reed, J. C. (1996). Elevated expression of Bcl-X and reduced Bak in primary colorectal adenocarcinomas. *Cancer Res.*, **56**, 2422–7.

Lakhani, S. R. and Ashworth, A. (2001). Microarray and histopathological analysis of tumors: the future and the past? *Nat. Rev. Cancer*, **1**, 151–7.

Lee, S. H., Shin, M. S., Kim, H. S. *et al.* (2001). Somatic mutations of TRAIL-receptor 1 and TRAIL-receptor 2 genes in non-Hodgkin's lymphoma. *Oncogene*, **20**, 399–403.

Letai, A., Bassik, M. C., Walensky, L. D., Sorcinelli, M. D., Weiler, S., and Korsmeyer, S. J. (2002). Distinct BH3 domains either sensitize or activate mitochondrial apoptosis, serving as prototype cancer therapeutics. *Cancer Cell*, **2**, 183–92.

Lewis, J., Devin, A., Miller, A. *et al.* (2000). Disruption of hsp90 function results in degradation of the death domain kinase, receptor-interacting protein (RIP), and blockage of tumor necrosis factor-induced nuclear factor-kappaB activation. *J. Biol. Chem.*, **275**, 10519–26.

Lorenzo, H. K., Susin, S. A., Penninger, J., and Kroemer, G. (1999). Apoptosis inducing factor (AIF): a phylogenetically old, caspase-independent effector of cell death. *Cell Death Differ.*, **6**, 516–24.

Mariadason, J. M., Corner, G. A., and Augenlicht, L. H. (2000). Genetic reprogramming in pathways of colonic cell maturation induced by short chain fatty acids: comparison with trichostatin A, sulindac, and curcumin and implications for chemoprevention of colon cancer. *Cancer Res.*, **60**, 4561–72.

Meijerink, J. P., Mensink, E. J., Wang, K. *et al.* (1998). Hematopoietic malignancies demonstrate loss-of-function mutations of BAX. *Blood*, **91**, 2991–7.

Mueller, C. M. and Scott, D. W. (2000). Distinct molecular mechanisms of Fas resistance in murine B lymphoma cells. *J. Immunol.*, **165**, 1854–62.

Muschen, M., Warskulat, U., and Beckmann, M. W. (2000). Defining CD95 as a tumor suppressor gene. *J. Mol. Med.*, **78**, 312–25.

Nimmanapalli, R. and Bhalla, K. (2002). Novel targeted therapies for Bcr-Abl-positive acute leukemias: beyond STI571. *Oncogene*, **21**, 8584–90.

Olson, M. and Kornbluth, S. (2001). Mitochondria in apoptosis and human disease. *Curr. Mol. Med.*, **1**, 91–122.

Ozoren, N., Fisher, M. J., Kim, K. *et al.* (2000). Homozygous deletion of the death receptor DR4 gene in a nasopharyngeal cancer cell line is associated with TRAIL resistance. *Int. J. Oncol.*, **16**, 917–25.

Pandey, P., Saleh, A., Nakazawa, A. *et al.* (2000). Negative regulation of cytochrome c-mediated oligomerization of Apaf-1 and activation of procaspase-9 by heat shock protein 90. *EMBO J.*, **19**, 4310–22.

Pelengaris, S., Rudolph, B., and Littlewood, T. (2000). Action of Myc in vivo – proliferation and apoptosis. *Curr. Opin. Genet. Dev.*, **10**, 100–5.

Perez, D. and White, E. (2003). E1A sensitizes cells to tumor necrosis factor alpha by downregulating c-FLIP S. *J. Virol.*, **77**, 2651–62.

Prendergast, G. C. (1999). Mechanisms of apoptosis by c-Myc. *Oncogene*, **18**, 2967–87.

Pruschy, M., Rocha, S., Zaugg, K. *et al.* (2001). Key targets for the execution of radiation-induced tumor cell apoptosis: the role of p53 and caspases. *Int. J. Radiat. Oncol. Biol. Phys.*, **49**, 561–7.

Puthalakath, H. and Strasser, A. (2002). Keeping killers on a tight leash: transcriptional and post-translational control of the pro-apoptotic activity of BH3-only proteins. *Cell Death Differ.*, **9**, 505–12.

Rampino, N., Yamamoto, H., Ionov, Y. *et al.* (1997). Somatic frameshift mutations in the BAX gene in colon cancers of the microsatellite mutator phenotype. *Science*, **275**, 967–9.

Ravagnan, L., Roumier, T., and Kroemer, G. (2002). Mitochondria, the killer organelles and their weapons. *J. Cell Physiol.*, **192**, 131–7.

Richmond, A. (2002). Nf-kappa B, chemokine gene transcription and tumor growth. *Nat. Rev. Immunol.*, **2**, 664–74.

Richon, V. M., Emiliani, S., Verdin, E. *et al.* (1998). A class of hybrid polar inducers of transformed cell differentiation inhibits histone deacetylases. *Proc. Natl. Acad. Sci. USA*, **95**, 3003–7.

Rossi, A., Kapahi, P., Natoli, G. *et al.* (2000). Anti-inflammatory cyclopentenone prostaglandins are direct inhibitors of IkappaB kinase. *Nature*, **403**, 103–8.

Rudner, J., Jendrossek, V., and Belka, C. (2002). New insights in the role of Bcl-2: Bcl-2 and the endoplasmic reticulum. *Apoptosis*, **7**, 441–7.

Sato, S., Fujita, N., and Tsuruo, T. (2000). Modulation of Akt kinase activity by binding to Hsp90. *Proc. Natl. Acad. Sci. USA*, **97**, 10832–7.

Seitz, S., Wassmuth, P., Fischer, J. *et al.* (2002). Mutation analysis and mRNA expression of trail-receptors in human breast cancer. *Int. J. Cancer*, **102**, 117–28.

Shi, Y. (2002). Mechanisms of caspase activation and inhibition during apoptosis. *Mol. Cell*, **9**, 459–70.

Shibata, M. A., Liu, M. L., Knudson, M. C. *et al.* (1999). Haploid loss of bax leads to accelerated mammary tumor development in C3(1)/SV40-TAg transgenic mice: reduction in protective apoptotic response at the preneoplastic stage. *EMBO J.*, **18**, 2692–701.

Shin, M. S., Kim, H. S., Lee, S. H. *et al.* (2001). Mutations of tumor necrosis factor-related apoptosis-inducing ligand receptor 1 (TRAIL-R1) and receptor 2 (TRAIL-R2) genes in metastatic breast cancers. *Cancer Res.*, **61**, 4942–6.

Simon, R., Radmacher, M. D., Dobbin, K., and McShane, L. M. (2003). Pitfalls in the use of DNA microarray data for diagnostic and prognostic classification. *J. Natl. Cancer Inst.*, **95**, 14–18.

Skorski, T. (2002). Oncogenic tyrosine kinases and the DNA-damage response. *Nat. Rev. Cancer*, **2**, 351–60.

Skorski, T., Bellacosa, A., Nieborowska-Skorska, M. *et al.* (1997). Transformation of hematopoietic cells by BCR/ABL requires activation of a PI-3k/Akt-dependent pathway. *EMBO J.*, **16**, 6151–61.

Smyth, M. J., Takeda, K., Hayakawa, Y., Peschon, J. J., van den Brink, M. R., and Yagita, H. (2003). Nature's TRAIL – on a path to cancer immunotherapy. *Immunity*, **18**, 1–6.

Soengas, M. S., Capodieci, P., Polsky, D. *et al.* (2001). Inactivation of the apoptosis effector Apaf-1 in malignant melanoma. *Nature*, **409**, 207–11.

Sporn, M. B., Suh, N., and Mangelsdorf, D. J. (2001). Prospects for prevention and treatment of cancer with selective PPARgamma modulators (SPARMs). *Trends Mol. Med.*, **7**, 395–400.

Teitz, T., Lahti, J. M., and Kidd, V. J. (2001). Aggressive childhood neuroblastomas do not express caspase-8: an important component of programmed cell death. *J. Mol. Med.*, **79**, 428–36.

Teitz, T., Wei, T., Valentine, M. B. *et al.* (2000). Caspase 8 is deleted or silenced preferentially in childhood neuroblastomas with amplification of MYCN. *Nat. Med.*, **6**, 529–35.

Tepper, C. G. and Seldin, M. F. (1999). Modulation of caspase-8 and FLICE-inhibitory protein expression as a potential mechanism of Epstein–Barr virus tumorigenesis in Burkitt's lymphoma. *Blood*, **94**, 1727–37.

Thome, M., Schneider, P., Hofmann, K. *et al.* (1997). Viral FLICE-inhibitory proteins (FLIPs) prevent apoptosis induced by death receptors. *Nature*, **386**, 517–21.

Timmer, T., de Vries, E. G., and de Jong, S. (2002). Fas receptor-mediated apoptosis: a clinical application? *J. Pathol.*, **196**, 125–34.

West, K. A., Sianna Castillo, S., and Dennis, P. A. (2002). Activation of the PI3K/Akt pathway and chemotherapeutic resistance. *Drug Resist. Update*, **5**, 234–48.

Wong, W. W., Dimitroulakos, J., Minden, M. D., and Penn, L. Z. (2002). HMG-CoA reductase inhibitors and the malignant cell: the statin family of drugs as triggers of tumor-specific apoptosis. *Leukemia*, **16**, 508–19.

Yamasaki, L. (1999). Balancing proliferation and apoptosis in vivo: the Goldilocks theory of E2F/DP action. *Biochim. Biophys. Acta*, **1423**, M9–15.

4

Neuronal cell death in human neurodegenerative diseases and their animal/cell models

Lee J. Martin, Zhiping Liu, Juan Troncoso, and Donald L. Price

Departments of Pathology, Division of Neuropathology, Neuroscience, and Neurology, Johns Hopkins University School of Medicine, Baltimore, Maryland, USA

4.1 Introduction

Cell death is important in the normal histogenesis of organs, steady state kinetics of healthy adult tissues, pathogenesis of tissue damage and disease, and disease therapy. Pathologists conceived the concept of cell death as a mechanism for disease to aid in diagnosis and therapy (Virchow, 1858). Developmental biologists then realized the essential role of cell death in tissue and organ development (Glücksmann, 1951; Lockshin and Williams, 1964; Saunders, 1966). This idea was at first received with skepticism, but now it is dogma that cell number in tissues is controlled precisely in developing and adult tissues. The absence of this precise control of cell number in tissues causes cancer (impaired apoptosis is a central step toward neoplasia). Pathologic stimuli can be extrinsic or intrinsic and can inactivate normal cell death networks or can cause abrupt or delayed cell death. Cell demise can occur as multiple types of death (Schweichel and Merker, 1973; Lockshin and Zakeri, 2002). It is compelling that a goal of human disease therapy is, on the one hand, to prevent cell death in neurologic disease and, on the other hand, to stimulate cell death in malignancy. Thus, the study of cell death is fundamental to human pathobiology and disease treatment. In this chapter, recent critical views on the contributions of the different forms of cell death to human neurodegenerative diseases and their animal and cell models will be presented.

4.2 Types of cell death

Cells can die in different ways. This basic idea has been pursued since the late 1800s (Lockshin and Zakeri, 2001). The death of cells has been classified generally

Apoptosis in Health and Disease: Clinical and Therapeutic Aspects, ed. Martin Holcik, Alex E. MacKenzie, Robert G. Korneluk, and Eric C. LaCasse. Published by Cambridge University Press. © Cambridge University Press 2004.

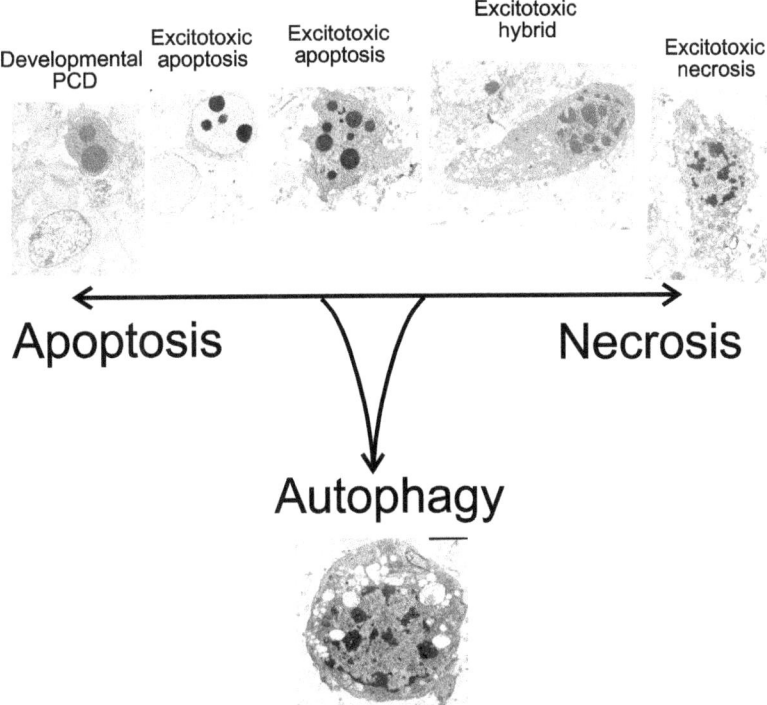

Figure 4.1   *Gallery of cell death*. Electron micrographs of striatal neurons at mid-stages of degeneration. Developmental PCD of neurons in the early postnatal brain is a "gold standard" for neuronal apoptosis. Glutamate receptor excitotoxicity induces classical apoptosis, hybrids of apoptosis and necrosis, classical necrosis, and autophagy. A spectrum of cell death morphologies can be identified that suggests a continuum for cell death.

as three distinct types – called apoptosis, necrosis, and autophagy. These forms of cellular degeneration were originally classified differently because microscopically they look different (Figure 4.1). The orderly dismantling of selective cells by apoptosis (derived from a Greek word for "dropping of leaves from trees") is distinct from the lytic destruction of necrosis. Apoptosis was later considered as an example of programmed cell death (PCD) – that is, a genetically driven form of cell suicide – but other non-apoptotic forms of genetically regulated PCD exist (Lockshin and Zakeri, 2002). Another road to cell death is autophagy. Autophagy is an intracellular catabolic process that occurs by lysosomal degradation of damaged or expendable organelles. However, some of the morphologic distinctions between different forms of cell death are now becoming blurred with the proposal that cell death could exist as a continuum with apoptosis and necrosis at opposite ends of this continuum (Figure 4.1) (Hirsch *et al.*, 1997; Leist *et al.*, 1997; Portera-Cailliau *et al.*, 1997a,b; Martin *et al.*, 1998). The molecular mechanisms of cell death are now becoming known (Orrenius *et al.*, 2003; Yuan *et al.*, 2003), further revealing the distinctiveness of different cell death processes as well as the potential overlap among different cell death forms.

4.2.1 Necrosis

Cell death resulting from cytoplasmic swelling and karyolysis has been classified traditionally as necrosis (Trump *et al.*, 1964). Necrosis results from rapid and severe failure to sustain cellular homeostasis, notably volume control, rather than from a specific molecular program within dying cells as in PCD (Majno and Joris, 1995; Trump and Berezesky, 1996). Nevertheless, with necrosis, specific signaling pathways still could be activated to cause cell death (Proskuryakov *et al.*, 2003). The process of necrosis involves damage to the structural and functional integrity of the cell plasma membrane and associated enzymes, for example Na^+, K^+ ATPase, rapid mitochondrial damage and energetic failure, and abrupt influx of ions (e.g. Na^+ and Ca^{2+}) and H_2O (Bonfoco *et al.*, 1995; Leist *et al.*, 1997; Golden *et al.*, 2001). Metabolic inhibition and oxidative stress by reactive oxygen species (ROS) are major culprits in the generation of necrosis. The structure of classic necrosis is distinct (Figure 4.1). The main features are swelling and degeneration of organelles, destruction of membrane integrity, random digestion of chromatin, and dissolution of the cell. The overall configuration of the moribund cell is maintained as it blends into the surrounding tissue parenchyma and induces an inflammatory reaction in vivo. In necrosis, dying cells do not bud to form discrete, membrane-bound fragments. The nuclear pyknosis and karyolysis appear as condensation of chromatin into many irregularly shaped, small clumps, sharply contrasting with the formation of few, uniformly dense and regularly shaped chromatin aggregates that occur in apoptosis. These differences in the cytoplasmic changes and condensation of nuclear chromatin in pure apoptosis and pure necrosis are very diagnostic.

4.2.2 Apoptosis

Apoptosis is a form of PCD because it is carried out by active, intrinsic transcription-dependent (Tata, 1966) or -independent mechanisms involving specific molecules (Table 4.1). Apoptosis should not be used as a synonym for PCD because non-apoptotic forms of PCD exist (Schwartz *et al.*, 1993; Amin *et al.*, 2000) and because apoptosis refers to a morphologic process. Apoptosis is only one example of PCD; it is critical for the normal growth and differentiation of organ systems in vertebrates and invertebrates (Glücksmann, 1951; Lockshin and Williams, 1964; Saunders, 1966; Tata, 1966). The structure of apoptosis is similar to the type I form of PCD described by Clarke (Clarke, 1990). In physiologic settings in adult tissues, apoptosis is a normal process, occurring continuously in populations of cells that

Table 4.1 *Molecular regulation of apoptosis*

Bcl-2 family				
Anti-apoptotic proteins	Pro-apoptotic proteins	Caspase family	IAP family	Tumor suppressor
Bcl-2	Bax	Apoptosis "initiators:" caspase-2, -8, -9, -10	NAIP	p53 p63 p73
Bcl-x$_L$	Bak	Apoptosis "executioners:" caspase-3, -6, -7	cIAP1	
Boo	Bcl-x$_S$ Bad Bid Bik Spike Noxa Puma	Cytokine processors: caspase-1, -4, -5, -11	cIAP2 XIAP	PTEN

undergo slow proliferation (e.g. liver and adrenal glands) or rapid proliferation (e.g. epithelium of intestinal crypts) (Wyllie *et al.*, 1980; Bursch *et al.*, 1990). Apoptosis is a normal event in the immune system when lymphocyte clones are deleted after an immune response (Nagata, 1999). Kerr and colleagues were the first to describe apoptosis in pathologic settings (Kerr *et al.*, 1972), but many descriptions were made prior to this time in studies of developing systems (Lockshin and Zakeri, 2001). The link between apoptosis and a Ca^{2+}-activated DNase that generated a DNA ladder was identified by Wyllie (1980).

Classical apoptosis has a distinctive structural appearance (Figure 4.1). The cell condenses and is dismantled in an organized way into small packages that can be consumed by nearby cells. Nuclear breakdown is orderly. The chromatin condenses into sharply delineated, uniformly dense masses that appear as crescents abutting the nuclear envelope or as smooth, round masses within the nucleus (Figure 4.1). The emergence of the apoptotic nuclear morphology may be independent of the degradation of chromosomal DNA (Sakahira *et al.*, 1999). Cytoplasmic breakdown is also orderly. The cytoplasm condenses (as reflected by a darkening of the cell in electron micrographs; Figure 4.1), and subsequently the cell shrinks in size, while the plasma membrane remains intact. Condensation of the cytoplasm can be associated with the formation of translucent cytoplasmic vacuoles. The origin of these clear vacuoles is still uncertain, but they may be derived from the endoplasmic reticulum (ER) or the Golgi apparatus. During the course of these events, it is believed that the mitochondria are required for apoptosis. Subsequently, the nuclear and plasma membranes become convoluted, and then the cell undergoes a process called budding. In this process, the nucleus, containing smooth, uniform masses of condensed chromatin, undergoes fragmentation in association with the condensed cytoplasm, forming cellular debris (called apoptotic bodies) composed of pieces of nucleus surrounded by cytoplasm with closely packed and apparently intact organelles. A double membrane surrounds some nuclear fragments, but others are not membrane bound. Apoptotic cells display surface markers (e.g. phosphatidylserine or sugars) for recognition by phagocytic cells. Phagocytosis of cellular debris by adjacent cells is the final phase of apoptosis in vivo.

It has been known that variants of classical apoptosis or non-classical apoptosis can occur during nervous system development (Pilar and Landmesser, 1976; Clarke, 1990) and also in pathophysiologic settings in the nervous system (Portera-Cailliau *et al.*, 1997a,b; Martin *et al.*, 1998). Axotomy and target deprivation in the mature nervous system can induce apoptosis in neurons that is similar structurally to, but not identical to, developmental PCD (Al-Abdulla and Martin, 1998; Al-Abdulla *et al.*, 1998; Martin *et al.*, 1999). Exposure to neuronal excitotoxins can induce non-classical forms of apoptosis (Portera-Cailliau *et al.*, 1997a,b). Types of cell death similar to those seen with excitotoxicity occur in pathologic cell death resulting from hypoxia–ischemia (Martin *et al.*, 2000a; Nakajima *et al.*, 2000; Northington *et al.*, 2001).

Cells can die by PCD carried out through mechanisms that are distinct from apoptosis (Schwartz *et al.*, 1993; Amin *et al.*, 2000). The structure of non-apoptotic PCD is similar to the type II or type III forms of cell death described by Clarke (Clarke, 1990). Interestingly, there is no inter-nucleosomal fragmentation of DNA in some forms of non-apoptotic PCD (Schwartz *et al.*, 1993; Amin *et al.*, 2000).

4.2.3 Autophagy

Autophagy is a mechanism whereby eukaryotic cells degrade their own cytoplasm and organelles (Klionsky and Emr, 2000). The degradation of organelles and long-lived proteins is carried out by the lysosomal system. Autophagy functions as a cell death mechanism and as a homeostatic non-lethal stress response mechanism for recycling proteins to protect cells from low supplies of nutrient. Autophagy is also called type II PCD (Clarke, 1990). A hallmark of autophagic cell death is the accumulation of autophagic vacuoles of lysosomal origin. Autophagy has been linked to developmental and pathologic conditions. Insect metamorphosis involves autophagy (Lockshin and Zakeri, 2001), and developing neuronal populations use autophagy as a PCD mechanism (Schweichel and Merker, 1973; Xue *et al.*, 1999).

Degeneration of Purkinje neurons in the mouse mutant *Lucher* may be a form of autophagy, thus possibly linking excitotoxic cell death, due to constitutive activation of the GluRδ2 glutamate receptor, to autophagy (Yue *et al.*, 2002). Autophagy may have roles in Alzheimer's disease (Cataldo *et al.*, 1994), Parkinson's disease (Anglade *et al.*, 1997), and Huntington's disease (Petersen *et al.*, 2001).

The molecular controls of autophagy appear to be common to eukaryotic cells from yeast to human, and it is believed that autophagy evolved before apoptosis (Yuan *et al.*, 2003). However, most of the work has been done on yeast, with detailed work on mammalian cells only now beginning (Mizushima *et al.*, 2002). Double-membrane autophagosomes for sequestration of cytoplasmic components are derived from the ER or the plasma membrane. Tor kinase, phosphatidylinositol 3 (PI3)-kinase, a family of cysteine proteases called autophagins, and death-associated proteins function in autophagy (Bursch, 2001; Inbal *et al.*, 2002). Autophagic and apoptotic cell death pathways interact. The product of the tumor suppressor gene Beclin1 (the human homolog of the yeast autophagy gene, APG6) interacts with Bcl-2 (Liange *et al.*, 1998). Autophagy can block apoptosis by the sequestration of mitochondria. When autophagic capacity is reduced, stressed cells die by apoptosis, whereas inhibition or blockade of molecules that function in apoptosis turn the cell death into autophagy (Ogier-Denis and Codogno, 2003). Thus, a continuum between autophagy and apoptosis exists (Figure 4.1).

4.2.4   The cell death continuum

Apoptosis and necrosis as well as apoptosis and autophagy may represent extremes in graded scales of cell death (Figure 4.1). Apoptosis can be induced by injurious stimuli of lesser amplitude than insults causing necrosis (Raffray and Cohen, 1997). Toxicologic studies of cultured non-neuronal cells have verified that stimulus intensity influences the mode of cell death, although the modes of cell death are still viewed as being mechanistically distinct (Lennon *et al.*, 1991; Fernandes and Cotter, 1994; Bonfoco *et al.*, 1995). The death of neurons is not always strictly via apoptosis or necrosis, according to a traditional binary classification of cell death, but also occurs as intermediate or hybrid forms with co-existing characteristics that lie along a structural continuum, with apoptosis and necrosis at the extremes (Portera-Cailliau *et al.*, 1997a,b). The neuronal cell death continuum is best revealed in in vivo excitotoxic paradigms (Portera-Cailliau *et al.*, 1997a,b) and in closely related insults such as hypoxia–ischemia in newborn rat brain (Nakajima *et al.*, 2000; Northington *et al.*, 2001). These hybrid cells undergo progressive compaction of chromatin into a few, discrete, large, irregularly shaped clumps (Figure 4.1). This morphology contrasts with the formation of few, uniformly shaped, dense, round masses in classic apoptosis and the formation of numerous, smaller, irregularly shaped chromatin clumps in classic necrosis. The cytoplasmic changes in hybrid cells have a basic pattern that appear to be more similar to necrosis than apoptosis, but which differ in severity.

The molecular mechanisms that drive this structural continuum are speculative. ATP levels (Leist *et al.*, 1997), intracellular Ca^{2+} levels (Trump and Berezesky, 1996), and mitochondrial permeability transition (Hirsch *et al.*, 1997) are likely to be involved. Moreover, the age or maturity of neurons and the subtype of glutamate receptor that is activated influence neuronal cell death along this continuum (Portera-Cailliau *et al.*, 1997a,b; Martin, 2001; Natale *et al.*, 2002). Hence, neuronal death induced by injury is not the same in the mature and immature brain and may not be identical in every neuron. In vitro systems with homogeneous populations of neurons given a constant insult are needed to dissect out the mechanisms of the neuronal cell death continuum more precisely. In cell culture systems, apoptosis mechanisms are different in immature neurons compared with mature neurons (Lesuisse and Martin, 2002).

The concept of the cell death continuum can be extended to a hypothetical cell death matrix (Figure 4.2). Therefore, when studying mechanisms of cell death in a system, it is important to consider that the presence of apoptosis, necrosis, autophagy, or non-apoptotic PCD may not be strictly "black and white." A matrix might be a useful tool for pathology in general and for delineating how drugs and other treatments for human disease (e.g. cancer) work. The specific contributions of the different forms of cell death, and the possible identification of previously unrecognized forms of cell death, need to be identified in human diseases and in their animal models. We need to identify better the relationships between mechanisms of cell death and the structure of dying cells in human pathology in developing and adult organs as well as in animal and cell models of toxicity in immature and mature cells. The concept of a cell death matrix could be important for understanding neuronal degeneration in a variety of pathophysiologic settings, and thus may be important for mechanism-based neuroprotective treatments in neurologic disorders in infants, children, and adults. If brain maturity dictates how neurons die (Martin *et al.*, 1998; Martin, 2001), then major forms of neuronal degeneration in adults may be mechanistically different from neuronal degeneration in infants and children; thus, therapeutic targets could be dissimilar in different age groups.

4.3 Molecular and cellular regulation of apoptosis

Apoptosis is a structurally and biochemically organized form of cell death. The basic machinery of apoptosis is conserved in yeast, hydra, nematode, fruitfly, zebrafish, mouse, and human (Ameisen, 2002). Much of our current understanding of the molecular mechanisms of apoptosis in mammalian cells is built on studies by Horvitz and colleagues of PCD in the nematode *Caenorhabditis elegans*. They

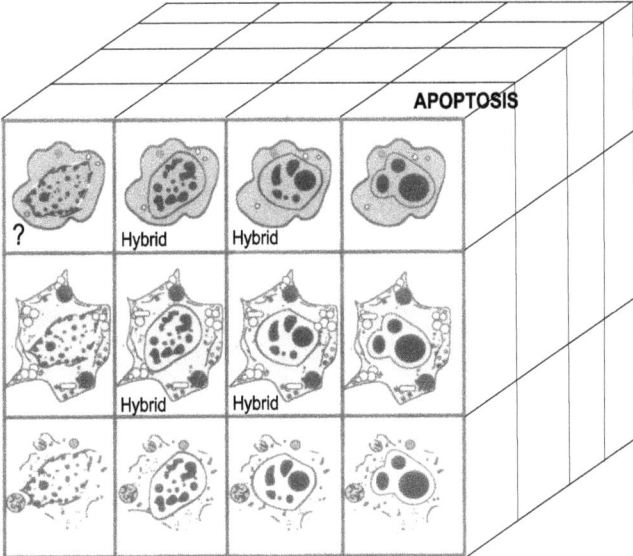

NECROSIS

Figure 4.2   *Cell death matrix.* This diagram summarizes in a three-dimensional format the concept of the apoptosis–necrosis continuum of cell death. The front matrix of the cube shows some of the numerous possible structures of neuronal cell death near or at the terminal stages of degeneration. Combining different nuclear morphologies and cytoplasmic morphologies generates a non-linear matrix of possible cell death structures. In the cell at the extreme upper right corner, nuclear and cytoplasmic morphologies combine to form an apoptotic neuron that is typical of naturally occurring PCD during nervous system development. This death is classical apoptosis. In contrast, in the cell at the extreme lower left corner, the merging of necrotic nuclear and necrotic cytoplasmic morphologies forms a typical necrotic neuron resulting from NMDA receptor excitotoxicity and cerebral ischemia. Between these two extremes, hybrids of cell death can be produced with varying contributions of apoptosis and necrosis. The typical apoptosis–necrosis hybrid cell death structure is best exemplified by neurons in the adult CNS dying from non-NMDA GluR-mediated excitotoxicity. Other apoptosis–necrosis hybrid forms of neuronal cell death might occur in chronic neurodegenerative disease such as ALS, PD, and HD. The death forms shown in the front matrix of the cube represent only a small number of the possible forms of cell death that can be envisioned to fill the empty cells of the matrix. Neuronal maturity and the subtypes of GluRs that are over-activated are known to influence where an injured/degenerating neuron falls within the matrix. The types and levels of DNA damage that are sustained by a cell might also influence the position of a degenerating cell within the death matrix. This concept may also be relevant to cell death in general, and thus may be applicable to cancer biology and the mechanisms of action of chemotherapies and radiation therapies.

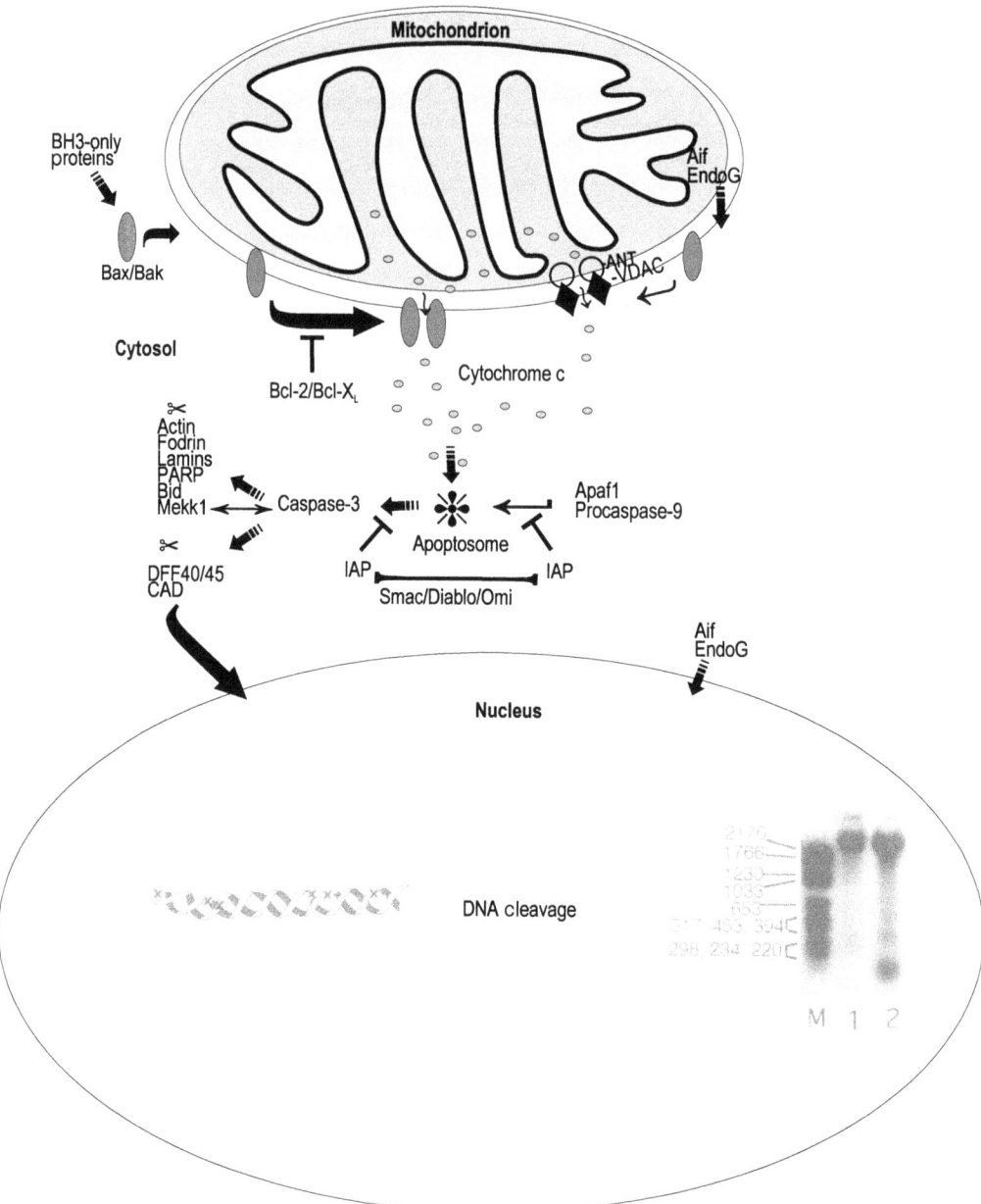

Figure 4.3   *Mitochrondrial regulation of apoptosis.* Bcl-2 family members regulate apoptosis by modulating the release of cytochrome c. Bax and Bak are pro-apoptotic. They physically interact
and form channels that are permeable to cytochrome c. BH3-only members (e.g. Bid,
Noxa, and Puma) are pro-apoptotic and can modulate the conformation of Bax. Bcl-2 and
Bcl-X$_L$ are anti-apoptotic and can block the function of Bax/Bak. The permeability transition
pore (PTP), formed by the interaction of the adenine nucleotide translocator (ANT)
and the voltage-dependent anion channel (VDAC) during the process of swelling, is a

pioneered the understanding of the genetic control of developmental cell death by showing that this death is regulated predominantly by three genes (*ced-3*, *ced-4*, and *ced-9*). Several families of apoptosis regulation genes (Table 4.1) have been identified in mammals. Apoptotic cell death is controlled by the Bcl-2 family (Merry and Korsmeyer, 1997; Cory and Adams, 2002), the caspase family of cysteine-containing, aspartate-specific proteases (Wolf and Green, 1999), the p53 gene family (Levrero *et al.*, 2000), death receptors (Nagata, 1999), and other apoptogenic factors including Ca^{2+}, cytochrome c, apoptosis-inducing factor (AIF), and second mitochondrial activator of caspases (Smac/DIABLO). Moreover, a family of inhibitor of apoptosis proteins (IAPs) actively blocks cell death (Liston *et al.*, 1996), and IAPs are inhibited by Omi (Hegde *et al.*, 2002).

Specific organelles have been identified as being critical for the apoptotic process, including mitochondria and the ER (Figures 4.3 and 4.4). In a seminal study by Wang and colleagues, it was discovered that the mitochondrion integrates death signals mediated by proteins in the Bcl-2 family and releases molecules residing in the mitochondrial intermembrane space, such as cytochrome c, that activate caspase proteases leading to internucleosomal cleavage of DNA (Li *et al.*, 1997). The finding that cytochrome c, first described in 1930 (Keilin, 1930), has a function in apoptosis, in addition to its more well-known role in oxidative phosphorylation, was astounding, although foreshadowing clues were available. The translocation of cytochrome c from mitochondria to the cytosol with concomitant reduced oxidative phosphorylation was described as the "cytochrome c effect" in γ-irradiated cancer cells (van Bekkum, 1957). The ER which regulates intracellular Ca^{2+} levels appears to participate in a loop with mitochondria to modulate mitochondrial permeability transition and cytochrome c release (Scorrano *et al.*, 2003).

Mitochondrial permeability transition is a mitochondrial state in which the proton-motive force is disrupted (Crompton, 1999; van Gurp *et al.*, 2003). This disruption involves the mitochondrial permeability transition pore (PTP). The PTP is a transmembrane channel formed at the contact sites between the inner

---

transmembrane channel that emerges at contact sites between the inner mitochondrial and the outer mitochondrial membranes. The PTP has a role in regulating mitochondrial membrane potential and the release of cytochrome c. In the cytosol, cytochrome c, Apaf1, and procaspase-9 interact to form the apoptosome that drives the activation of caspase-3. The family of inhibitors of apoptosis (IAPs) block this process. The IAPs are inhibited by mitochondrially derived Smac, Diablo, and Omi. Caspase-3 cleaves many substrate proteins, some of which are endonucleases that translocate to the nucleus to cleave DNA into internucleosomal fragments (see DNA gel at lower right, showing molecular weight standards [M], control brain tissue [lane 1], and brain tissue undergoing apoptosis [lane 2]). Aif and endonuclease G are mitochondrially released proteins with nuclease activity that can translocate to the nucleus. See text for detailed descriptions.

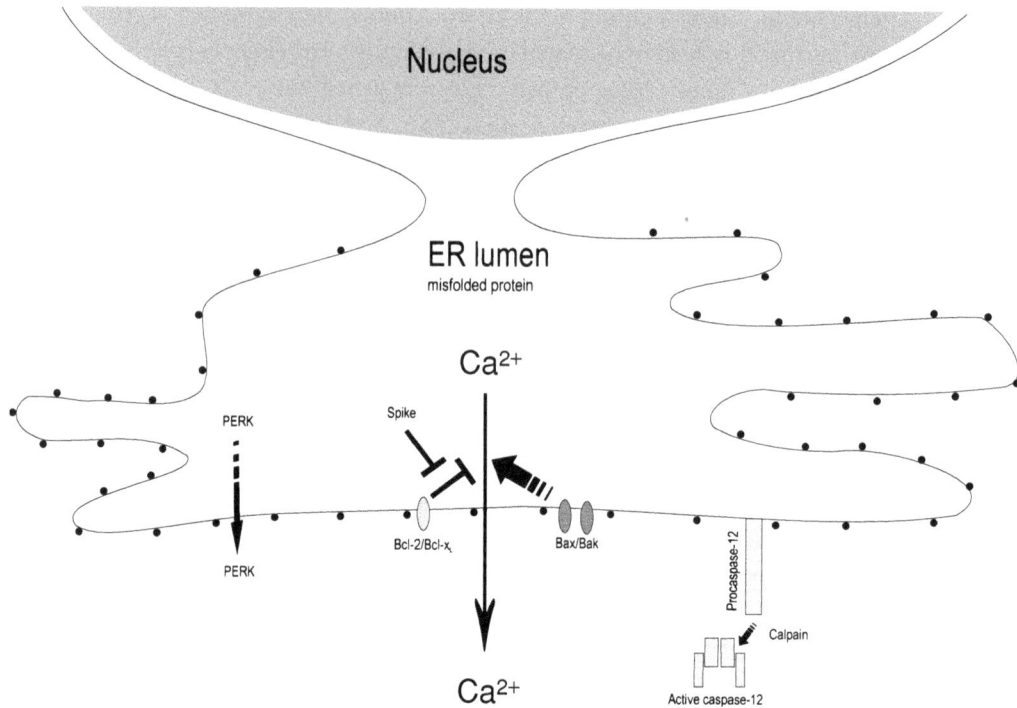

Figure 4.4   *The endoplasmic reticulum (ER) functions in apoptosis.* Under conditions of ER stress, such as conditions resulting in protein misfolding, Bax and Bak regulate the release of ER Ca^{2+} into the cytosol. Bcl-2 and Bcl-X_L can block the release of ER Ca^{2+}. The BH3-only protein Spike inhibits Bcl-2/Bcl-X_L. Procaspase-12 is localized to the ER and is cleaved by calpain into the active form in response to prolonged stress. The elF2α kinase known as PERK is a transmembrane protein resident in the ER. PERK functions to couple ER stress signals to translation inhibition. ER stress increases the activity of PERK, which then phosphorylates elF2α, resulting in reduced translation and cell cycle arrest.

mitochondrial membrane and the outer mitochondrial membrane. The components of the PTP are the voltage-dependent anion channel (VDAC) in the outer mitochondrial membrane and the adenine nucleotide translocator (ANT) in the inner mitochondrial membrane (Crompton, 1999). The VDAC makes the inner mitochondrial membrane permeable to most small molecules <5 kilodaltons (kD) for free exchange of respiratory chain substrates. The ANT mediates the exchange of ADP for ATP. During normal mitochondrial function, the intermembrane space separates the outer and inner mitochondrial membranes and the VDAC and the ANT do not interact. When mitochondrial permeability is activated by the formation of the PTP, the inner mitochondrial membrane loses its integrity, and oxidative phosphorylation is uncoupled. When this occurs, oxidation of metabolites by O_2 proceeds with electron flux not coupled to proton pumping, resulting in dissipation of the transmembrane proton gradient and ATP production and the production of ROS (van Grup *et al.*, 2003). Bcl-2 family members modulate mitochondrial permeability transition.

4.3.1 Bcl-2 family

The *bcl-2* proto-oncogene family is a group of apoptosis regulatory genes encoding for proteins defined by at least one conserved Bcl homology domain (BH1–BH4 can be present) that function in protein–protein interactions (Merry and Korsmeyer, 1997; Cory and Adams, 2002). At present, this family encodes for more than 20 proteins. Some of the products of these genes (e.g. Bcl-2, Bcl-x_L, and Boo) are anti-apoptotic, whereas other gene products (e.g. Bax, Bcl-x_S, Bad, Bak, Bid, Bim, Bik, Spike, Noxa, and Puma) are pro-apoptotic (Table 4.1). Because Bcl-x_L possesses structural similarities to the pore-forming subunit of diphtheria toxin (Much-more *et al.*, 1999), Bcl-2 family members appear to function by conformation-induced insertion into the outer mitochondrial membrane to form channels or pores that release apoptogenic factors (Figure 4.3). These proteins have differential tissue distributions and subcellular localizations. The tissue distributions of Bcl-2, Bcl-x_L, Bcl-x_S, Bax, Bak, and Bad are widespread, whereas the expression of Boo is highly restricted (Song *et al.*, 1999). The subcellular distributions of Bax, Bak, and Bad in healthy adult rodent central nervous system (CNS) tissue (Martin *et al.*, 2003) are consistent with in vitro studies of non-neuronal cells (Wolter *et al.*, 1997; Nechushtan *et al.*, 2001). Bax, Bad, and Bcl-2 reside primarily in the cytosol, whereas Bak resides primarily in mitochondria. Bcl-2 family members can form homodimers or heterodimers and higher order multimers with other family members. Bax forms homodimers or heterodimers with Bak, Bcl-2, or Bcl-x_L. When Bax and Bak are present in excess, the anti-apoptotic activity of Bcl-2 is antagonized. The formation of Bax homo-oligomers promotes apoptosis through the formation of channels that release cytochrome c, whereas Bax heterodimerization with either Bcl-2 or Bcl-x_L prevents apoptosis by blocking the release of cytochrome c. In mammalian cells, cytochrome c triggers the assembly of the cytoplasmic apoptosome (a protein complex of Apaf1, cytochrome c, and procaspase-9) which is the engine that drives caspase-3 activation (Li *et al.*, 1997). Release of apoptogenic proteins from mitochondria may occur through mechanisms that involve the formation of membrane channels comprised of Bax (Antonsson *et al.*, 1997), Bax and the adenine nucleotide translocator (Marzo *et al.*, 1998), and the voltage-dependent anion channel (Shimizu *et al.*, 2000). Bcl-2 and Bcl-x_L block the release of cytochrome c and AIF (Kluck *et al.*, 1997; Yang *et al.*, 1997; Susin *et al.*, 1999) from mitochondria and thus the activation of caspase-3 (Liu *et al.*, 1996; Li *et al.*, 1997). The blockade of cytochrome c release from mitochondria by Bcl-2 and Bcl-x_L (Liu *et al.*, 1996; Vander Heiden *et al.*, 1997) is caused by the inhibition of Bax channel-forming activity in the outer mitochondrial membrane (Antonsson *et al.*, 1997) or by the modulation of mitochondrial membrane potential and volume homeostasis (Vander Heiden *et al.*, 1997). Bcl-x_L also has anti-apoptotic activity by interacting with Apaf-1 and caspase-9 and inhibiting the Apaf-1-mediated maturation of caspase-9 (Hu *et al.*, 1998). Boo can inhibit Bak- and Bik-induced apoptosis (but not Bax-induced cell death) possibly through heterodimerization and by interactions with Apaf-1 and caspase-9 (Song *et al.*, 1999). Bax and Bak double knock-out cells are completely resistant to mitochondrial cytochrome c release during apoptosis (Wei *et al.*, 2001). BH3-only proteins such as Bim and Bid appear to induce a conformational change in Bax that allow it to form pores in the outer mitochondrial membrane (Letai *et al.*, 2002).

Although many studies have focused on how Bcl-2 family members regulate apoptosis at the mitochondrial level, it is now becoming evident that ER stress can initiate apoptosis (Figure 4.4). This finding is very relevant to neurodegeneration, where protein misfolding may be important to pathogenesis (Bucciantini *et al.*, 2002). The ER functions to fold proteins and, when this capacity is compromised, an unfolded protein response (UPR) is engaged. The UPR can lead to a return to homeostasis or to cell death. Studies have found Bcl-2 localized to the ER (Lithgow *et al.*, 1994). Over-expression of Bcl-2 and Bcl-x_L can block thapsigargin-induced apoptosis (Srivastava *et al.*, 1999). Bak and Bax also operate in the ER and function in the activation of ER-specific caspase-12 (Zong *et al.*, 2003). Cells lacking Bax and Bak are resistant to ER stress-induced apoptosis (Wei *et al.*, 2001). A recently identified pro-apoptotic Bcl-2 family member called Spike is found exclusively in the ER and functions to inhibit the formation of a complex between Bcl-x_L and Bcl-2-associated protein-31 (Mund *et al.*, 2003).

Protein phosphorylation regulates the functions of Bcl-2 family members. Following serine phosphorylation, Bcl-2 losses its anti-apoptotic activity, possibly because its antioxidant function is inactivated (Haldar *et al.*, 1995). In addition to interacting with homologous proteins, Bcl-2 can associate with non-homologous proteins, including the protein kinase Raf-1 (Wang *et al.*, 1996). Bcl-2 is thought to target Raf-1 to mitochondrial membranes, allowing this kinase to phosphorylate Bad at serine residues. The phosphatidylinositol 3-kinase (PI3-K)–Akt pathway also regulates the function of Bad (Datta *et al.*, 1997; del Peso *et al.*, 1997) and caspase-9 (Cardone *et al.*, 1998) through phosphorylation. In the presence of trophic factors, Bad is phosphorylated. Phosphorylated Bad translocates to the cytosol and interacts with soluble protein 14-3-3 and, when bound to protein 14-3-3, Bad is unable to interact with Bcl-2 and Bcl-x_L, thereby promoting survival (Zha *et al.*, 1996). Conversely, when Bad is dephosphorylated by calcineurin (Wang *et al.*, 1999), it dissociates from protein 14-3-3 in the cytosol and translocates to the mitochondria where it expresses pro-apoptotic activity. Non-phosphorylated Bad heterodimerizes with membrane-associated Bcl-2 and Bcl-x_L, thereby displacing

Bax from Bax–Bcl-2 and Bax–Bcl-x_L dimers and promotes cell death (Yang et al., 1995). The phosphorylation status of Bad helps to regulate glucokinase activity, thereby linking glucose metabolism to apoptosis (Danial et al., 2003).

4.3.2 Caspases

Caspases (cysteinyl aspartate-specific proteinases) are cysteine proteases that have a near absolute requirement for aspartate in the P_1 position of the peptide bond. Fourteen members have been identified (Wolf and Green, 1999). Caspases exist as constitutively expressed pro-enzymes (30–50 kD) in healthy cells. The protein contains three domains: an amino-terminal pro-domain, a large subunit (∼20 kD), and a small subunit (∼10 kD). Caspases are activated through regulated proteolysis of pro-enzyme with "initiator" caspases activating "executioner" caspases (Table 4.1), although some caspase pro-enzymes (e.g. caspase-9) have low activity without processing (Stennicke et al., 1999). Other caspase family members function in inflammation by processing cytokines (Table 4.1). Activation of caspases involves proteolytic processing between domains, and then association of large and small subunits to form a heterodimer with both subunits contributing to the catalytic site. Two heterodimers associate to form a tetramer that has two catalytic sites that function independently. Active caspases have many target proteins (Schwartz and Milligan, 1996) that are cleaved during regulated and organized cell death. Caspases cleave nuclear proteins (e.g. PARP, DNA–PK, heteronuclear ribonucleoproteins, transcription factors, or lamins), cytoskeletal proteins (e.g. actin and fodrin), and cytosolic proteins (e.g. other caspases, protein kinases, Bid, and DNases).

In cell models of apoptosis based on in vitro experiments using human cell lines, activation of caspase-3 occurs when caspase-9 pro-enzyme (also known as Apaf-3) is bound by Apaf-1 in a process initiated by cytochrome c (identified as Apaf-2) and either ATP or dATP (Li et al., 1997). Cytosolic ATP or dATP are required co-factors for cytochrome c-induced caspase activation. Apaf-1, a 130 kD protein, serves as a docking protein for procaspase-9 (Apaf-3) and cytochrome c (Li et al., 1997). Apaf-1 becomes activated when ATP is bound and hydrolyzed, with the hydrolysis of ATP and the binding of cytochrome c promoting Apaf-1 oligomerization (Zou et al., 1999). This oligomeric complex recruits and activates procaspase-9 (forming the apoptosome) that disassociates from the complex and becomes available to activate caspase-3. Once activated, caspase-3 cleaves a protein with DNase activity (DFF-45), and this cleavage activates a pathway leading to the internucleosomal fragmentation of genomic DNA (Liu et al., 1997).

So far, three caspase-related signaling pathways have been identified that can lead to apoptosis (Li et al., 1997; Li et al., 1998), but there seems to be room for considerable overlap in these pathways. The intrinsic mitochondria-mediated pathway is controlled by Bcl-2 family proteins. It is regulated by cytochrome c release from mitochondria, promoting the activation of caspase-9 through Apaf-1 and then caspase-3 activation. This pathway is also regulated by Smac and Omi. The extrinsic death receptor pathway involves the activation of cell-surface death receptors, including Fas and tumor necrosis factor receptor, leading to the formation of the death-inducible signaling complex (DISC) and caspase-8 activation that, in turn, cleaves and activates downstream caspases such as caspase-3, -6, and -7. Caspase-8 can also cleave Bid, leading to the translocation, oligomerization, and insertion of Bax or Bak into the mitochondrial membrane. Another pathway involves the activation of caspase-2 by DNA damage as a pre-mitochondrial signal (Robertson *et al.*, 2002).

Not all forms of apoptotic cell death are caspase dependent (Beresford *et al.*, 2001; Fan *et al.*, 2003). The serine protease granzyme A (GrA) mediates a caspase-independent apoptotic pathway (Beresford *et al.*, 2001). GrA is delivered to target cells through Ca^{2+}-dependent, perforin-generated pores and activates a DNase (GrA–DNase, GAAD). GAAD activity is inhibited by a specific inhibitor, known as the SET complex, which is located in the ER and comprised of the nucleosome assembly protein SET, an inhibitor of protein phosphatase 2A, apurinic endonuclease-1, and a high mobility group protein (a non-histone DNA-binding protein that induces alterations in DNA architecture). GrA cleaves components of the SET complex to release activated GAAD that translocates to the nucleus to induce single-strand DNA nicks and apoptosis (Fan *et al.*, 2003).

4.3.3 IAP family

To prevent unwanted apoptosis in normal cells, the activity of pro-apoptotic proteins must be placed in check. Apoptosis is regulated by the IAP family, in addition to Bcl-2, Bcl-x_L, and Boo (Liston *et al.*, 1996; LaCasse *et al.*, 1998; Holcik, 2002). In mammals, this family includes X-chromosome-linked IAP (XIAP), cIAP1, cIAP2, NAIP (neuronal apoptosis inhibitory protein), Survivin, Livin, Ts-IAP, and Apollon. These proteins are characterized by one or more baculoviral IAP repeats consisting of a zinc finger domain of ~70–80 amino acids (Holcik, 2002). The main identified anti-apoptotic function of IAPs is the suppression of caspase activity (Deveraux *et al.*, 1998). Procaspase-9 and procaspase-3 are major targets of IAPs. IAPs reversibly interact directly with caspases to block substrate cleavage. However, IAPs do not prevent caspase-8-induced proteolytic activation of procaspase-3. IAPs can also block apoptosis by reciprocal interactions with the nuclear transcription factor, NF6B (LaCasse *et al.*, 1998). NAIP is expressed throughout the CNS in neurons (Xu *et al.*, 1997). The importance of the IAP gene family in neurodegeneration is underscored by the finding that NAIP is deleted partially in a significant proportion of children with spinal muscular atrophy (Roy *et al.*, 1995).

Proteins exist that inhibit mammalian IAPs. A murine mitochondrial protein called Smac (second mitochondria-derived activator of caspases) and its human ortholog DIABLO (direct IAP-binding protein with low pI) inactivate the anti-apoptotic actions of IAPs and thus exert pro-apoptotic actions (Du *et al.*, 2000; Verhagen *et al.*, 2000). These IAP inhibitors are 23 kD mitochondrial proteins (derived from 29 kD precursor proteins processed in the mitochondria) that are released from the intermembrane space and sequester IAPs. Omi/Htra2 is another mitochondrial protein that exerts pro-apoptotic activity by inhibiting IAPs (Verhagen *et al.*, 2002).

4.3.4 AIF

AIF is a mammalian cell mitochondrial protein identified as a flavoprotein (Susin *et al.*, 1999). AIF has an N-terminal mitochondrial localization signal and, after import into the inner mitochondrial space, the mitochondrial localization is cleaved off to generate a mature protein of 57 kD. Under normal physiologic conditions, AIF might function as a free radical scavenger targeting hydrogen peroxide (Klein *et al.*, 2002) or in redox cycling with NAD(P)H (Mate *et al.*, 2002). With apoptotic stimuli, AIF translocates to the nucleus (Susin *et al.*, 1999). Over-expression of AIF induces cardinal features of apoptosis, including chromatin condensation, high molecular weight DNA fragmentation, and loss of mitochondrial transmembrane potential (Susin *et al.*, 1999).

4.3.5 p53/p63/p73 family and other tumor suppressors

Tumor suppressors limit cell proliferation and induce apoptosis. Apoptosis can be induced by the tumor suppressor protein p53 and related DNA-binding proteins identified as p73, and p63 (Levrero *et al.*, 2000). p53, p73, and p63 function in growth arrest and/or apoptosis. They commit to death cells that have sustained DNA damage from free radicals, irradiation, and other genotoxic stresses (Levrero *et al.*, 2000). p53 and p73 have similar oligomerization and transactivation properties. p73 exists as a group of full-length isoforms (including p73α and p73β) and as truncated isoforms that lack the transactivation domain (ΔN-p73). p53 is the most well studied of this family of proteins.

p53 is a short-lived protein with a half-life of ~5–20 minutes in most types of cells studied. p53 rapidly accumulates several fold in response to DNA damage. This rapid regulation is mediated by post-translational modification such as phosphorylation and acetylation as well as intracellular redox state (Giaccia and Kastan, 1998). The elevation in p53 protein levels occurs through stabilization and prevention of degradation. p53 is degraded rapidly in a ubiquitination-dependent proteosomal pathway (Maki *et al.*, 1996; Chang *et al.*, 1998). Mdm2 (murine double minute 2; the human homolog is Hdm2) has a crucial role in this degradation pathway

(Shieh *et al.*, 1997). Mdm2 functions in a feedback loop to limit the duration or magnitude of the p53 response to DNA damage. Expression of the Mdm2 gene is controlled by p53 (Shieh *et al.*, 1997). Mdm2 binds to the N-terminal transcriptional activation domain of p53 and regulates the transactivator activity and stability of p53 by direct association. Mdm2 has ubiquitin ligase activity for p53 through the ubiquitin-conjugating enzyme E2. Stabilization of p53 is achieved through phosphorylation of serine[15], resulting in inhibition of the formation of Mdm2–p53 complexes. Activated p53 binds the promoters of several genes encoding proteins associated with growth control and cell cycle checkpoints (e.g. p21, Gadd45, and Mdm2) and apoptosis (e.g. Bax, Bcl-2, Bcl-x_L, and Fas). The BH3-only proteins Puma and Noxa are critical mediators of p53-mediated apoptosis (Villunger *et al.*, 2003).

In sympathetic ganglion neurons, p53 and p73 regulate cell survival. p53 has a critical apoptotic role in response to neurotrophin withdrawal (Aloyz *et al.*, 1998). p53 deficiency protects against neuronal apoptosis induced by axotomy (Martin *et al.*, 2001; Martin and Liu, 2002). p53-mediated neuronal apoptosis can be blocked by the ΔN-p73 isoform by direct binding and inactivation of p53 (Pozniak *et al.*, 2000).

Phosphatase and tensin homolog deleted on chromosome 10 (PTEN) is another tumor suppressor that is relevant to the CNS. PTEN is lipid phosphatase present in human, mouse, and rat brain (Sano *et al.*, 1999; Lachyankar *et al.*, 2000; Martin *et al.*, unpublished observations). Antisense knockdown of PTEN expression induces death of differentiating precursor cells in the subventricular zone (Lachyankar *et al.*, 2000). Mice lacking PTEN die in utero and show cephalic overgrowth (DiCristofano *et al.*, 1998). The role of PTEN in the adult nervous system is less clear, with possible roles in neuronal growth and maintenance (Kwon *et al.*, 2001).

4.3.6 Death receptors

Cell death by apoptosis can also be initiated at the cell membrane by surface death receptors of the tumor necrosis factor (TNF) receptor family. Fas (CD95/Apo-1) and the 75-kD neurotrophin receptor (p75[NTR]) are members of the TNF receptor family (Nagata, 1999). Apoptosis through Fas is independent of new RNA or protein synthesis. The signal for apoptosis is initiated at the cell surface by aggregation (trimerization) of Fas. This activation of Fas is induced by the binding of the multivalent Fas ligand (FasL), a member of the TNF–cytokine family. FasL is expressed on activated T-cells and natural killer cells. Clustering of Fas by FasL recruits Fas-associated death domain (FADD), a cytoplasmic adapter molecule that functions in the activation of the caspase-8–Bid pathway, thus forming the "death-induced signaling complex" (DISC) (Li *et al.*, 1998). In this pathway, Bid (a pro-apoptotic family member that is a substrate for caspase-8) is cleaved in the cytosol, and then truncated Bid translocates to mitochondria, thereby functioning as a transducer of Fas apoptotic signals at the cell plasma membrane to mitochondria (Li *et al.*, 1998). Bid translocation from the cytosol to mitochondrial membranes is associated with a conformational change in Bax (that is prevented by Bcl-2 and Bcl-x_L) and is accompanied by the release of cytochrome c from mitochondria (Desagher *et al.*, 1999).

Apoptosis can be mediated by p75^{NTR} (Kaplan and Miller, 1997). Activation of p75^{NTR} occurs through the binding of nerve growth factor. When p75^{NTR} is activated without Trk receptors, neurotrophin binding induces homodimer formation and activates an apoptotic cascade. p75^{NTR} activation leads to the generation of ceramide through sphingomyelin hydrolysis (Casaccia-Bonnefil *et al.*, 1996). Ceramide production is associated with the activation of Jun N-terminal kinase (JNK) that phosphorylates and activates c-Jun and other transcription factors. p75 mediates hippocampal neuron death in response to neurotrophin withdrawal, involving cytochrome c, Apaf1, and caspases-9, -6, and -3 (but not caspase-8), and thus is different from Fas-mediated cell death (Troy *et al.*, 2002).

4.3.7 Mitogen-activated protein kinase (MAPK) pathway

The MAPK pathways functions in the regulation of cellular proliferation, differentiation, survival, and death (Cardone *et al.*, 1997; English *et al.*, 1999), although in post-mitotic cells (such as neurons) the functions of this pathway are much less understood. Historically, MAPK networks exist as three kinase modules. The Raf-MEKK-MEK-Erk module functions in the transduction of cell type-specific growth and differentiation signals from tyrosine kinase receptors and G-protein-coupled receptors; however, it has become realized recently that this module may also function in cell death. Raf1 is the predominant activator of the Erk module and functions in mitogenesis, cell cycle progression, and apoptosis in non-neuronal cells. The p38 MAPK module mediates inflammatory and stress responses to cytokines. The JNK module has been given functions that include the transduction of stress signals such as oxidation and DNA damage, but there now appears to be functional overlap between the JNK and Erk modules.

The duality of the MAPK pathway in regulating cell death or survival has been realized previously. In non-neuronal cells, basal constitutive activity of Erk functions in cell survival, whereas a transient upregulation of Erk induces apoptosis (Ishikawa and Kitamura, 1999). Depending on the cell type and context, Erk signaling can participate in neuronal survival (Hetman *et al.*, 1999) or neuronal death (Satoh *et al.*, 2000; Stanciu *et al.*, 2000). Erk activation by neurotrophin is important for the survival of rat cortical neurons in vitro (Hetman *et al.*, 1999). However, activation of Erk42/44 might contribute to neuronal apoptosis in some in vitro models of neurotoxicity (Stanciu *et al.*, 2000), although the mechanisms are not known. The MAPK pathway has a survival function in unstressed mature neurons but not in unstressed immature neurons (Lesuisse and Martin, 2002). This finding is consistent with the observation that immature neurons do not require constitutive phosphorylation of Erk42/44 for survival. Therefore, Erk42/44 phosphorylation is less important as a survival signal in unstressed immature neurons compared with mature neurons. However, in mature and immature neurons undergoing genotoxic stress, the MAPK pathway assumes a pro-apoptotic function (Lesuisse and Martin, 2002). When activated, Erk42/44 translocates to the nucleus to phosphorylate target proteins, including transcription factors (Elk, Sap, and Sp1) and histone proteins (English *et al.*, 1999), many of which are involved in apoptosis. This nuclear translocation of active Erk42/44 may function as a pre-apoptotic signal or a transient, compensatory survival signal in mature neurons. There is precedence for Erk activation functioning pro-apoptotically in mesangial cells (Ishikawa and Kitamura, 1999) and neuronal cells (Satoh *et al.*, 2000; Stanciu *et al.*, 2000) after oxidative stress. Blocking Erk activation by MEK inhibition protects cortical neurons against apoptosis (Lesuisse and Martin, 2002), supporting a pro-apoptotic role for Erk activation in mature neurons.

MEKK1 could account for the pro-apoptotic activity of the MAPK pathway. Most of the work on MEKK1 function has been done on non-neuronal cells, although studies of MEKK1 in neuronal apoptosis are emerging (Lesuisse and Martin, 2002). This upstream kinase in the MAPK cascade also has dual functions. Full-length MEKK1 promotes cell survival and has anti-apoptotic actions through the activation of NFκB, whereas MEKK1 C-terminal fragments promote apoptosis (Cardone *et al.*, 1997; Deak *et al.*, 1998; Widmann *et al.*, 1998). Caspase-3 seems to act as a switch to convert MEKK1 from a survival signal to a pro-apoptotic effector. MEKK1 fragments have constitutive kinase activity that can activate caspases, comprising a positive feedback loop for driving apoptosis (Cardone *et al.*, 1997). MEKK1 activity is required for apoptosis in non-neuronal cells following DNA damage (Widmann *et al.*, 1998) and Fas ligation (Deak *et al.*, 1998). Cleavage of MEKK1 in monkey kidney cells after an apoptotic stimulus results in changes in the intracellular distribution of MEKK1 (insoluble particulate to soluble diffuse cytoplasmic; Deak *et al.*, 1998).

MEKK1 appears to have a role to play in neuronal apoptosis induced by DNA damage (Lesuisse and Martin, 2002). This role might be different in immature and mature neurons stimulated to undergo apoptosis as well as in unstressed normal neurons. For instance, cleaved MEKK1 levels are constitutively higher in mature neurons than in immature neurons, suggesting that C-terminal fragments of MEKK1 are inactive as pro-apoptotic kinases in mature neurons. Major C-terminal fragments of MEKK1 are not present constitutively in immature neurons. However, during DNA damage-induced apoptosis of immature cortical neurons, the formation of C-terminal fragments of MEKK1 coincides with caspase-3 cleavage and decreased full-length MEKK1 (Lesuisse and Martin, 2002). In contrast, during DNA damage-induced apoptosis of mature cortical neurons, levels of major C-terminal fragments of MEKK1 do not appear to change, but levels of full-length (survival) MEKK1 decline (Lesuisse and Martin, 2002), suggesting different mechanisms for MEKK1 processing and stabilization in mature and immature neurons. The loss of full-length MEKK1 may be sufficient for engaging apoptosis in mature neurons.

4.4   Regulation of developmental PCD in the nervous system

As anticipated from the work of Horvitz and others in nematodes, cell death proteins control developmental PCD in the mammalian nervous system. The numbers of neurons appear increased (presumably resulting from deficient apoptosis) in some CNS regions in transgenic mice that over-express the *bcl-2* gene (Martinou *et al.*, 1994; Farlie *et al.*, 1995; Bonfanti *et al.*, 1996), mice with *bax* gene inactivation (Deckwerth *et al.*, 1996), and mice with gene deletions in caspase-3 (Kuida *et al.*, 1998), caspase-9 (Hakem *et al.*, 1998; Kuida *et al.*, 1998), and Apaf-1 (Yoshio *et al.*, 1998). Homozygous deficiencies in caspase-3, caspase-9, and Apaf-1 are embryonic lethal and cause cerebral malformations. Most $p73^{-/-}$ mice die within the first postnatal month and present with cerebral dysgenesis in the form of hippocampal hypertrophy (Yang *et al.*, 2000). Bax, Bcl-2, and p53 homozygous deficient mice all survive to adulthood, but Bcl-2-deficient mice show progressive degeneration of motor neurons after the PCD period during early postnatal development (Michaelidis *et al.*, 1996).

4.5   Cell death in human neurodegenerative diseases

4.5.1   Alzheimer's disease (AD)

AD is the most common cause of dementia occurring in middle and late life (Katzman, 1993). Population-based surveys estimate that AD affects 7–10% of individuals over 65 years of age and possibly 50–60% of people over 85 years of age (McKhann *et al.*, 1984; Evans *et al.*, 1989). AD now affects about 2% of the population, or about 4 million people in the USA (Olshansky *et al.*, 1993) and ~12 million people worldwide. The prevalence of AD is increasing proportionally to increased life expectancy (estimates predict that ~25% of the population will be over 65 years of age in the year 2050).

Most cases of AD have unknown etiologies and are called "sporadic" and have a late onset; however, some cases, particularly those with early onset, are familial and are inherited as autosomal dominant disorders linked to mutations in the gene that encodes the amyloid precursor protein (Chartier-Harlin *et al.*, 1991; Goate *et al.*, 1991; Naruse *et al.*, 1991) or genes that encode for presenilin proteins (Campion *et al.*, 1995; Sherrington *et al.*, 1995). For late-onset sporadic cases, a variety of risk factors have been identified in addition to old age (Kalaria, 2003). The apolipoprotein E (ApoE) allele is a susceptibility locus, with the ApoE4 type showing dose-dependent contributions (Roses, 1996). Cardiovascular disease and head trauma are additional risk factors for AD (Katzman, 1993).

Neurons in the neocortex, hippocampus, and basal forebrain are selectively vulnerable in AD (Whitehouse *et al.*, 1982; Gomez-Isla *et al.*, 1996; Mouton *et al.*, 1998; West *et al.*, 2000; Pelvig *et al.*, 2003). The mechanisms that cause the profound degeneration and loss of neurons in AD are not known, and existing information is incomplete. An abnormal processing or modification of the amyloid precursor protein (APP) and the cytoskeletal protein tau (a microtubule-associated protein) is involved in the pathogenesis (Hardy and Selkoe, 2002), resulting in amyloid (Aβ) deposits and neurofibrillary changes consisting of paired helical filaments, neurofibrillary tangles (NFT), and dystrophic neurites (Kosik *et al.*, 1986) (Figure 4.5). Cortical and hippocampal neuronal degeneration could be the consequence of a combination of variety of mechanisms including perturbations in protein metabolism, excitotoxicity, oxidative stress, and inflammation. The possible specific mechanisms for neuronal degeneration in AD may involve dysfunction of *N*-methyl-D-aspartate (NMDA) receptors (Sze *et al.*, 2001; Kemp and McKernan, 2002), dysregulation of Ca^{2+} homeostasis (Mattson *et al.*, 1992), defects in synapses (DeKosky and Scheff, 1990; Terry *et al.*, 1991; Martin *et al.*, 1994; Sze *et al.*, 1997; Selkoe, 2002), abnormalities in the metabolism of APP and presenilin (PS) proteins, toxic actions of Aβ protein derived from APP (Yankner *et al.*, 1989; Younkin, 1995), and cytoskeletal pathology (Fath *et al.*, 2002; Rapoport *et al.*, 2002).

The classification of the neurodegeneration in AD is surprisingly still not clear. Some results suggest that this neuronal death could be apoptosis based on in situ DNA fragmentation/damage assays (Smale *et al.*, 1995; Anderson *et al.*, 1996; Adamec *et al.*, 1999); however, this DNA damage is not exclusively indicative of apoptosis (Stadelmann *et al.*, 1999). Dying neurons are found in cortical and subcortical regions in the AD brain (Figure 4.5). Other studies indicate that apoptosis does not have a major role in the neuronal degeneration of AD (Lucassen *et al.*, 1997). Experiments on changes in the levels of proteins in the Bcl-2 family in AD postmortem brains are difficult to interpret, with studies showing upregulation of both anti-apoptotic and pro-apoptotic proteins or no changes (Su *et al.*, 2001; Anderson *et al.*, 1996; Kitamura *et al.*, 1998). The processes might involve caspase-3 activation as determined by the immunohistochemical detection of cleaved caspase-3 (Stadelmann *et al.*, 1999; Su *et al.*, 2001; Gastard *et al.*, 2003) and caspase

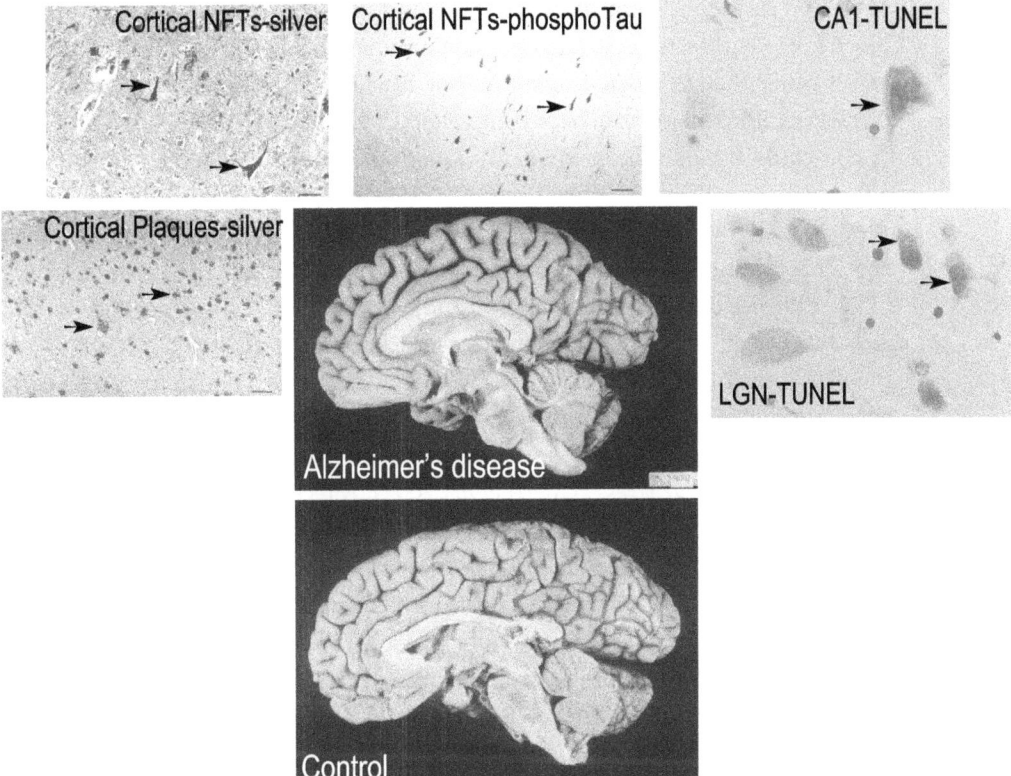

Figure 4.5　*Neurodegeneration in AD*. Mid-sagittal views (center pictures) of the brains from an 85-year-old individual with AD and an 86-year-old normal individual. The cerebral cortex in the person with AD is atrophic, as indicated by the widening of the sulci and narrowing of the gyri. The cerebral cortex in the control is normal with broad gyri and narrow sulci. Senile plaques and neurofibrillary tangles (NFT) are brain lesions that are formed in patients with AD (images clockwise from left). Numerous senile plaques and abnormal extracellular deposits of Aβ amyloid protein are formed in the brain (arrows). Scale bar = 200 μm. Neurofibrillary tangles, which are abnormal intracellular aggregates of protein (arrows), are formed in pyramidal neurons. Scale bar = 50 μm. Neurofibrillary tangles are composed of hyperphosphorylated tau proteins (arrows). Scale bar = 100 μm. Cell death assays, such as TUNEL (brown nuclear staining), show that neurons die in the hippocampus CA1 and in subcortical regions such as the thalamic lateral geniculate nucleus (LGN).
(For a colour version of this figure, please see www.cambridge.org/9780521159449.)

gene expression (Pompl *et al.*, 2003). Cleaved caspase-3 (Gastard *et al.*, 2003) and caspase-9 (Rohn *et al.*, 2002) have been found in NFT-bearing neurons in AD. However, other groups have not found evidence for the accumulation of cleaved caspase-3 in neurons in the AD brain (Selznick *et al.*, 1999), but changes seen in early and late AD may differ (Gastard *et al.*, 2003). Despite these findings, immunodetection of cleaved caspase-3 is not always equivalent to caspase-3 activation as determined biochemically (Martin *et al.*, 2003). Furthermore, caspase-3 may function in processes other than cell death, including neuronal differentiation, migration, and plasticity (Shimohama *et al.*, 2001; Yan *et al.*, 2001).

p53 could have a role in the neuronal degeneration of AD. The levels of p53 are elevated in AD postmortem brain as determined by immunoblotting (de la Monte *et al.*, 1997, 1998). Immunolocalization studies are conflicting, as results show p53 accumulation in degenerating neurons and in glia (de la Monte *et al.*, 1997, 1998), although other evidence indicates that p53 is upregulated in glial cells but not in neurons in the AD brain (Kitamura *et al.*, 1997). Interestingly, p53 has been found in neurites (de la Monte *et al.*, 1997), and it may have a local transcription-independent function at synapses (Gilman *et al.*, 2003). However, fibroblasts from sporadic AD patients have impaired p53-mediated cell death in response to oxidative stress (Uberti *et al.*, 2002). A recent genetic study has found no association between the common proline/arginine polymorphism at codon 72 in the p53 gene and the presence of AD (Rosenmann *et al.*, 2003).

Death receptors might also have a role in the neuronal degeneration of AD, but the evidence is not yet clear. Immunoblot studies of Fas have reported elevations (de la Monte *et al.*, 1997, 1998) or no changes (Ferrer *et al.*, 2001; Su *et al.*, 2003) in AD postmortem brain. The levels of FasL appear to be increased in AD cortical samples and in cortical neuron cultures exposed to $A\beta_{1-42}$ (Su *et al.*, 2003) but are decreased in AD hippocampal samples (Ferrer *et al.*, 2001). Fas and FasL are expressed in tangle-bearing and non-tangle-bearing neurons without relationship to DNA fragmentation (Ferrer *et al.*, 2001).

Progressively slow changes in cytoskeletal dynamics can be related to neuronal cell death in AD. A striking feature of the AD brain is the formation of intracellular aggregates comprised of abnormally phosphorylated tau (Kosik *et al.*, 1986) (Figure 4.5). Activation of the MAPK cascade might have a role in tau phosphorylation and neuronal cell death in AD (Holzer *et al.*, 2001). A protein kinase known as cyclin-dependent kinase 5 (Cdk5) could mediate this neuronal death. Cdk5 is activated by its regulatory protein p35 which is truncated to a more potent activator p25 by proteases in AD leading to cdk5 activation, tau hyperphosphorylation, and apoptosis (Patrick *et al.*, 1999). In mice, Cdk5 deregulation by p25 causes tau hyperphosphorylation and possible neurofibrillary degeneration (Cruz *et al.*, 2003).

Neuronal cell death in AD occurs over a lengthy period. There are many gaps in our knowledge of the mechanisms of AD. When considering the pathologic classification of the primary neurodegeneration in AD, it appears safe to conclude that based on morphology it is not classical apoptosis or necrosis. Autophagy could have a role in this neuronal cell death (Stadelmann *et al.*, 1999; Cataldo *et al.*, 1994). It might be more useful to consider neurofibrillary cell death separately from classical

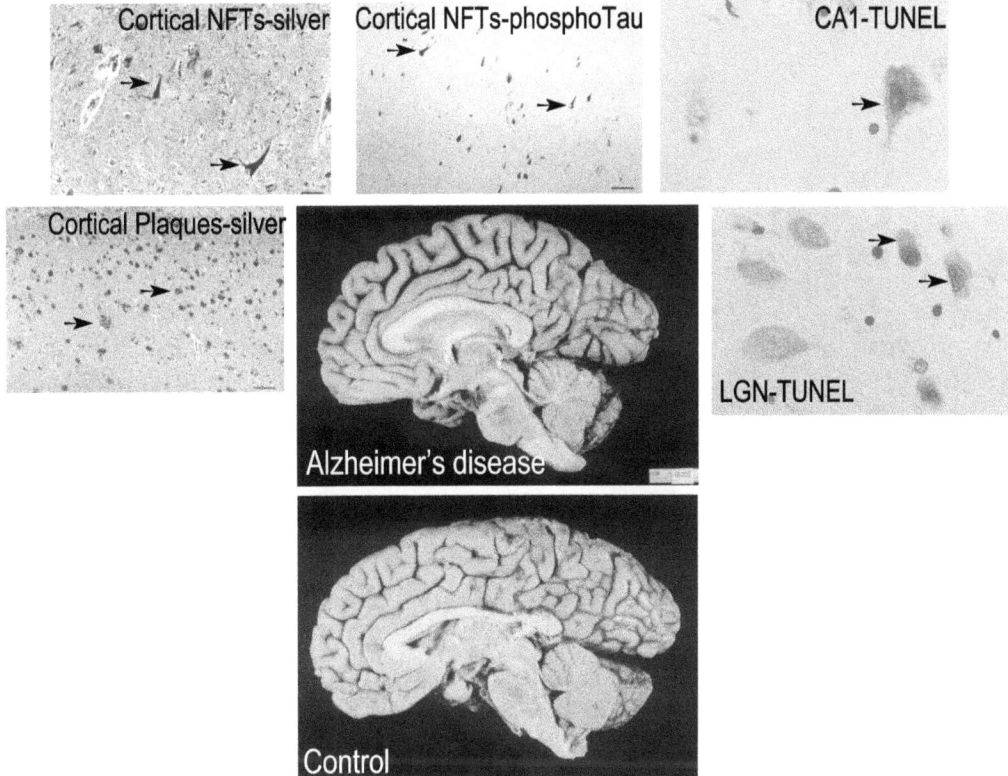

Plate 1   *Neurodegeneration in AD*. (See Figure 4.5, page 117.)

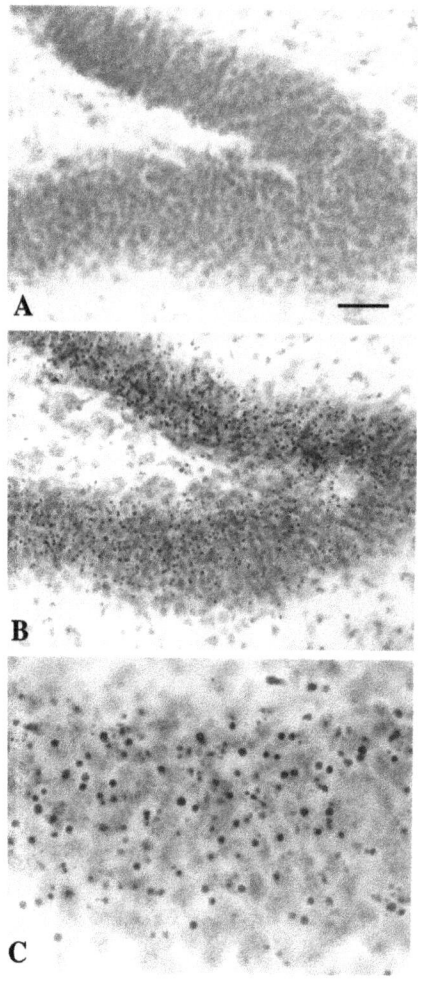

Plate 2   *Animal model of hippocampal neuron degeneration.* (See Figure 4.6, page 123.)

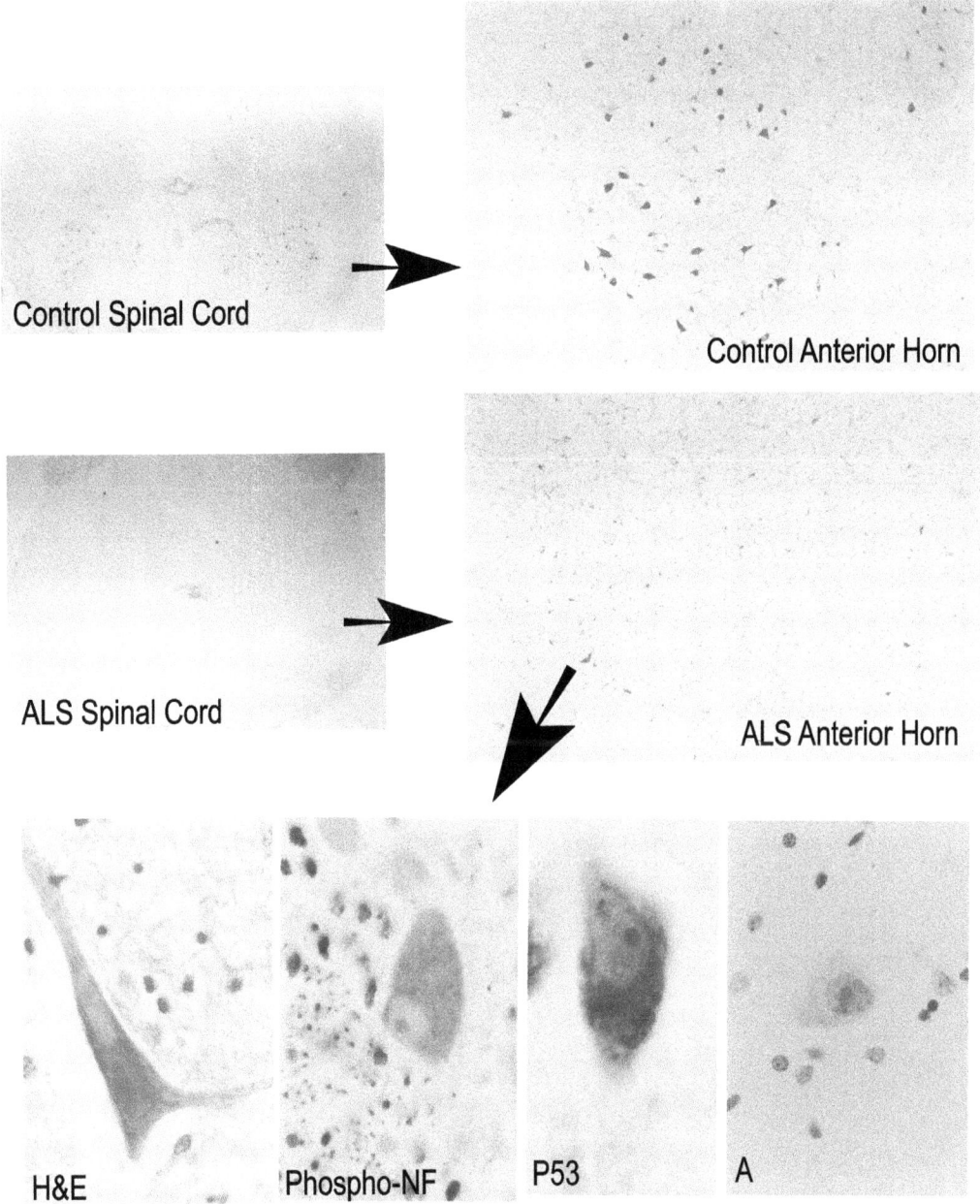

Plate 3 *Motor neurons in spinal cord degenerate in individuals with ALS.* (See Figure 4.7, page 125.)

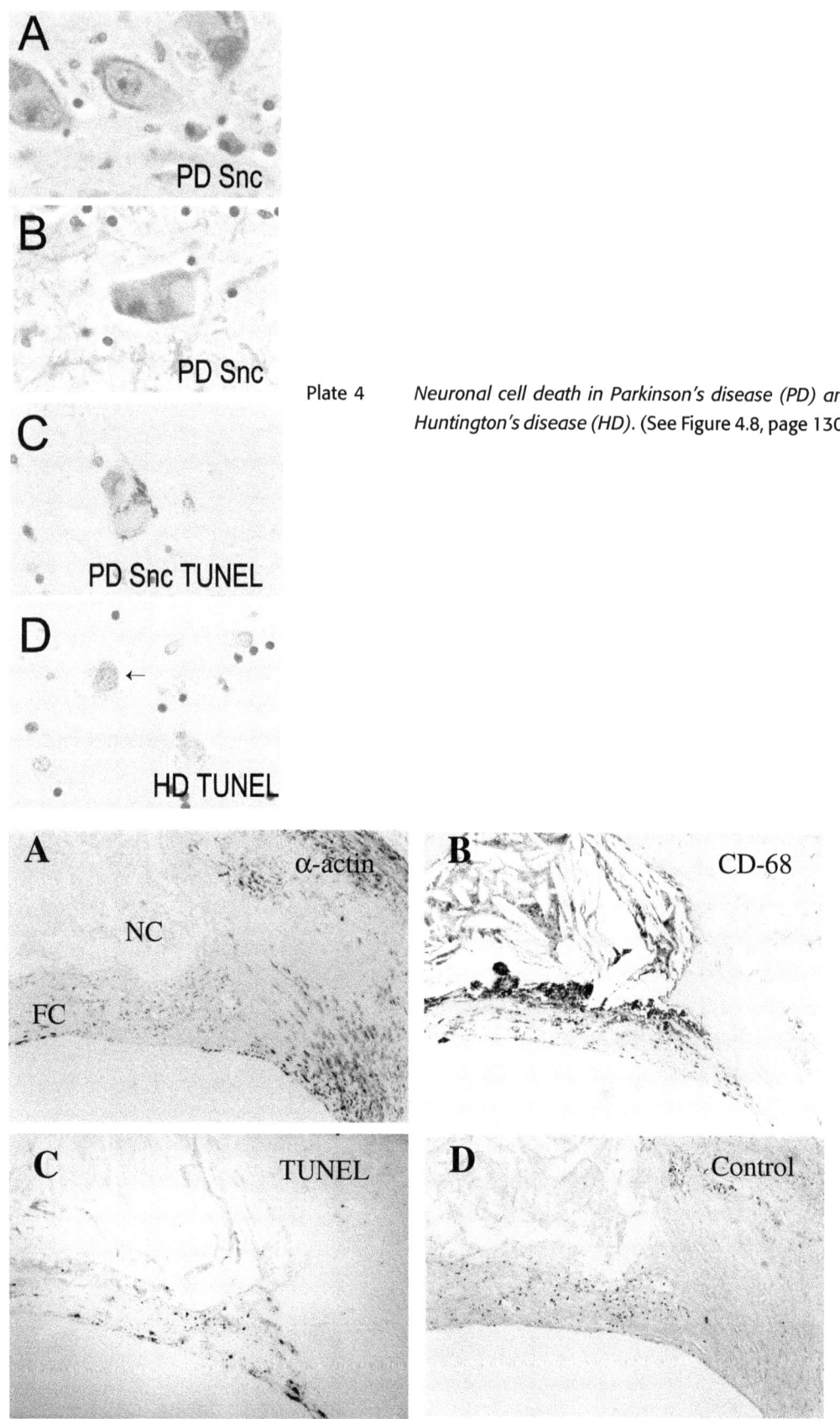

Plate 4   *Neuronal cell death in Parkinson's disease (PD) and Huntington's disease (HD). (See Figure 4.8, page 130.)*

Plate 5   *Typical vulnerable atherosclerotic plaque, prone to rupture. (See Figure 5.1, page 163.)*

apoptosis and necrosis, but they may have some overlap in mechanisms. It could be of interest to decipher how neurons with neurofibrillary degeneration escape classical apoptosis and necrosis.

4.5.1.1 Neuronal cell death in transgenic (Tg) mouse models of AD

Animal experiments will provide critical insight into the mechanisms of neurodegeneration in AD. However, most human *APP* and/or *PS1* Tg mice show substantial Aβ deposits in the hippocampus and cortex, but do not develop significant neuronal loss (Irizarry *et al.*, 1997; Wong *et al.*, 2002). Evidence for caspase-3 activation has not been found in APPswe Tg mice (Selznick *et al.*, 1999). Analysis of a different Tg mouse line (APP23) showed that Aβ deposition is accompanied by a modest loss of CA1 neurons but no cortical neuron loss, despite an Aβ burden similar to that seen in CA1 (Calhoun *et al.*, 1998). None of these studies has found the formation of NFT; however, a recent study of triple-Tg mice harboring mutant PS1, APP, and tau transgenes is encouraging (Oddo *et al.*, 2003). These mice have intraneuronal accumulations of Aβ and phosphorylated tau, but neuronal loss was not reported (Oddo *et al.*, 2003). In Tg mice over-expressing Aβ protein, extracellular deposition of Aβ and neuronal cell death are observed (LaFerla *et al.*, 1996). The neuronal cell death mechanism in these mice may involve p53 (LaFerla *et al.*, 1996), consistent with in vitro findings on Aβ toxicity (Zhang *et al.*, 2002).

4.5.1.2 In vitro models of cortical and hippocampal neuron degeneration and interactions between APP, Aβ, tau, and caspases

Aβ has long been thought to be a primary cause of AD. Extracellular application of Aβ can induce apoptosis (Loo *et al.*, 1993) or necrosis (Behl *et al.*, 1994) in neuronal cell culture. Extracellularly applied $A\beta_{25-53}$ or $A\beta_{1-40}$ causes mitochondrial dysfunction in primary cortical neurons (Casley *et al.*, 2002), enhanced production of free radicals, intracellular Ca^{2+} destabilization, and DNA damage (Duker *et al.*, 2001; Kuperstein and Yavin, 2002). Studies of the specific intracellular signaling pathways that are activated by Aβ to trigger cell death are only now appearing in the literature. In studies of neurons exposed to extracellular Aβ, the induction of apoptosis involves the Fas–JNK pathway (Morishima *et al.*, 2001), the $p75^{NTR}$–JNK pathway (Yaar *et al.*, 2002), the MAPK pathway (Kuperstein and Yavin, 2002), and caspase-12 (Nakagawa *et al.*, 2000). Neuronal cultures exposed to Aβ can be protected from neurotoxicity by caspase-8 inhibition and expression of dominant–negative FADD, both components of the Fas pathway (Ethell *et al.*, 2002). Some experiments have implicated p53 in the hippocampal neuron death triggered by extracellular Aβ (Culmsee *et al.*, 2001; Chan *et al.*, 2002), but other studies with Aβ-treated cortical neurons reveal a p53-independent mechanism for cell death involving E2F1 (Giovanni *et al.*, 2000). It is noteworthy that the results of in vitro experiments using extracellular application of Aβ are likely to be dependent on Aβ exposure concentration. When human neuron primary cultures are treated with Aβ at concentrations closer to physiologic levels for up to 3 days, evidence for apoptosis is scarce; however, Bcl-2 levels are downregulated and Bax levels are upregulated (Paradis *et al.*, 1997). Thus, extracellular Aβ at physiologic concentrations might render neurons more sensitive to cell stress rather than kill them outright. Intracellular $Aβ_{1-42}$ exposure (as little as 1 pM) is, in contrast, toxic to human cortical neurons, and this toxicity requires de novo protein synthesis, Bax, p53, and caspases, indicating neuronal cell death by apoptosis (Zhang *et al.*, 2002).

Experiments have been performed to link APP, presenilins, and tau to cell death. Over-expression and intracellular accumulation of APP activates caspase-3 (Uetsuki *et al.*, 1999). APP is a target of caspase-3 (Weidemann *et al.*, 1999), and APP cleavage by caspase-3 or caspase-6 may promote Aβ formation (Gervais *et al.*, 1999; LeBlanc *et al.*, 1999). Thus, increased production of Aβ may be a consequence of neuronal apoptosis. Presenilin proteins are also substrates for caspase-3 (Kim *et al.*, 1997). Presenilin proteins can influence mitochondrial regulation of apoptosis, such as Bax activation and cytochrome c release, through interactions with $Bcl-X_L$ (Passer *et al.*, 1999). Studies have reported that over-expression of wild-type or mutant human presenilin-1 or presenilin-2 does not enhance apoptosis in neurons (Bursztajn *et al.*, 1998; Gamliel *et al.*, 2003). Other work indicates that the presenilin-1 mutation sensitizes neurons to DNA damage-induced apoptosis (Chan *et al.*, 2002). Interesting links between Aβ, tau, and cell death are emerging. Dysregulation of cell cycle protein kinases (e.g. Cdk5), GSK3, and MAPK could play a role in cytoskeletal perturbations leading to cortical neuron degeneration because tau is a substrate for these kinases that can be activated by Aβ (Takashima *et al.*, 1993; Alvarez *et al.*, 2001). Recent work shows that tau is directly involved in the formation of dystrophic neurons in response to fibrillar Aβ (Rapoport *et al.*, 2002). More in vitro work needs to be done on the basic mechanisms of cortical neuron degeneration and on Aβ neurotoxicity mechanisms under basal conditions and in the presence of FAD-related and tau gene mutations.

4.5.1.3   Animal model of neuronal death induced by target deprivation

Populations of neurons in the AD brain, notably subcortical neurons (Figure 4.5), are believed to undergo cytoskeletal changes and retrograde degeneration as a result of target deprivation (Martin *et al.*, 1998). Specific examples of neuronal groups that undergo retrograde degenerative changes include neurons in the basal forebrain magnocellular complex, raphae, locus coeruleus, and thalamus (Figure 4.5). Animal models of axotomy and target deprivation are useful for gaining insight into the mechanisms of progressive neuronal injury and retrograde degeneration. Ablation of the visual neocortex induces retrograde neuronal degeneration in the dorsal lateral geniculate nucleus (dLGN) of the thalamus (Lashley, 1941; Barron *et al.*,

1973). The geniculocortical projection neurons die by a morphologic process that is unequivocal apoptosis (Al-Abdulla *et al.*, 1998; Al-Abdulla and Martin, 1998). This cell death is preceded by an accumulation of apparently active mitochondria in the perikaryon and might emerge with oxidative damage to genomic DNA of the highly vulnerable projection neurons (Al-Abdulla and Martin, 1998). Apoptosis of geniculocortical projection neurons requires the presence of the *Bax* gene and is modified by a functional *p53* gene, further supporting a role for apoptosis in this neurodegeneration and the possible role of DNA damage as a trigger (Martin *et al.*, 2001).

The molecular regulation of target deprivation-induced neuronal apoptosis has been studied by Martin *et al.*, 2003. The synchronized apoptosis of dLGN projection neurons occurs in association with differential subcellular changes in pro-apoptotic molecules. Bax increases in mitochondria within 1 day after target deprivation, while Bak increases later, after about 4 days. Few studies have demonstrated such a rapid (within 1 day) redistribution of Bax during apoptosis of neurons or of cells in general. Very few studies have addressed the role of Bak in neuronal apoptosis. An in vitro study of neonatal peripheral nervous system neurons deprived of NGF failed to reveal a change in Bak during apoptosis (Putcha *et al.*, 1999). However, experiments on non-neuronal cells suggest that Bax and Bak are essentially interchangeable (Wei *et al.*, 2001). In central neurons, Bax and Bak show an increased level and subcellular redistribution during target deprivation-induced apoptosis, but, interestingly, the timing of the changes is different for these two pro-apoptotic molecules (Martin *et al.*, 2003). Bax may be a rapid response protein, whereas Bak may be a delayed response protein. Bax and Bak may have coordinated hierarchical or independent functions during neuronal apoptosis. The finding that Bax translocates to mitochondria and then coalesces with Bak into mitochondria-associated clusters during apoptosis in non-neuronal cells (Nechushtan *et al.*, 2001) supports the idea of a coordinated participation of these two molecules. Bak may, therefore, be involved in reinforcing the cell death process after Bax engages the process in CNS neurons.

4.5.1.4 Animal models of cytoskeletal pathology-induced neuronal death

A major form of neurodegeneration in AD is related to cytoskeletal pathology (Kosik *et al.*, 1986). NFT are composed primarily of abnormally hyperphosphorylated tau molecules (Figure 4.5). Mutations in the *tau* gene have been identified in several neurodegenerative diseases, including fronto-temporal dementia with Parkinsonism (collectively called tauopathies) linked to chromosome 17 (Froelich *et al.*, 1997; Poorkaj *et al.*, 1998). Tg mice expressing mutant human tau protein (P301L) form abnormal tau-containing filaments in the brain (Gotz *et al.*, 2001a,b). These filaments have similarities to the NFT occurring in AD and fronto-temporal dementia, but the abundance is comparatively low compared with the human diseases. When mutant tau Tg mice are crossed with mutant *APP* mice and when mutant tau Tg mice receive intracerebral injection of $A\beta_{42}$, neurofibrillary degeneration in the brain is enhanced (Gotz *et al.*, 2001a, b; Lewis *et al.*, 2001). However, evidence for neuronal cell death in these animals has not yet been reported.

Cytoskeletal destabilization induces apoptosis. Taxol, vincristine, nocodazole, and colchicine are well-known examples of microtubule depolymerizing agents that induce apoptosis. Colchicine binds tightly to the β-tubulin subunit of the β/β-tubulin heterodimer, thereby decreasing the soluble tubulin pool and inhibiting microtubule assembly (Uppuluri *et al.*, 1993). In the CNS, the neurotoxicity of colchicine is highly cell-type selective, causing prominent degeneration of granule neurons in the dentate gyrus of the hippocampus (Figure 4.6) without affecting pyramidal neurons (Goldschmidt and Steward, 1980). The structure of this degeneration is classically apoptotic (Liu *et al.*, 2001) (Figure 4.6) and involves caspase-3 activation (Kim *et al.*, 2002) and tau hyperphosphorylation (Fath *et al.*, 2002). The detailed signaling mechanisms of this cytoskeletal damage-induced apoptosis in neurons remain to be explored.

4.5.2   Amyotrophic lateral sclerosis (ALS)

ALS is a fatal neurologic disease in humans that cannot be cured or treated effectively (Rowland and Shneider, 2001). More than 5000 people in the USA are diagnosed with ALS each year, and, in parts of the UK, one in $\sim$500 deaths is caused by some form of ALS (Nicholson *et al.*, 2000). ALS is characterized clinically by progressive weakness, muscle atrophy, and eventual paralysis and death within 3–5 years after symptoms begin. ALS is characterized neuropathologically by progressive degeneration and elimination of upper and lower motor neurons (MNs) in the spinal cord, brainstem, and cerebral cortex (Martin *et al.*, 2000b; Rowland and Shneider, 2001; Sathasivam *et al.*, 2001). It is still not understood why MNs are selectively vulnerable in ALS. The molecular pathogenesis of most forms of ALS is still understood poorly. Two forms of ALS exist: idiopathic (sporadic) and heritable (familial). The majority of ALS cases are sporadic with no known genetic component. Familial forms of ALS (FALS) are autosomal dominant and make up $\sim$10% or less of all ALS cases. A subset of FALS cases ($\sim$5–10%) has mutations in the superoxide dismutase 1 (SOD1) gene (Deng *et al.*, 1993; Rosen *et al.*, 1993). Recent work has shown that reduced expression of vascular endothelial growth factor is a risk factor for contracting ALS (Lambrechts *et al.*, 2003).

Many theories have been proposed for the possible causes of MN death in ALS. Focus is on glutamate excitotoxicity, autoimmunity, inflammation, neurofilament abnormalities, and genetic causes involving mutant SOD1. Glutamate excitotoxicity, resulting from loss or downregulation of astroglial glutamate transporter,

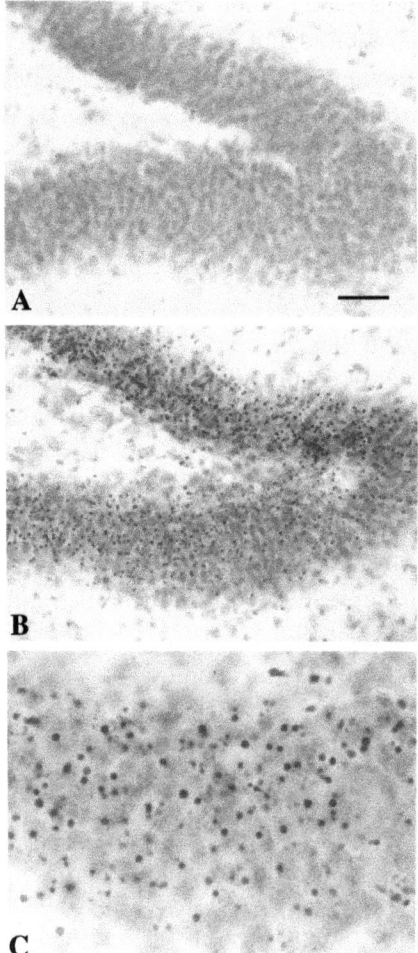

Figure 4.6   *Animal model of hippocampal neuron degeneration.* Intraventricular injection of colchicine induces large-scale apoptosis of neurons in the dentate gyrus. Images show the dentate gyrus granule cell layer in cresyl violet-stained sections of adult rat brain. (A) Control without apoptosis. (B) (low mag) and, (C) (high mag). At 48 hours after colchicine injection, copious apoptosis (dark round degenerating profiles) is seen.
(For a colour version of this figure, please see www.cambridge.org/9780521159449.)

may be a primary mechanism for MN degeneration in ALS (Rothstein *et al.*, 1992; Rothstein *et al.*, 1995). However, there appears to be a lack of disease-specific and region-specific defects in glutamate transporter subtypes, and these changes may be secondary (Rowland and Shneider, 2001), possibly related to hypoxia (Martin *et al.*, 1998). Autoimmunity may also have a role in the pathogenesis of ALS, but this theory is controversial as well (Rowland and Shneider, 2001). Activation of inflammatory cells or other non-neuronal cells may participate in the degeneration of MNs, suggesting that MN death is not cell autonomous (Raoul *et al.*, 2002). The involvement of SOD1 is so far the best lead, but the basis for the selective vulnerability of MNs in the presence of mutant SOD1 (mSOD1) is still not clear, because SOD1 is widely expressed in cells throughout the body. Moreover, the role of SOD1 in the pathogenesis of the vast majority of ALS cases is not clear. Oxidative stress, once considered to be a consequence of the disease, is now pushing to the forefront as a cause of the disease, and peroxynitrite ($ONOO^-$), formed from superoxide and nitric oxide, might be a major culprit in the disease process (Beckman *et al.*, 1993, 2001).

The degeneration of MNs in ALS may be a form of apoptosis (Martin, 1999; Sathasivam *et al.*, 2001). This idea is based on morphologic and biochemical findings. A staging scheme for the structural progression of MN death has been devised (Martin, 1999). This staging arrangement reveals that MN degeneration in ALS may be a non-classical form of apoptosis. After an initial chromatolytic stage, MNs progressively accumulate neurofilament and undergo attrition of the cell body and dendrites that culminates in a residual MN of only ~20% of normal diameter (Figure 4.7). During somatodendritic attrition, both the cytoplasm and the nucleus become condensed and dark, consistent with apoptosis (Figure 4.7). The nuclear condensation in ALS MNs differs from classical apoptosis because the chromatin is not discretely organized into uniformly round, dense clumps as in animal models of neuronal apoptosis. The terminal transferase-biotinUTP nick-end labeling (TUNEL) method was used to identify when nuclear DNA fragmentation commences during the staging of MN degeneration in ALS. DNA fragmentation is detected in MNs at the somatodendritic attrition and apoptotic stages of neuronal death but not in MNs in the chromatolytic stage of degeneration. DNA fragmentation is also found in subsets of pyramidal neurons in ALS motor cortex, but not somatosensory cortex. Immunoblotting experiments of subcellular fractions have shown that Bax and Bak protein levels are increased and that Bcl-2 protein level is decreased in selectively vulnerable motor regions, but the levels of Bcl-X_L are unchanged (Martin, 1999). The biochemical activity of caspase-3 activity is increased selectively in ALS MN regions (Martin, 1999). A DNA fragmentation factor endonuclease is activated, and internucleosomal fragmentation of genomic DNA can be found (Martin, 1999). These findings lead to the hypothesis that an inappropriate re-emergence of a PCD mechanism, involving cytosol-to-membrane and membrane-to-cytosol redistributions of cell death proteins, participates in the pathogenesis of MN-degeneration ALS (Martin, 1999). However, precise structure–molecular correlations are needed to validate this MN-degeneration staging scheme, and the protein expression data lack resolution for MN-specific abnormalities, despite the use of a micropunch tissue sampling method for anterior horn and motor cortex. There is non-neuronal cell contamination in the samples. The

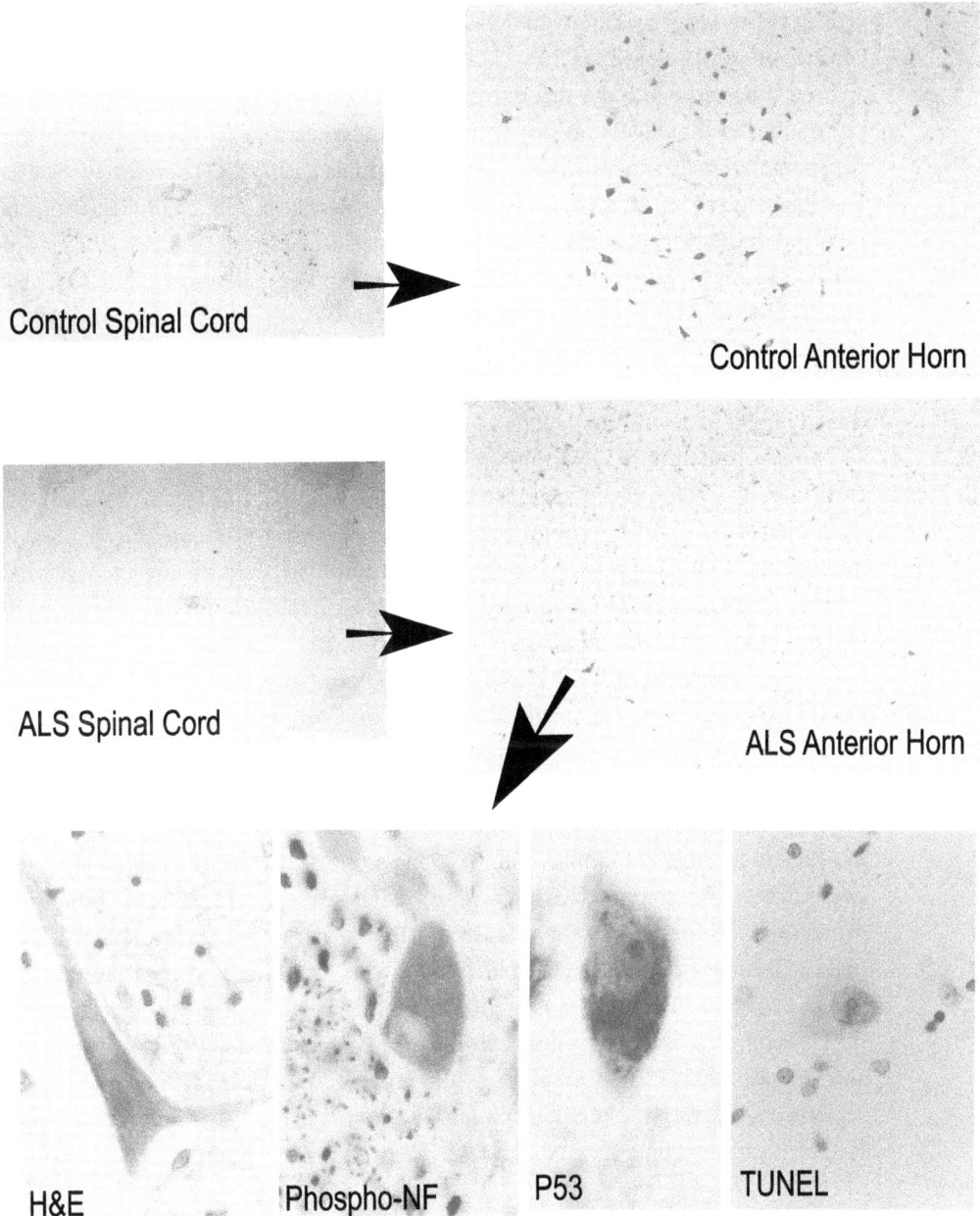

Figure 4.7   *Motor neurons in spinal cord degenerate in individuals with ALS.* In normal control individuals, the anterior horns of the spinal cord contain many large, multipolar motor neurons. In ALS cases, the anterior horn is depleted of large neurons and the remaining neurons are atrophic. These attritional chromatolytic motor neurons display a dark condensed nucleus as seen in H&E-stained sections and accumulate phosphorylated neurofilament (brown cytoplasmic staining). p53 accumulates in the nucleus of ALS motor neurons (brown labeling). Cell death assays (e.g. TUNEL) identify subsets of motor neurons in the process of DNA fragmentation (brown nuclear staining).
(For a colour version of this figure, please see www.cambridge.org/9780521159449.)

resolution and MN specificity in these experiments need to be enhanced using laser capture microdissection (LCM).

Even if MNs in ALS die through some form of apoptosis, this outcome might be too late in the degenerative process for antagonism of apoptotic pathways to have any major therapeutic benefit for individuals with ALS. Therefore, it is vital to explore upstream mechanisms for MN degeneration. DNA damage could be involved in the pathogenesis of ALS (Bradley and Krasin, 1982). A key DNA lesion is 8-hydroxydeoxyguanosine (OHdG). OHdG is a mutagenic lesion. OHdG adducts are elevated in ALS CNS tissue extracts (Ferrante *et al.*, 1997; Fitzmaurice *et al.*, 1996), but the contribution of MNs to the OHdG pool was not identified. Subsequent immunolocalization experiments have shown that OHdG lesions occur directly in ALS MN (Martin, 2001).

The accumulation of DNA damage in MNs could be a stimulus for MN degeneration in ALS. This idea is supported by the finding that p53 is activated in MNs in human ALS. By immunoblotting, p53 levels increase in vulnerable regions in individuals with ALS (Martin, 2000). p53 accumulates specifically in ALS MNs (Figure 4.7). This p53 is functionally active because it is phosphorylated at serine[392] and has increased DNA-binding activity (Martin, 2000, 2001).

The accumulation of DNA damage in ALS could also signify perturbations in DNA repair processes. The number of studies on DNA repair in ALS is few. Repair enzyme gene mutations have been identified in ALS (Olkowski, 1998), but other studies have not identified prominent contributions of the mutations (Hayward *et al.*, 1999; Tomkins *et al.*, 2000). Biochemical studies have examined DNA repair in ALS skin fibroblasts (Tandan *et al.*, 1987), lymphocytes (Robison *et al.*, 1993), and postmortem frontal cortex (Kisby *et al.*, 1997). These results are also conflicting. The expression and function of the class II apurinic/apyrimidinic endonuclease (APE) in ALS CNS has been studied (Shaikh and Martin, 2002). This protein is interesting to ALS because APE functions as a redox factor (redox factor-1, Ref-1) that facilitates the DNA binding of transcription factors through redox modulation (Xanthoudakis and Curran, 1992). APE protein levels and repair activity are increased in ALS MN regions, supporting the possibility that DNA damage is an upstream mechanism for MN degeneration.

4.5.2.1 MN degeneration in the mutant SOD1 mouse model of ALS

MNs in mice harboring mutant human SOD1 undergo prominent degeneration. However, the type of cell death that these neurons undergo and the mechanisms for this degeneration are controversial. Structurally, the degeneration of MNs in mSOD1 mice does not match an apoptotic profile (Wong *et al.*, 1995; Ikonomidou *et al.*, 1996; Migheli *et al.*, 1999; Bendotti *et al.*, 2001). No evidence for DNA fragmentation has been found (Migheli *et al.*, 1999). Nevertheless, G93A mice over-expressing Bcl-2 (Kostic *et al.*, 1997) or treated with caspase inhibitor (Friedlander *et al.*, 1997; Li *et al.*, 2000) had modest, but significant, delays in the onset of disease and prolonged survival. The localizations of p53 are abnormal in mouse ALS (de Aguilar *et al.*, 2000). However, deletion of p53 does not protect against the disease in ALS mice (Kuntz *et al.*, 2000; Prudlo *et al.*, 2000), revealing that the MN degeneration in these mice is independent of p53. These experiments do not rule out the participation of other p53 family members (Levrero *et al.*, 2000). While p63 functions in morphogenesis, p73 also functions in cell survival/death and is expressed in nervous tissue, importantly in MNs (Martin *et al.*, unpublished observations). Cells deficient in p53 can die through the functional redundancy of p73 (Jost *et al.*, 1997; White and Prives, 1999). Tissue homogenate studies have identified some biochemical features of apoptosis in mutant ALS mice, including changes in Bax, Bid, and caspases (Vukosavic *et al.*, 1999; Guegan *et al.*, 2001), but these experiments lack cellular resolution for MN-specific events because whole spinal cords were studied. In spinal cord, MNs form the minority of cells. MN degeneration in ALS mice may also involve inflammatory processes driven by microglia and nitric oxide-mediated cell death (Raoul *et al.*, 2002). Several issues still need to be resolved on the basic classification of the degeneration of MNs in ALS mice. Studies need to be conducted on the relationships between cell death structure and cell biochemistry using LCM to resolve the apparent mismatch between apoptotic structure and biochemistry in the MNs of ALS mice. Moreover, the possible role of p73 needs to be evaluated, as well as the autonomy of the MN death process.

4.5.2.2   MN death mechanisms in nerve injury models

Axonal injury paradigms are very useful for delineating the responses of MNs to axonal perturbations and for discovering cell death mechanisms in MNs and therapeutic targets. Several factors influence the progression of axotomy-induced MN degeneration and the likelihood of subsequent neuronal death or survival, including the age of the animal at the time of injury, the location of axonal trauma in relation to the cell body, and the species (Prestige, 1970; Lieberman, 1971; Torvik, 1976). For example, transection of the seventh cranial nerve or sciatic nerve in newborn rodents causes loss of MNs in the facial nucleus or lumbar cord, respectively; in contrast, a similar lesion in adults produces no discernible loss of MNs (Prestige, 1970). Axotomy-induced degeneration of MNs in the immature CNS resembles apoptosis, based on morphologic evidence in mouse (Romanes, 1946) and chick (O'Connor and Wyttenbach, 1974; Chu-Wang and Oppenheim, 1978). An apoptotic mechanism is supported by the finding that Bcl-2 over-expression ameliorates the MN death in newborn mice in response to facial nerve transection (Dubois-Dauphin *et al.*, 1994) or sciatic nerve transection (Farlie *et al.*, 1995).

However, *bcl-2* transgenic mice have more MNs than wild-type mice (Farlie *et al.*, 1995). Differential changes in IAP levels do not seem to account for the age-related susceptibility of MNs to axotomy (McPhail *et al.*, 2003), but IAPs have a key role in glial cell line-derived neurotrophic factor (GDNF)-mediated protection of MNs in vivo (Perrelet *et al.*, 2002).

In adult animals, sciatic nerve avulsion (SNA) reliably causes loss of MNs in lumbar spinal cord by an unequivocal apoptotic process (Martin *et al.*, 1999). MN apoptosis can be staged based on structure. MNs appear normal and then pass through chromatolysis, somatodendritic attrition, end-stage apoptosis, and phagocytosis. This MN death requires Bax and is modulated by p53 because the apoptosis is blocked in mice deficient in Bax and functional p53 (Martin and Liu, 2002). Single-cell gel electrophoresis (comet assay), immunocytochemistry, and immunogold electron microscopy have been used to measure molecular changes in MNs during the progression of apoptosis (Martin and Liu, 2002). Injured MNs accumulate single-strand breaks in DNA by 5 days, and p53 accumulates in nuclei of MNs destined to undergo apoptosis. p53 is functionally activated by 4–5 days post-lesion, as revealed by serine^{392}-phosphorylated p53. By 7 days post-lesion, Bax translocates to mitochondria, cytochrome c accumulates in the cytoplasm, and caspase-3 is activated (Martin and Liu, 2002). These results suggest that MN apoptosis in the adult spinal cord is controlled by upstream mechanisms involving DNA damage and activation of p53 and downstream mechanisms involving Bax, cytochrome c, and caspase-3. The details of the biochemical mechanisms of apoptosis in MNs are still unknown.

4.5.3 Parkinson's disease (PD)

PD is a chronically progressive, age-related, fatal neurologic disease in humans (Olanow and Tatton, 1999). It affects at least 1% of the population over 55 years of age. In the USA, ~1 million people are affected, placing PD as the second most common neurodegenerative disease with an adult onset (after AD). Progressive resting tremor, rigidity, bradykinesia, gait disturbance, and postural instability characterize PD clinically. The disease progression is associated with dementia and autonomic dysfunction. Progressive degeneration and elimination of dopamine neurons in the substantia nigra compacta (SNc) and the formation of intracytoplasmic inclusions known as Lewy bodies are the neuropathologic hallmarks of PD. The molecular pathogenesis of PD is still not understood. Several candidate genes have been screened (including tyrosine hydroxylase, dopamine receptors, glutathione peroxidase, catalase, and superoxide dismutases), but no linkages have been identified. At least two forms of PD exist: idiopathic and heritable (familial). The majority of PD cases are idiopathic with no known genetic component. The cause of idiopathic PD is unknown. Epidemiologic studies reveal several risk factors for developing idiopathic PD in addition to aging. Pesticides, herbicides, well water, and industrial chemicals are possible neurotoxic agents related to PD. About 5–10% of PD patients have a familial form with an autosomal dominant pattern of inheritance. Mutations in genes have been identified in rare familial forms of PD. These gene mutations occur in two enzymes of the ubiquitin–proteosome system (parkin and ubiquitin C-terminal hydrolyase L1) and in a protein of little known function called α-synuclein.

There is little doubt about the loss of dopamine neurons in PD, yet an important unresolved basic question is how SNc neurons die. The mechanism by which nigral neurons degenerate in human PD is not well understood and requires further study for the development of effective treatments. Some studies report that apoptotic PCD contributes to the neurodegeneration in human PD (Tatton *et al.*, 1998; Hartmann *et al.*, 2000; Tatton *et al.*, 2003), but other studies caution about such claims (Jellinger, 1999; Wüllner *et al.*, 1999). Some of this uncertainty is due to how degenerating nigral neurons appear and how PCD is defined. Nigral neurons in human PD do not degenerate with the morphology consistent with classical apoptosis. However, non-apoptotic forms of PCD exist (Schwartz *et al.*, 1993; Klionsky and Emr, 2000). The pathologic process in PD may involve autophagy (Anglade *et al.*, 1997). The reports on the incidence of dying cells in the PD SNc with DNA fragmentation based on in situ end-labeling methods are conflicting. Some studies report no labeling (Jellinger, 1999), but other experiments (including the authors, Figure 4.8) have found nuclear DNA fragmentation (Tompkins *et al.*, 1997; Tatton *et al.*, 1998). Some uncertainty in these results is due to the use of detection systems that do not distinguish between DNA strand breaks associated with apoptosis or necrosis and to arbitrary morphologic interpretations of chromatin condensation (Martin *et al.*, 1998; Jellinger, 1999).

The ongoing pathology in the SNc of idiopathic PD cases is not a simple picture. H&E sections reveal atrophy, degeneration, and loss of large, multipolar, melanin-containing neurons. Inflammatory cells laden with the debris of dead SNc neurons are conspicuous. In TUNEL preparations, subsets of large, melanin-containing, atrophic neurons have DNA fragmentation (Figure 4.8). Other neurons without melanin are TUNEL[+]; glial and macrophage-like cells are TUNEL[+], as well. Studies need to be done using high-resolution techniques (e.g. LCM) with unambiguous specificity for nigral neurons to demonstrate quantitatively the activation of apoptotic mechanisms directly in SNc neurons. Currently, there have been no reports of nigral neuron death in α-synuclein Tg mice.

4.5.3.1 Nigral neuron death mechanisms in MPTP and 6-OHDA models of PD

Mitochondrial poisons destroy dopaminergic neurons in cell culture and in animal models. This degeneration may be a form of apoptosis; however, the mechanisms

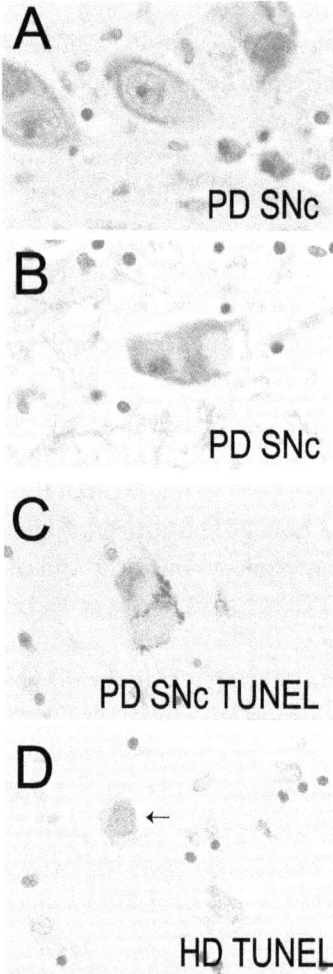

Figure 4.8   *Neuronal cell death in Parkinson's disease (PD) and Huntington's disease (HD).* In PD cases, dopaminergic neurons in the substantia nigra pars compacta (SNc) can be found at different stages of degeneration in H&E-stained sections. As in ALS (Martin, 1999), the neurodegeneration in PD can be staged as chromatolytic (A), attritional (B), and apoptotic. Cell death assays (e.g. TUNEL) show nuclear DNA fragmentation during the attritional stages (C, brown labeling in the shrunken nucleus). In HD, subsets of striatal neurons (D, arrow) show DNA fragmentation.
(For a colour version of this figure, please see www.cambridge.org/9780521159449.)

have not been defined clearly. Caspase inhibitors attenuate MPP$^+$ toxicity in primary cultures of mesencephalic dopamine neurons (Bilsland *et al.*, 2002), but the role of Bax in the death of dopamine neurons in vitro is unclear (Hartmann *et al.*, 2001). MPTP (1-methyl-4-phenyl-1, 2 ,3, 6-tetrahydropyridine) toxicity in nigral dopamine neurons in vivo is attenuated by the over-expression of Bcl-2 (Offen *et al.*,

1998; Yang et al., 1998), Bax gene ablation (Vila et al., 2001), and p53 gene deletion (Trimmer et al., 1996). However, the mechanisms for this Bax- and p53-control of nigral neuron death have not been revealed clearly. PC12 cells exposed to 6-OHDA undergo caspase-mediated apoptosis, possibly through p53- and Bax-dependent mechanisms (Blum et al., 1997). Previous studies have reported apoptosis of SNc neurons after 6-OHDA (6-hydroxy dopamine) poisoning in rats (He et al., 2000; Zuch et al., 2000) by mechanisms that are not fully understood. It has been found that neuronal apoptosis-inhibitory protein (NAIP) can modulate nigral neuron degeneration induced by 6-OHDA (Crocker et al., 2001).

4.5.3.2 Regulation of cell death by synucleins

α-synuclein has a role in cell death but the precise pro-apoptotic versus anti-apoptotic roles in specific types of neurons remain to be better delineated. Wild-type α-synuclein can protect neurons from apoptosis by the inhibition of caspase-3 (Alves da Costa et al., 2002). Mutations in α-synuclein can abolish its inhibitory modulation of caspase-3 (Alves da Costa et al., 2002). Aggregation of wild-type and PD-related mutated α-synuclein is associated with enhanced cell death (Giasson et al., 2000). This process could be due to intrinsic pro-apoptotic properties of α-synuclein or to loss of an anti-apoptotic function. The latter possibility is supported by evidence that α-synuclein lowers the sensitivity of neurons to apoptosis driven by p53 (Alves da Costa et al., 2002) and activates MAPK survival pathways (Iwata et al., 2001). The functions of α-synuclein might be context and neuron type dependent. For example, α-synuclein is neuroprotective in non-dopaminergic cortical neurons but is pro-apoptotic in dopaminergic neurons, with toxicity requiring dopamine and the production of ROS (Xu et al., 2002).

4.5.4 Huntington's disease (HD)

HD is another chronically progressive, age-related, fatal neurologic disease in humans (Harper, 1996); it begins in mid-adulthood and is an autosomal dominant disease. Individuals with HD display motor (chorea), cognitive, and psychiatric abnormalities that progressively increase in severity with a typical duration of ~18 years. At end-stage disease, these individuals show extensive atrophy of the caudate nucleus, putamen, and frontal cortex (Vonsattel et al., 1985). The initial and primary population of neurons that is vulnerable in HD is the medium-sized spiny GABAergic/peptidergic neurons in the striatum, and 95% of this population can be lost (Vonsattel and DiFiglia, 1998). HD is caused by an expansion of a polyglutamine tract (CAG trinucleotide repeat) in the N-terminus of the huntingtin gene that encodes a 350 kD protein of unknown function (The Huntington's Disease Collaborative Research Group, 1993). Recent studies have implicated the formation of intranuclear aggregates of truncated polyglutamine-containing fragments in the pathogenesis of CAG trinucleotide diseases. However, it is still not known how this mutation leads to the selective degeneration of striatal neurons, particularly in light of the finding that huntingtin (Htt) is expressed throughout the body.

The degeneration of striatal neurons in HD is profound and there is no doubt that these neurons die (Vonsattel *et al.*, 1985; Vonsattel and DiFiglia, 1998; Figure 4.8). The evidence for a contribution of apoptosis in HD is controversial. Some studies have argued for a role of apoptosis in human HD because subsets of striatal neurons become dark and condensed or contain fragmented DNA as detected by the TUNEL method (Vonsattel and DiFiglia, 1998; Figure 4.8). However, these assays are not definitive for apoptosis (Martin *et al.*, 1998). Other studies using electron microscopy have revealed that the structure of striatal neuron degeneration in HD is distinct from apoptosis (Turmaine *et al.*, 2000).

Tg mouse models of HD expressing N-terminal Htt fragments have confirmed that polyglutamine expansion in Htt causes a neurologic disease phenotype. Mice expressing smaller N-terminal mutant Htt develop more progressive phenotypes (Davies *et al.*, 1997; Schilling *et al.*, 1999) than mice expressing full-length mutant Htt (Reddy *et al.*, 1998; Hodgson *et al.*, 1999). No major neuronal cell loss has been found in the brains of Tg HD mice with severe phenotypes and early death at 3–6 months of age. However, Tg HD mice expressing full-length mutant Htt with a shorter repeat than 150 glutamines or at a lower expression level than that of endogenous Htt develop neurodegeneration at hypokinetic/akinetic stages (Reddy *et al.*, 1998; Hodgson *et al.*, 1999; Turmaine *et al.*, 2000). This neurodegeneration appears as pyknotic eosinophilic cells in H&E-stained sections of striatum, CA1 hippocampus, neocortex, and cerebellum (Reddy *et al.*, 1998) or as dark shrunken neurons by EM (Hodgson *et al.*, 1999; Turmaine *et al.*, 2000), similar in regional distributions, cellular vulnerabilities, and morphology to hypoxic–ischemic brain damage (Martin *et al.*, 1998). Evidence for apoptosis (using ultrastructural and DNA fragmentation as end-points) has not been found in three lines of symptomatic Tg HD mice (Turmaine *et al.*, 2000). However, it is evident that different lines of Tg HD mice harboring mutant Htt can display very different neurodegenerative responses (Yu *et al.*, 2003). The specific roles of Htt and apoptosis in these neurodegenerative changes still need to be examined.

An exciting link between striatal neurodegeneration and apoptosis regulators has been identified recently (Jones *et al.*, 2003). The mouse mutant mnd2 (motor neuron degeneration 2) exhibits muscle wasting and neurodegeneration, and death by 40 days of age (Rathke-Hartlieb *et al.*, 2002). This mutant was described originally as a model of MN degeneration, but the striatal neurodegeneration occurs earlier and is more profound than the MN degeneration. This neurodegeneration and lethality in mnd2 mice results from a defect in mitochondrial

Omi (Jones *et al.*, 2003), but is not prevented by Bcl-2 (Rathke-Hartlieb *et al.*, 2002).

In vitro experiments are beginning to reveal relationships between mutant Htt and cell death. Studies have found that Htt acts in the nucleus to induce apoptosis independent of the formation of intranuclear and cytoplasmic inclusions (Saudou *et al.*, 1998; Kim *et al.*, 1999). p53 has been implicated in the neurodegeneration of HD through its interaction with Htt (Steffan *et al.*, 2000). An Htt-interacting protein (Hip1) has been identified that has pro-apoptotic activity through the activation of caspase-3 (Hackam *et al.*, 2000). Hip1 interacts with another protein called Hip1 protein interactor (Hippi) with the formation of Hippi–Hip1 heterodimers recruiting and activating caspase-8 through a novel mechanism without activating the plasma membrane death receptor (Gervais *et al.*, 2002).

4.5.5   Excitotoxic cell death

Neuronal death can be induced through excitotoxicity. This basic observation was made originally in 1957 (Lucas and Newhouse, 1957), formulated into a concept by Olney (Olney, 1971), and then extended mechanistically by Choi (Choi, 1992). This concept has fundamental importance to a variety of acute neurologic insults, such as cerebral ischemia and trauma and possibly chronic neurodegenerative diseases, particularly HD, ALS, AD, and PD (Martin, 2001). This pathologic neurodegeneration is mediated by excessive activation of glutamate-gated ion channel receptors as well as voltage-dependent ion channels. The excessive interaction of ligand with subtypes of glutamate receptors causes changes in intracellular ion concentrations, pH, protein phosphorylation, and energy metabolism (Choi, 1992; Lipton and Rosenberg, 1994). An increase in cytosolic free Ca^{2+} causes an activation in Ca^{2+}-sensitive proteases, protein kinases/phosphatases, and phospholipases when glutamate receptors are stimulated. The precise mechanisms of excitotoxic cell death are still being intensively examined, with the hope of identifying therapeutic targets for neurologic/neurodegenerative disorders with excitotoxic components. Both in vitro and in vivo data are discordant with regard to whether excitotoxic neuronal death is apoptotic or necrotic or perhaps even a peculiar form of cell death that is unique to excitotoxicity.

The contribution of apoptotic mechanisms to excitotoxic death of neurons has been examined in neuronal culture. However, these studies have provided conflicting results. Excitotoxicity results in an activation of endonucleases and subsequent internucleosomal DNA fragmentation in cultures of cortical neurons (Kure *et al.*, 1991; Gwag *et al.*, 1997) and cerebellar granule cells (Ankarcrona *et al.*, 1995; Simonian *et al.*, 1996). Other studies have not found internucleosomal DNA fragmentation in cerebellar granule cell cultures (Dessi *et al.*, 1993). Excitotoxic cell death in vitro has been shown to be prevented (Kure *et al.*, 1991) or unaffected (Dessi *et al.*, 1993; Simonian *et al.*, 1996; Gwag *et al.*, 1997) by inhibitors of RNA or protein synthesis and sensitive (Kure *et al.*, 1991; Simonian *et al.*, 1996) or insensitive (Dessi *et al.*, 1993) to the endonuclease inhibitor aurintricarboxylic acid. In primary culture of mouse cortical cells, kainic acid (KA) exposure induced an increase in Bax protein, and *bax* gene deficiency significantly protected cells against KA receptor toxicity (Xiang *et al.*, 1998). However, NMDA (N-methyl-D-aspartate) receptor toxicity in mouse cerebellar granule neurons (Miller *et al.*, 1997) and mouse cortical cells (Dargusch *et al.*, 2001) was not Bax related. In vitro studies of caspase activation in excitotoxic cell death are also conflicting. Glutamate (100 μM) stimulation of mouse cortical cells did not cause an increase in caspase activity (Johnson *et al.*, 1999), but NMDA-treated rat cortical cells showed an increased caspase activity (Tenneti and Lipton, 2000). In cerebellar granule neurons, glutamate (100 μM–1 mM) did not activate caspase activity and adenoviral-mediated expression of IAPs did not influence excitotoxic cell death (Simons *et al.*, 1999). These conflicting results may be due to the finding that activation of different subtypes of glutamate receptors appears to activate different cell death pathways (Portera-Cailliau *et al.*, 1997b).

The in vivo morphologic characteristics of excitotoxicity in many neurons include somatodendritic swelling, chromatin condensation into irregular clumps, and organelle damage (Olney, 1971; van Lookeren Campagne *et al.*, 1995; Portera-Cailliau *et al.*, 1997a,b), features that are thought to be typical of cellular necrosis; however, in other neurons, excitotoxicity causes cytologic features more reminiscent of apoptosis (van Lookeren Campagne *et al.*, 1995; Portera-Cailliau *et al.*, 1997a,b). Excitotoxic degeneration of CA3 neurons in response to KA is increased in NAIP-deleted mice, further supporting a contribution of apoptosis (Holcik *et al.*, 1999). Excitotoxic neurodegeneration in vivo has been shown to be either sensitive (Schreiber *et al.*, 1993) or insensitive (Leppin *et al.*, 1992) to protein synthesis inhibition; therefore, a role for de novo protein synthesis in the expression of a PCD cascade in excitotoxicity is uncertain.

The precise mechanisms of excitotoxic neuronal apoptosis in vivo have not been identified specifically. Neurons in the immature rodent CNS undergo massive apoptosis in response to glutamate receptor excitotoxicity (Portera-Cailliau *et al.*, 1997a,b). Apoptosis is much more prominent after excitotoxic injury in the immature brain compared with the mature brain (Portera-Cailliau *et al.*, 1997b). Intrastriatal administration of KA in newborn rodents causes copious apoptosis of striatal neurons (Portera-Cailliau *et al.*, 1997a,b), serving as an unequivocal model of apoptosis in neurons that are selectively vulnerable in HD. This apoptosis has been verified structurally with light microscopy and EM and by immunolocalization of cleaved caspase-3 (Lok and Martin, 2002). Ubiquitous apoptosis is observed at 24 hours after the insult. DNA degradation by internucleosomal fragmentation further confirms the presence of apoptosis. Excitotoxic neuronal apoptosis is associated with rapid (within 2 hours after neurotoxin exposure) translocation of Bax and cleaved caspase-3 to the mitochondria (Lok and Martin, 2002). Moreover, this study revealed that the ratio of mitochondrial Bax to soluble Bax in normal developing striatum changes prominently with brain maturation. Newborn rat striatum has a much greater proportion of Bax in the mitochondrial fraction with lower levels of soluble Bax. Mature rat striatum has a much larger proportion of Bax in the soluble fraction and low amounts of Bax in the mitochondrial fraction. With brain maturation, there is a linear decrease in the ratio of mitochondrial Bax to soluble Bax. This developmental subcellular redistribution of Bax might be related to the observation that immature neurons exhibit a more robust classical apoptosis response compared with adult neurons after brain damage (Martin, 2001).

ACKNOWLEDGEMENTS

This work was supported by grants from the US Public Health Service, NIH-NINDS (NS34100), NIH-NIA (AG16282), and the Department of Defense US Army Medical Research and Materiel Command (DAMD17-99-1-9553).

REFERENCES

Adamec, E., Vonsattel, J. P., and Nixon, R. A. (1999). DNA strand breaks in Alzheimer's disease. *Brain Res.*, **849**, 67–77.

Al-Abdulla, N. A. and Martin, L. J. (1998). Apoptosis of retrogradely degenerating neurons occurs in association with the accumulation of perikaryal mitochondria and oxidative damage to the nucleus. *Am. J. Pathol.*, **153**, 447–56.

Al-Abdulla, N. A., Portera-Cailliau, C., and Martin, L. J. (1998). Occipital cortex ablation in adult rat causes retrograde neuronal death in the lateral geniculate nucleus that resembles apoptosis. *Neuroscience*, **86**, 191–209.

Aloyz, R. S., Bamji, S. X., Pozniak, C. D. *et al.* (1998). p53 is essential for developmental neuron death regulated by the TrkA and p75 neurotrophin receptors. *J. Cell Biol.*, **143**, 1691–703.

Alvarez, A., Munoz, J. P., and Maccioni, R. B. (2001). A cdk5-p35 stable complex is involved in the β-amyloid-induced deregulation of cdk5 activity in hippocampal neurons. *Exp. Cell Res.*, **264**, 266–74.

Alves da Costa, C., Paitel, E., Vincent, B., and Checler, F. (2002). Alpha-synuclein lowers p53-dependent apoptotic of neuronal cell: abolishment by 6-hydroxydopamine and implication for Parkinson's disease. *J. Biol. Chem.*, **277**, 50980–4.

Ameisen, J. C. (2002). On the origin, evolution, and nature of programmed cell death: a timeline of four billion years. *Cell Death Diff.*, **9**, 367–93.

Amin, F., Bowen, I. D., Szegedi, Z., Mihalik, R., and Szende, B. (2000). Apoptotic and non-apoptotic modes of programmed cell death in MCF-7 human breast carcinoma cells. *Cell Biol. Intl.*, **24**, 253–60.

Anderson, A. J., Su, J. H., and Cotman, C. W. (1996). DNA damage and apoptosis in Alzheimer's disease: colocalization with c-jun immunoreactivity, relationship to brain area, and effect of postmortem delay. *J. Neurosci.*, **16**, 1710–19.

Anglade, P., Vyas, S., Javoy-Agid, F. *et al.* (1997). Apoptosis and autophagy in nigral neurons of patients with Parkinson's disease. *Histol. Histopathol.*, **12**, 25–31.

Ankarcrona, M., Dypbukt, J. M., Bonfoco, E. *et al.* (1995). Glutamate-induced neuronal death: a succession of necrosis or apoptosis depending on mitochondrial function. *Neuron*, **15**, 961–73.

Antonsson, B., Conti, F., Ciavatta, A. *et al.* (1997). Inhibition of bax channel-forming activity by bcl-2. *Science*, **277**, 370–2.

Barron, K. D., Means, E. D., and Larsen, E. (1973). Ultrastructure of retrograde degeneration in thalamus of rat. I. Neuronal somata and dendrites. *J. Neuropathol. Exp. Neurol.*, **32**, 218–44.

Beckman, J. S., Carson, M., Smith, C. D., and Koppenol, W. H. (1993). ALS, SOD and peroxynitrite. *Nature*, **364**, 548.

Beckman, J. S., Estóvez, A. G., Crow, J. P., and Barbeito, L. (2001). Superoxide dismutase and the death of motoneurons in ALS. *Trends Neurosci.*, **24** (suppl), S15–S20.

Behl, C., Davis, J. B., Klier, F. G., and Schubert, D. (1994). Amyloid beta peptide induces necrosis rather than apoptosis. *Brain Res.*, **645**, 253–64.

Bendotti, C., Calvaresi, N., Chiveri, L. *et al.* (2001). Early vacuolization and mitochondrial damage in motor neurons of FALS mice are not associated with apoptosis or with changes in cytochrome oxidase histochemical reactivity. *J. Neurol. Sci.*, **191**, 25–33.

Beresford, P. J., Zhang, D., Oh, D. Y. *et al.* (2001). Granzyme A activates an endoplasmic reticulum-associated caspase-independent nuclease to induce single-stranded DNA nicks. *J. Biol. Chem.*, **276**, 43285–93.

Bilsland, J., Roy, S., Xanthoudakis, S. *et al.* (2002). Caspase inhibitors attenuate 1-methyl-4-phenylpyridinum toxicity in primary cultures of mesencephalic dopaminergic neurons. *J. Neurosci.*, **22**, 2637–49.

Blum, D., Wu, Y., Nissou, M-F., Arnaud, S., Benabid, A.-L., and Verna, J.-M. (1997). p53 and Bax activation in 6-hydroxydopamine-induced apoptosis in PC12 cells. *Brain Res.*, **751**, 139–42.

Bonfanti, L., Strettoi, E., Chierzi, S. *et al.* (1996). Protection of retinal ganglion cells from natural and axotomy-induced cell death in neonatal transgenic mice overexpressing bcl-2. *J. Neurosci.*, **16**, 4186–94.

Bonfoco, E., Krainc, D., Ankarcrona, M., Nicotera, P., and Lipton, S. A. (1995). Apoptosis and necrosis: two distinct events induced, respectively, by mild and intense insults with N-methyl-D-aspartate or nitric oxide/superoxide in cortical cell culture. *Proc. Natl. Acad. Sci. USA*, **92**, 7162–6.

Bradley, W. G. and Krasin, F. (1982). A new hypothesis of the etiology of amyotrophic lateral sclerosis. The DNA hypothesis. *Arch. Neurol.*, **39**, 677–80.

Bucciantini, M., Giannoni, E., Chiti, F. *et al.* (2002). Inherent toxicity of aggregates implies a common mechanism for misfolding diseases. *Nature*, **416**, 507–11.

Bursch, W. (2001). The autophagosomal-lysosomal compartment in programmed cell death. *Cell Death Diff.*, **8**, 569–81.

Bursch, W., Paffe, S., Putz, B., Barthel, G., and Schulte-Hermann, R. (1990). Determination of the length of the histological stages of apoptosis in normal liver and in altered hepatic foci of rats. *Carcinogenesis*, **11**, 847–53.

Bursztajn, S., DeSouza, R., McPhie, D. L. *et al.* (1998). Overexpression in neurons of human presenilin-1 or a presenilin-1 familial Alzheimer disease mutant does not enhance apoptosis. *J. Neurosci.*, **18**, 9790–9.

Calhoun, M. E., Wiederhold, K. H., Abramowski, D. *et al.* (1998). Neuron loss in APP transgenic mice. *Nature*, **395**, 755–6.

Campion, D., Flaman, J. M., Brice, A. *et al.* (1995). Mutations of the presenilin 1 gene in families with early-onset Alzheimer's disease. *Hum. Mol. Genet.*, **4**, 2373–7.

Cardone, M. H., Roy, N., Stennicke, H. R. *et al.* (1998). Regulation of cell death protease caspase-9 by phosphorylation. *Science*, **282**, 1318–21.

Cardone, M. H., Salvesen, G. S., Widmann, C., Johnson, G., and Frisch, S. M. (1997). The regulation of anoikis: MEKK-1 activation requires cleavage by caspases. *Cell*, **90**, 315–23.

Casaccia-Bonnefil, P., Carter, B. D., Dobrowsky, R. T., and Chao, M. V. (1996). Death of oligodendrocytes mediated by the interaction of nerve growth factor with its receptor p75. *Nature*, **383**, 716–19.

Casley, C. S., Land, J. M., Sharpe, M. A., Clark, J. B., Duchen, M. R., and Canevari, L. (2002). β-amyloid fragment 25–35 causes mitochondrial dysfunction in primary cortical neurons. *Neurobiol. Dis.*, **10**, 258–67.

Cataldo, A. M., Hamilton, D. J., and Nixon, R. A. (1994). Lysosomal abnormalities in degenerating neurons link neuronal compromise to senile plaque development in Alzheimer's disease. *Brain Res.*, **640**, 68–80.

Chan, S. L., Culmsee, C., Haughey, N., Klapper, W., and Mattson, M. P. (2002). Presenilin-1 mutations sensitize neurons to DNA damage-induced death by a mechanism involving perturbed calcium homeostasis and activation of calpains and caspase-12. *Neurobiol. Dis.*, **11**, 2–19.

Chang, Y.-C., Lee, Y.-S., Tejima, T. *et al.* (1998). Mdm-2 and bax, downstream mediators of the p53 response, are degraded by the ubiquitin-proteasome pathway. *Cell Growth Diff.*, **9**, 79–84.

Chartier-Harlin, M.-C., Crawford, F., Houlden, H. *et al.* (1991). Early-onset Alzheimer's disease caused by mutations at codon 717 of the β-amyloid precursor protein gene. *Nature*, **353**, 844–6.

Choi, D. W. (1992). Excitotoxic cell death. *J. Neurobiol.*, **23**, 1261–76.

Chu-Wang, I.-W. and Oppenheim, R. W. (1978). Cell death of motoneurons in the chick embryo spinal cord. I. A light and electron microscopic study of naturally occurring and induced cell loss during development. *J. Comp. Neurol.*, **177**, 33–58.

Clarke, P. G. H. (1990). Developmental cell death: morphological diversity and multiple mechanisms. *Anat. Embryol.*, **181**, 195–213.

Cory, S. and Adams, J. M. (2002). The bcl-2 family: regulators of the cellular life-or-death switch. *Nat. Rev.*, **2**, 647–56.

Crocker, S. J., Wigle, N., Liston, P. *et al.* (2001). NAIP protects the nigrostriatal dopamine pathway in an intrastriatal 6-OHDA rat model of Parkinson's disease. *Eur. J. Neurosci.*, **14**, 391–400.

Crompton, M. (1999). The mitochondrial permeability transition pore and its role in cell death. *Biochem. J.*, **341**, 233–49.

Cruz, J. C., Tseng, H. C., Goldman, J. A., Shih, H., and Tsai, L.-H. (2003). Aberrant Cdk5 activation by p25 triggers pathological events leading to neurodegeneration and neurofibrillary tangles. *Neuron*, **40**, 471–83.

Culmsee, C., Zhu, X., Yu, Q. S. *et al.* (2001). A synthetic inhibitor of p53 protects neurons against death induced by ischemic and excitotoxic insults, and amyloid beta-peptide. *J. Neurochem.*, **77**, 220–8.

Danial, N. N., Gramm, C. F., Scorrano, L. *et al.* (2003). Bad and glucokinase reside in a mitochondrial complex that integrates glycolysis and apoptosis. *Nature*, **424**, 952–6.

Dargusch, R., Piasecki, D., Tan, S., Liu, Y., and Schubert, D. (2001). The role of Bax in glutamate-induced nerve cell death. *J. Neurochem.*, **76**, 295–301.

Datta, S. R., Dudek, H., Tao, X. *et al.* (1997). Akt phosphorylation of Bad couples survival signals to the cell-intrinsic death machinery. *Cell*, **91**, 231–41.

Davies, S. W., Turmaine, M., Cozens, B. A. *et al.* (1997). Formation of neuronal intranuclear inclusions underlies the neurological dysfunction in mice transgenic for the HD mutation. *Cell*, **90**, 537–48.

de Aguilar, J. L. G., Gorden, J. W., Rene, F. *et al.* (2000). Alteration of the cl-x/Bax ratio in a transgenic mouse model of amyotrophic lateral sclerosis: evidence for the implication of the p53 signaling pathway. *Neurobiol. Dis.*, **7**, 406–15.

Deak, J. C., Cross, J. V., Lewis, M. *et al.* (1998). Fas-induced proteolytic activation and intracellular redistribution of the stress-signaling kinase MEKK1. *Proc. Natl. Acad. Sci. USA*, **95**, 5595–600.

Deckwerth, T. L., Elliott, J. L., Knudson, C. M., Johnson, E. M., Snider, W. D., and Korsmeyer, S. J. (1996). Bax is required for neuronal death after trophic factor deprivation and during development. *Neuron*, **17**, 401–11.

DeKosky, S. T. and Scheff, S. W. (1990). Synapse loss in frontal cortex biopsies in Alzheimer's disease: correlation with cognitive severity. *Ann. Neurol.*, **27**, 457–64.

de la Monte, S., Sohn, Y. K., and Wands, J. R. (1997). Correlates of p53- and Fas (CD95)-mediated apoptosis in Alzheimer's disease. *J. Neurol. Sci.*, **152**, 73–83.

de la Monte, S. M., Sohn, Y. K., Ganju, N., and Wands, J. R. (1998). p53- and CD95-associated apoptosis in neurodegenerative diseases. *Lab. Invest.*, **78**, 401–11.

del Peso, L., Gonzlez-Garcia, M., Page, C., Herrera, R., and Nuez, G. (1997). Interleukin-3-induced phosphorylation of bad through the protein kinase Akt. *Science*, **278**, 687–9.

Deng, H.-X., Hentati, A., Tainer, J. A. *et al.* (1993). Amyotrophic lateral sclerosis and structural defects in Cu,Zn superoxide dismutase. *Science*, **261**, 1047–51.

Desagher, S., Osen-Sand, A., Nichols, A. *et al.* (1999). Bid-induced conformational change of bax is responsible for mitochondrial cytochrome c release during apoptosis. *J. Cell Biol.*, **144**, 891–901.

Dessi, F., Charriaut-Marlangue, C., Khrestchatisky, M., and Ben-Ari, Y. (1993). Glutamate-induced neuronal death is not a programmed cell death in cerebellar culture. *J. Neurochem.*, **60**, 1953–5.

Deveraux, Q. L., Roy, N., Stennicke, H. R. *et al.* (1998). IAPs block apoptotic events induced by caspase-8 and cytochrome *c* by direct inhibition of distinct caspases. *EMBO J.*, **17**, 2215–23.

Di Cristofano, A., Pesce, B., Cordon-Cardo, C., and Pandolfi, P. P. (1998). Pten is essential for embryonic development and tumor suppression. *Nat. Genet.*, **19**, 348–55.

Du, C., Fang, M., Li, Y., Li, L., and Wang, X. (2000). Smac, a mitochondrial protein that promotes cytochrome c-dependent caspase activation by eliminating IAP inhibition. *Cell*, **102**, 33–42.

Dubois-Dauphin, M., Frankowski, H., Tsujimoto, Y., Huarte, J., and Martinou, J.-C. (1994). Neonatal motoneurons overexpressing the bcl-2 protooncogene in transgenic mice are protected from axotomy-induced cell death. *Proc. Natl. Acad. Sci. USA*, **91**, 3309–13.

Duker, N. J., Sperling, J., Soprano, K. J., Druin, D. P., Davis, A., and Ashworth, R. (2001). β-amyloid protein induces the formation of purine dimers in cellular DNA. *J. Cell Biochem.*, **81**, 393–400.

English, J., Pearson, G., Wilsbacher, J. *et al.* (1999). New insights into the control of MAP kinase pathways. *Exp. Cell Res.*, **253**, 255–70.

Ethell, D. W., Kinloch, R., and Green, D. R. (2002). Metalloproteinase shedding of Fas ligand regulates beta-amyloid neurotoxicity. *Curr. Biol.*, **12**, 1595–600.

Evans, D. A., Funkenstein, H. H., Albert, M. S. *et al.* (1989). Prevalence of Alzheimer's disease in a community population of older persons: higher than previously reported. JAMA, **262**, 2551–6.

Fan, Z., Beresford, P. J., Oh, D. Y., Zhang, D., and Lieberman, J. (2003). Tumor suppressor NM23-H1 is a granzyme A-activated DNase during TL-mediated apoptosis, and the nucleosome assembly protein SET is its inhibitor. *Cell*, **112**, 659–72.

Farlie, P. G., Dringen, R., Rees, S. M., Kannourakis, G., and Bernard, O. (1995). Bcl-2 transgene expression can protect neurons against developmental and induced cell death. *Proc. Natl. Acad. Sci. USA*, **92**, 4397–401.

Fath, T., Eidenmuller, J., and Brandt, R. (2002). Tau-mediated cytotoxicity in a pseudohyper-phosphorylation model of Alzheimer's disease. *J. Neurosci.*, **22**, 9733–41.

Fernandes, R. S. and Cotter, T. G. (1994). Apoptosis or necrosis: intracellular levels of glutathione influence mode of cell death. *Biochem. Pharmacol.*, **48**, 675–81.

Ferrante, R. J., Browne, S. E., Shinobu, L. A. *et al.* (1997). Evidence of increased oxidative damage in both sporadic and familial amyotrophic lateral sclerosis. *J. Neurochem.*, **69**, 2064–74.

Ferrer, I., Puig, B., Krupinski, J., Carmona, M., and Blanco, R. (2001). Fas and Fas ligand expression in Alzheimer's disease. *Acta. Neuropathol.*, **102**, 121–31.

Fitzmaurice, P. S., Shaw, I. C., Kleiner, H. E. *et al.* (1996). Evidence for DNA damage in amyotrophic lateral sclerosis. *Muscle. Nerve.*, **19**, 797–8.

Friedlander, R. M., Brown, R. H., Gagliardini, V., Wang, J., and Yuan, J. (1997). Inhibition of ICE slows ALS in mice. *Nature*, **388**, 31.

Froelich, S., Basun, H., Forsell, C. *et al.* (1997). Mapping of a disease locus for familial rapidly progressive frontotemporal dementia to chromosome 17q12–21. *Am. J. Med. Genet.*, **74**, 380–5.

Gamliel, A., Teicher, C., Hartmann, T., Beyreuther, K., and Stein, R. (2003). Overexpression of wild-type presenilin 2 or its familial Alzheimer's disease-associated mutant does not induce or increase susceptibility to apoptosis in different cells. *Neuroscience*, **117**, 119–28.

Gastard, M. C., Troncoso, J. C., and Koliatsos, V. E. (2003). Caspase activation in the limbic cortex of subjects with early Alzheimer's disease. *Ann. Neurol.*, **54**, 393–8.

Gervais, F. G., Singaraja, R., Xanthoudakis, S. *et al.* (2002). Recruitment and activation of caspase-8 by the Huntingtin-interacting protein Hip-1 and a novel partner Hippi. *Nat. Cell Biol.*, **4**, 95–105.

Gervais, F. G., Xu, D., Robertson, G. S. *et al.* (1999). Involvement of caspases in proteolytic cleavage of Alzheimer's amyloid-beta precursor protein and amyloidogenic A beta peptide formation. *Cell*, **97**, 395–406.

Giaccia, A. J. and Kastan, M. B. (1998). The complexity of p53 modulation: emerging patterns from divergent signals. *Genes Develop.*, **12**, 2973–83.

Giasson, B. I., Duda, J. E., Murray, I. V. J. *et al.* (2000). Oxidative damage linked to neuro-degeneration by selective alpha-synuclein nitration in synucleinopathy lesions. *Science*, **290**, 985–9.

Gilman, C. P., Chan, S. L., Guo, Z., Zhu, X., Greig, N., and Mattson, M. P. (2003). p53 is present in synapses where it mediates mitochondrial dysfunction and synaptic degeneration in response to DNA damage, and oxidative and excitotoxic insults. *Neuromolecular Med.*, **3**, 159–72.

Giovanni, A., Keramaris, E., Morris, E. J. *et al.* (2000). E2F1 mediates death of β-amyloid-treated cortical neurons in a manner independent of p53 and dependent on Bax and caspase 3. *J. Biol. Chem.*, **275**, 11553–60.

Glücksmann, A. (1951). Cell deaths in normal vertebrate ontogeny. *Biol. Rev.*, **26**, 59–86.

Goate, A., Chartier-Harlin, M.-C., Mullan, M. *et al.* (1991). Segregation of a missense mutation in the amyloid precursor protein gene with familial Alzheimer's disease. *Nature*, **349**, 704–6.

Golden, W. C., Brambrink, A. M., Traystman, R. J., and Martin, L. J. (2001). Failure to sustain recovery of Na,K ATPase function is a possible mechanism for striatal neurodegeneration in hypoxic-ischemic newborn piglets. *Mol. Brain Res.*, **88**, 94–102.

Goldschmidt, R. B. and Steward, O. (1980). Preferential neurotoxicity of colchicine for granule cells of the dentate gyrus of adult rat. *Proc. Natl. Acad. Sci. USA*, **77**, 3047–51.

Gotz, J., Chen, F., Barmettler, R., and Nitsch, R. M. (2001a). Tau filament formation in transgenic mice expressing P301L tau. *J. Biol. Chem.*, **276**, 529–34.

Gotz, J., Chen, F., van Dorpe, J., and Nitsch, R. M. (2001b). Formation of neurofibrillary tangles in P301l tau transgenic mice induced by Abeta 42 fibrils. *Science*, **293**, 1491–5.

Gomez-Isla, T., Price, J. L., McKeel, D. W. Jr., Morris, J. C., Growdon, J. H., and Hyman, B. T. (1996). Profound loss of layer II entorhinal cortex neurons occurs in very mild Alzheimer's disease. *J. Neurosci.*, **16**, 4491–500.

Guegan, C., Vila, M., Rosoklija, G., Hays, A. P., and Przeborski, S. (2001). Recruitment of the mitochondrial-dependent apoptotic pathway in amyotrophic lateral sclerosis. *J. Neurosci.*, **21**, 6569–76.

Gwag, B. J., Koh, J. Y., DeMaro, J. A., Ying, H. S., Jacquin, M., and Choi, D. W. (1997). Slowly triggered excitotoxicity occurs by necrosis in cortical cultures. *Neuroscience*, **77**, 393–401.

Hackam, A. S., Yassa, A. S., Singaraja, R. *et al.* (2000). Huntingtin interacting protein 1 induces apoptosis via a novel caspase-dependent death effector domain. *J. Biol. Chem.*, **275**, 41299–308.

Hakem, R., Hakem, A., Duncan, G. S. *et al.* (1998). Differential requirement for caspase 9 in apoptotic pathways in vivo. *Cell*, **94**, 339–52.

Haldar, S., Jena, N., and Croce, C. M. (1995). Inactivation of Bcl-2 by phosphorylation. *Proc. Natl. Acad. Sci. USA*, **92**, 4507–11.

Hardy, J. and Selkoe, D. J. (2002). The amyloid hypothesis of Alzheimer's disease: progress and problems on the road to therapeutics. *Science*, **297**, 353–6.

Harper, P. S. (1996). *Huntington's Disease*. London: W. B. Saunders.

Hartmann, A., Hunot, S., Michel, P. P. *et al.* (2000). Caspase-3: a vulnerability factor and final effector in apoptotic death of dopaminergic neurons in Parkinson's disease. *Proc. Natl. Acad. Sci.*, **97**, 2875–80.

Hartmann, A., Michel, P. P., Troadec, J.-D. *et al.* (2001). Is Bax a mitochondrial mediator in apoptotic death of dopaminergic neurons in Parkinson's disease? *J. Neurochem.*, **76**, 1785–93.

Hayward, C., Colville, S., Swingler, R. J., and Brock, D. J. H. (1999). Molecular genetic analysis of the APEX nuclease gene in amyotrophic lateral sclerosis. *Neurology*, **52**, 1899–901.

He, Y., Lee, T., and Leong, S. K. (2000). 6-hydroxydopamine induced apoptosis of dopaminergic cells in the rat substantia nigra. *Brain Res.*, **858**, 163–6.

Hegde, R., Srinivasula, S. M., Zhang, Z. *et al.* (2002). Identification of Omi/HtrA2 as a mitochondrial apoptotic serine protease that disrupts inhibitor of apoptosis protein-caspase interaction. *J. Biol. Chem.*, **277**, 432–8.

Hetman, M., Kanning, K., Cavanaugh, J. E., and Xia, Z. (1999). Neuroprotection by brain-derived neurotrophic factor is mediated by extracellular signal-regulated kinase and phosphatidyl-inositol 3-kinase. *J. Biol. Chem.*, **2744**, 22569–80.

Hirsch, T., Marchetti, P., Susin, S. A. *et al.* (1997). The apoptosis-necrosis paradox. Apoptogenic proteases activated after mitochondrial permeability transition determine the mode of cell death. *Oncogene*, **15**, 1573–81.

Hodgson, J. G., Agopyan, N., Gutekunst, C. A. *et al.* (1999). A YAC mouse model for Huntington's disease with full-length mutant huntingtin, cytoplasmic toxicity, and selective striatal neurodegeneration. *Neuron*, **23**, 181–92.

Holcik, M. (2002). The IAP proteins. *Trends Gen.*, **18**, 537–8.

Holcik, M., Thompson, C. S., Yaraghi, Z., Lefebvre, C. A., MacKenzie, A. E., and Korneluk, R. G. (1999). The hippocampal neurons of neuronal apoptosis inhibitory protein 1 (NAIP1)-deleted mice display increased vulnerability to kainic acid-induced injury. *Proc. Natl. Acad. Sci. USA*, **97**, 2286–90.

Holzer, M., Gartner, U., Klinz, F. J., Narz, F., Heumann, R., and Arendt, T. (2001). Activation of mitogen-activated protein kinase cascade and phosphorylation of cytoskeletal proteins after neurone-specific activation of p21ras. I. Mitogen-activated protein kinase cascade. *Neuroscience*, **105**, 1031–40.

Hu, Y., Benedict, M. A., Wu, D., Inohara, N., and Núñez, G. (1998). Bcl-x_L interacts with Apaf-1 and inhibits Apaf-1-dependent caspase-9 activation. *Proc. Natl. Acad. Sci. USA*, **95**, 4386–91.

The Huntington's Disease Collaborative Research Group (1993). A novel gene containing a trinucleotide repeat that is expanded and unstable on Huntington's disease chromosomes. *Cell*, **72**, 971–83.

Ikonomidou, C., Qin, Y. Q., Labruyere, J., and Olney, J. W. (1996). Motor neuron degeneration induced by excitotoxin agonists has features in common with those seen in the SOD-1 transgenic mouse model of amyotrophic lateral sclerosis. *J. Neuropathol. Exp. Neurol.*, **55**, 211–44.

Inbal, B., Bialik, S., Sabanay, I., Shani, G., and Kimchi, A. (2002). DAP kinase and DRP-1 mediate membrane blebbing and the formation of autophagic vesicles during programmed cell death. *J. Cell Biol.*, **157**, 455–68.

Irizarry, M. C., McNamara, M., Fedorchak, K., Hsiao, K., and Hyman, B. T. (1997). APPSw transgenic mice develop age-related A beta deposits and neuropil abnormalities, but no neuronal loss in CA1. *J. Neuropathol. Exp. Neurol.*, **56**, 965–73.

Ishikawa, Y. and Kitamura, M. (1999). Dual potential of extracellular signal-regulated kinase for the control of cell survival. *Biochem. Biophys. Res. Comm.*, **2644**, 696–701.

Iwata, A., Maruyama, M., Kanazawa, I., and Nukina, N. (2001). α-synuclein affects the MAPK pathway and accelerates cell death. *J. Biol. Chem.*, **276**, 45320–9.

Jellinger, K. A. (1999). Is there apoptosis in Lewy body disease? *Acta. Neuropathol.*, **97**, 413–15.

Johnson, M. D., Kinoshita, Y., Xiang, H., Ghatan, S., and Morrison, R. S. (1999). Contribution of p53-dependent caspase activation to neuronal cell death declines with neuronal maturation. *J. Neurosci.*, **19**, 2996–3006.

Jones, J. M., Datta, P., Srinivasula, S. M. *et al.* (2003). Loss of Omi mitochondrial protease activity causes the neuromuscular disorder of mnd2 mutant mice. *Nature*, **425**, 721–7.

Jost, C. A., Marin, M. C., and Kaelin, W. G. (1997). p73 is a human p53-related protein that can induce apoptosis. *Nature*, **389**, 191–4.

Kalaria, R. N. (2003). Dementia comes of age in the developing world. *Lancet*, **361**, 888–9.

Kaplan, D. R. and Miller, F. D. (1997). Signal transduction by the neurotrophin receptors. *Curr. Opin. Cell Biol.*, **9**, 213–21.

Katzman, R. (1993). Education and the prevalence of dementia and Alzheimer's disease. *Neurology*, **43**, 13–20.

Keilin, D. (1930). Cytochrome and intracellular oxidase. *Proc. R. Soc. Lond.*, **106**, 418–44.

Kemp, J. A. and McKernan, R. M. (2002). NMDA receptor pathways as drug targets. *Nat. Neurosci.* (suppl), **5**, 1039–42.

Kerr, J. F. R., Wyllie, A. H., and Currie, A. R. (1972). Apoptosis: a basic biological phenomenon with wide-ranging implications in tissue kinetics. *Br. J. Cancer*, **26**, 239–57.

Kim, J. A., Mitsukawa, K., Yamada, M. K., Nishiyama, N., Matsuki, N., and Ikegaya, Y. (2002). Cytoskeleton disruption causes apoptotic degeneration of dentate granule cells in hippocampal slice cultures. *Neuropharmacology*, **42**, 1109–18.

Kim, M., Lee, H. S., LaForet, G. *et al.* (1999). Mutant huntingtin expression in clonal striatal cells: dissociation of inclusion formation and neuronal survival by caspase inhibition. *J. Neurosci.*, **19**, 964–73.

Kim, T.-W., Pettingell, W. H., Jung, Y.-K., Kovacs, D. M., and Tazi, R. E. (1997). Alternative cleavage of Alzheimer-associated presenilins during apoptosis by a caspase-3 family protein. *Science*, **277**, 373–6.

Kisby, G. E., Miline, J., and Sweatt, C. (1997). Evidence of reduced DNA repair in amyotrophic lateral sclerosis brain tissue. *NeuroReport*, **8**, 1337–40.

Kitamura, Y., Shimohama, S., Kamoshima, W., Matsuoka, Y., Nomura, Y., and Taniguchi, T. (1997). Changes of p53 in the brains of patients with Alzheimer's disease. *Biochem. Biophys. Res. Comm.*, **232**, 418–21.

Kitamura, Y., Shimohama, S., Kamoshima, W. *et al.* (1998). Alteration of proteins regulating apoptosis, Bcl-2, Bcl-x, Bax, Bak, Bad, ICH-1 and CPP32, in Alzheimer's disease. *Brain Res.*, **780**, 260–9.

Klein, J. A., Longo-Guess, C. M., Rossmann, M. P. *et al.* (2002). The harlequin mouse mutation downregulates apoptosis-inducing factor. *Nature*, **419**, 367–74.

Klionsky, D. J. and Emr, S. D. (2000). Autophagy as a regulated pathway of cellular degradation. *Science*, **290**, 1717–21.

Kluck, R. M., Bossy-Wetzel, E., Green, D. R., and Newmeyer, D. D. (1997). The release of cytochrome c from mitochondria: a primary site for bcl-2 regulation of apoptosis. *Science*, **275**, 1132–6.

Kosik, K. S., Joachim, C. L., and Selkoe, D. J. (1986). Microtubule-associated protein tau is a major antigenic component of paired helical filaments in Alzheimer's disease. *Proc. Natl. Acad. Sci. USA*, **83**, 4044–8.

Kostic, V., Jackson-Lewis, V., de Bilbao, F., Dubois-Dauphin, M., and Przedborski, S. (1997). Bcl-2: prolonging life in a transgenic mouse model of familial amyotrophic lateral sclerosis, *Science*, **277**, 559–62.

Kuida, K., Haydar, T. F., Kuan, C.-Y. *et al.* (1998). Reduced apoptosis and cytochrome c-mediated caspase activation in mice lacking caspase-9. *Cell*, **94**, 325–37.

Kuntz, C., Kinoshita, Y., Beal, M. F., Donehower, L. A., and Morrison, R. S. (2000). Absence of p53: no effect in a transgenic mouse model of familial amyotrophic lateral sclerosis. *Exp. Neurol.*, **165**, 184–90.

Kuperstein, F. and Yavin, E. (2002). ERK activation and nuclear translocation in amyloid-beta peptide- and iron-stressed neuronal cell cultures. *Eur. J. Neurosci.*, **16**, 44–54.

Kure, S., Tominaga, T., Yoshimoto, T., Tada, K., and Narisawa, K. (1991). Glutamate triggers internucleosomal DNA cleavage in neuronal cells. *Biochem. Biophys. Res. Commun.*, **179**, 39–45.

Kwon, C. H., Zhu, X., Zhang, J. *et al.* (2001). Pten regulates neuronal soma size: a mouse model of Lhermitte–Duclos disease. *Nat. Genet.*, **29**, 410–11.

LaCasse, E. C., Baird, S., Korneluk, R. G., and MacKenzie, A. E. (1998). The inhibitors of apoptosis (IAPs) and their emerging role in cancer. *Oncogene*, **17**, 3247–59.

Lachyankar, M. B., Condon, P. J., Quesenberry, P. J. *et al.* (2000). A role for nuclear PTEN in neuronal differentiation. *J. Neurosci.*, **20**, 1404–13.

LaFerla, F. M., Hall, C. K., Ngo, L., and Jay, G. (1996). Extracellular deposition of β-amyloid upon p53-dependent neuronal cell death in transgenic mice. *J. Clin. Invest.*, **98**, 1626–32.

Lambrechts, D., Storkebaum, E., Morimoto, M. *et al.* (2003). VEGF is a modifier of amyotrophic lateral sclerosis in mice and humans and protects motoneurons against ischemic death. *Nat. Gen.*, **34**, 383–94.

Lashley, K. S. (1941). Thalamo-cortical connections of the rat's brain. *J. Comp. Neurol.*, **75**, 67–121.

LeBlanc, A., Liu, H., Goodyer, C., Bergeron, C., and Hammond, J. (1999). Caspase-6 role in apoptosis of human neurons, amyloidogenesis, and Alzheimer's disease. *J. Biol. Chem.*, **274**, 23426–36.

Leist, M., Single, B., Castoldi, A. F., Kühnle, S., and Nicotera, P. (1997). Intracellular adenosine triphosphate (ATP) concentration: a switch in the decision between apoptosis and necrosis. *J. Exp. Med.*, **185**, 1481–6.

Lennon, S. V., Martin, S. J., and Cotter, T. G. (1991). Dose-dependent induction of apoptosis in human tumour cell lines by widely diverging stimuli. *Cell Prolif.*, **24**, 203–14.

Leppin, C., Finiels-Marlier, F., Crawley, J. N., Montpied, P., and Paul, S. M. (1992). Failure of a protein synthesis inhibitor to modify glutamate receptor-mediated neurotoxicity in vivo. *Brain Res.*, **581**, 168–70.

Lesuisse, C. and Martin, L. J. (2002). Immature and mature neurons engage different apoptotic mechanisms involving caspase-3 and the mitogen-activated protein kinase pathway. *J. Cereb. Blood Flow Metabol.*, **22**, 935–50.

Letai, A., Bassik, M. C., Walensky, L. D., Sorcinelli, M. D., Weiler, S., and Korsmeyer, S. J. (2002). Distinct BH3 domains either sensitize or activate mitochondrial apoptosis, serving as prototype cancer therapeutics. *Cancer Cell*, **2**, 183–92.

Levrero, M., De Laurenzi, V., Costanzo, A. *et al.* (2000). The p53/p63/p73 family of transcription factors: overlapping and distinct functions. *J. Cell Sci.*, **113**, 1661–70.

Lewis, J., Dickson, D. W., Lin, W. L. *et al.* (2001). Enhanced neurofibrillary degeneration in transgenic mice expressing mutant tau and APP. *Science*, **293**, 1487–91.

Li, H., Zhu, H., Xu, C.-J., and Yuan, J. (1998). Cleavage of Bid by caspase 8 mediates the mitochondrial damage in the Fas pathway of apoptosis. *Cell*, **94**, 491–501.

Li, M., Ona, V. O., Guegan, C. *et al.* (2000). Functional role of caspase-1 and caspase-3 in an ALS transgenic mouse model. *Science*, **288**, 283–4.

Li, P., Nijhawan, D., Budihardjo, I. *et al.* (1997). Cytochrome c and dATP-dependent formation of Apaf-1/caspase-9 complex initiates an apoptotic protease cascade. *Cell*, **91**, 479–89.

Liange, X. H., Kleeman, L. K., Jiang, H. H. *et al.* (1998). Protection against fatal Sindbis virus encephalitis by beclin, a novel Bcl-2-interacting protein. *J. Virol.*, **72**, 8586–96.

Lieberman, A. R. (1971). The axon reaction: a review of the principal features of perikaryal responses to axon injury. *Int. Rev. Neurobiol.*, **14**, 49–124.

Lipton, S. A. and Rosenberg, P. A. (1994). Excitatory amino acids as a final common pathway for neurologic disorders. *N. Engl. J. Med.*, **330**, 613–22.

Liston, P., Roy, N., Tamai, K. *et al.* (1996). Suppression of apoptosis in mammalian cells by NAIP and a related family of IAP genes. *Nature*, **379**, 349–53.

Lithgow, T., van Driel, R., Bertram, J. F., and Strasser, A. (1994). The protein product of the oncogene *bcl-2* is a component of the nuclear envelope, the endoplasmic reticulum, and the outer mitochondrial membrane. *Cell Growth Differ.*, **5**, 411–17.

Liu, X., Kim, C. N., Yang, J., Jemmerson, R., and Wang, X. (1996). Induction of apoptotic program in cell-free extracts: requirement for dATP and cytochrome c. *Cell*, **86**, 147–57.

Liu, X., Zou, H., Slaughter, C., and Wang, X. (1997). DFF, a heterodimeric protein that functions downstream of caspase-3 to trigger DNA fragmentation during apoptosis. *Cell*, **89**, 175–84.

Liu, Z., Gastard, M., Verina, T., Bora, S., Mouton, P. R., and Koliatsos, V. E. (2001). Estrogens modulate experimentally induced apoptosis of granule cells in the adult hippocampus. *J. Comp. Neurol.*, **441**, 1–8.

Lockshin, R. A. and Williams, C. M. (1964). Programmed cell death: II. Endocrine potentiation of the breakdown of the intersegmental muscles of silkmoths. *J. Insect. Physiol.*, **10**, 643–9.

Lockshin, R. A. and Zakeri, Z. (2001). Programmed cell death and apoptosis: origins of the theory. *Nat. Rev. Mol. Cell Biol.*, **2**, 545–50.

(2002). Caspase-independent cell deaths. *Curr. Opin. Cell Biol.*, **14**, 727–33.

Lok, J. and Martin, L. J. (2002). Rapid subcellular redistribution of Bax precedes caspase-3 and endonuclease activation during excitotoxic neuronal apoptosis in rat brain. *J. Neurotrauma*, **19**, 815–28.

Loo, D. T., Copani. A., Pike, C. J., Whittemore, E. R., Walencewicz, A. J., and Cotman C. W. (1993). Apoptosis is induced by β-amyloid in cultured central nervous system neurons. *Proc. Natl. Acad. Sci. USA*, **90**, 7951–5.

Lucas, D. R. and Newhouse, J. P. (1957). The toxic effect of sodium L-glutamate on the inner layers of the retina. *Arch. Ophthal.*, **58**, 193–201.

Lucassen, P. J., Chung, W. C. J., Kamphorst, W., and Swaab D. F. (1997). DNA damage distribution in the human brain as shown by in situ end labeling: area-specific differences in aging and Alzheimer's disease in the absence of apoptotic morphology. *J. Neuropathol. Exp. Neurol.*, **56**, 887–900.

Majno, G. and Joris, I. (1995). Apoptosis, oncosis, and necrosis. An overview of cell death. *Am. J. Pathol.*, **146**, 3–15.

Maki, C. G., Huibregtse, J. M., and Howley, P. M. (1996). In vivo ubiquitination and proteasome-mediated degradation of p53. *Cancer Res.*, **56**, 2649–54.

Martin, L. J. (1999). Neuronal death in amyotrophic lateral sclerosis is apoptosis: possible contribution of a programmed cell death mechanism. *J. Neuropathol. Exp. Neurol.*, **58**, 459–71.

(2000). p53 is abnormally elevated and active in the CNS of patients with amyotrophic lateral sclerosis. *Neurobiol. Dis.*, **7**, 613–22.

(2001). Neuronal cell death in nervous system development, disease, and injury. *Int. J. Mol. Med.*, **7**, 455–78.

Martin, L. J. and Liu, Z. (2002). Injury-induced spinal motor neuron apoptosis is preceded by DNA single-strand breaks and is p53- and Bax-dependent. *J. Neurobiol.*, **50**, 181–97.

Martin, L. J., Al-Abdulla, N. A., Brambrink, A. M., Kirsch, J. R., Sieber, F. E. and Portera-Cailliau, C. (1998). Neurodegeneration in excitotoxicity, global cerebral ischemia, and target deprivation: a perspective on the contributions of apoptosis and necrosis. *Brain Res. Bull.*, **46**, 281–309.

Martin, L. J., Kaiser, A., and Price, A. C. (1999). Motor neuron degeneration after sciatic nerve avulsion in adult rat evolves with oxidative stress and is apoptosis. *J. Neurobiol.*, **40**, 185–201.

Martin, L. J., Kaiser, A., Yu, J. W., Natale, J. E., and Al-Abdulla, N. A. (2001). Injury-induced apoptosis of neurons in adult brain is mediated by p53-dependent and p53-independent pathways and requires Bax. *J. Comp. Neurol.*, **433**, 299–311.

Martin, L. J., Pardo, C. A., Cork, L. C., and Price, D. L. (1994). Synaptic pathology and glial responses to neuronal injury precede the formation of senile plaques and amyloid deposits in the aging cerebral cortex. *Am. J. Pathol.*, **145**, 1358–81.

Martin, L. J., Price, A. C., McClendon, K. B. *et al.* (2003). Early events of target deprivation/axotomy-induced neuronal apoptosis in vivo: oxidative stress, DNA damage, p53 phosphorylation and subcellular redistribution of death proteins. *J. Neurochem.*, **85**, 234–47.

Martin, L. J., Sieber, F. E., and Traystman, R. J. (2000a). Apoptosis and necrosis occur in separate neuronal populations in hippocampus and cerebellum after ischemia and are associated with alterations in metabotropic glutamate receptor signaling pathways. *J. Cereb. Blood Flow Metab.*, **20**, 153–67.

Martin, L. J., Price, A. C., Kaiser, A., Shaikh A. Y., and Liu, Z. (2000b). Mechanisms for neuronal degeneration in amyotrophic lateral sclerosis and in models of motor neuron death. *Int. J. Mol. Med.*, **5**, 3–13.

Martinou, J.-C., Dubois-Dauphin, M., Staple, J. K. *et al.* (1994). Overexpression of bcl-2 in transgenic mice protects neurons from naturally occurring cell death and experimental ischemia. *Neuron*, **13**, 1017–30.

Marzo, I., Brenner, C., Zamzami, N. *et al.* (1998). Bax and adenine nucleotide translocator cooperate in the mitochondrial control of apoptosis. *Science*, **281**, 2027–31.

Mate, M. J., Ortiz-Lombardia, M., Boitel, B. *et al.* (2002). The crystal structure of the mouse apoptosis-inducing factor AIF. *Nature Struct. Biol.*, **9**, 442–6.

Mattson, M. P., Cheng, B., Davis, D., Bryant, K., Lieberburg, I., and Rydel, R. E. (1992). β-amyloid peptides destabilize calcium homeostasis and render human cortical neurons vulnerable to excitotoxicity. *J. Neurosci.*, **12**, 376–89.

McKhann, G., Drachman, D., Folstein, M., Katzman, R., Price, D., and Stadlan, E. M. (1984). Clinical diagnosis of Alzheimer's disease: report of the NINCDS-ADRDA work group under the auspices of the Department of Health and Human Services task force on Alzheimer's disease. *Neurology*, **34**, 939–44.

McPhail, L. T., Vanderluit, J. L., McBride, C. B. *et al.* (2003). Endogenous expression of inhibitor of apoptosis proteins in facial motoneurons of neonatal and adult rats following axotomy. *Neuroscience*, **117**, 567–75.

Merry, D. E. and Korsmeyer, S. J. (1997). Bcl-2 gene family in the nervous system. *Ann. Rev. Neurosci.*, **20**, 245–67.

Michaelidis, T. M., Sendtner, M., Cooper, J. D. *et al.* (1996). Inactivation of bcl-2 results in progressive degeneration of motoneurons, sympathetic and sensory neurons during early postnatal development. *Neuron*, **17**, 75–89.

Migheli, A., Atzori, C., Piva, R. *et al.* (1999). Lack of apoptosis in mice with ALS. *Nature Med.*, **5**, 966–7.

Miller, T. M., Moulder, K. L., Knudson, C. M. *et al.* (1997). Bax deletion further orders the cell death pathway in cerebellar granule cells and suggests a caspase-independent pathway to cell death. *J. Cell Biol.*, **139**, 205–17.

Mizushima, N., Ohsumi, Y., and Yoshimori, T. (2002). Autophagosome formation in mammalian cells. *Cell Struct. Funct.*, **27**, 421–9.

Morishima, Y., Gotoh, Y., Zieg, J. *et al.* (2001). β-amyloid induces neuronal apoptosis via a mechanism that involves the c-Jun N-terminal kinase pathway and the induction of Fas ligand. *J. Neurosci.*, **21**, 7551–60.

Mouton, P. R., Martin, L. J., Calhoun, M. E., Dal Forno, G., and Price, D. L. (1998). Cognitive decline strongly correlates with cortical atrophy in Alzheimer's disease. *Neurobiol. Aging*, **19**, 371–7.

Muchmore, S. W., Sattler, M., Liang, H. *et al.* (1999). X-ray and NMR structure of human Bcl-xL, an inhibitor of programmed cell death. *Nature*, **381**, 335–41.

Mund, T., Gewies, A., Schoenfield, N., Bauer, M. K. A., and Grimm, S. (2003). Spike, a novel BH3-only protein, regulates apoptosis at the endoplasmic reticulum. *FASEB J.* (February 19). 10.1096/fj.02-0657fje.

Nagata, S. (1999). Fas ligand-induced apoptosis. *Annu. Rev. Genet.*, **33**, 29–55.

Nakagawa, T., Zhu, H., Morishima, N. *et al.* (2000). Caspase-12 mediates endoplasmic reticulum-specific apoptosis and cytotoxicity by amyloid-beta. *Nature*, **403**, 98–103.

Nakajima, W., Ishida, A., Lange, M. S. *et al.* (2000). Apoptosis has a prolonged role in the neurodegeneration after hypoxic ischemia in the newborn rat. *J. Neurosci.*, **20**, 7994–8004.

Naruse, S., Igarashi, S., Kobayashi, H. *et al.* (1991). Mis-sense mutation Val→Ile in exon 17 of amyloid precursor protein gene in Japanese familial Alzheimer's disease. *Lancet.*, **337**, 978–9.

Natale, J. E., Cheng, Y., and Martin, L. J. (2002). Thalamic neuron apoptosis emerges rapidly after cortical damage in immature mice. *Neuroscience*, **112**, 665–76.

Nechushtan, A., Smith, C. L., Lamensdorf, I., Yoon, S.-H., and Youle, R. J. (2001). Bax and Bak co-alesce into novel mitochondria-associated clusters during apoptosis. *J. Cell Biol.*, **153**, 1265–76.

Nicholson, S. J., Witherden, A. S., Hafezparast, M., Martin, J. E., and Fisher, E. M. C. (2000). Mice, the motor system, and human motor neuron pathology. *Mamm. Genome*, **11**, 1041–52.

Northington, F. J., Ferriero, D. M., Graham, E. M., Traystman, R. J., and Martin, L. J. (2001). Early neurodegeneration after hypoxia-ischemia in neonatal rat is necrosis while delayed neuronal death is apoptosis. *Neurobiol. Dis.*, **8**, 207–19.

O'Connor, T. M. and Wyttenbach, C. R. (1974). Cell death in the embryonic chick spinal cord. *J. Cell Biol.*, **60**, 448–59.

Oddo, S., Caccamo, A., Shepherd, J. D. *et al.* (2003). Triple-transgenic model of Alzheimer's disease with plaques and tangles: intracellular Aβ and synaptic dysfunction. *Neuron*, **39**, 409–21.

Offen, D., Beart, P. M., Cheung, N. S. *et al.* (1998). Transgenic mice expressing human Bcl-2 in their neurons are resistant to 6-hydroxydopamine and 1-methyl-4-phenyl-1,2,3,6-tetrahydropyridine neurotoxicity. *Proc. Natl. Acad. Sci.*, **95**, 5789–94.

Ogier-Denis, E. and Codogno, P. (2003). Autophagy: a barrier or an adaptive response to cancer. *Biochim. Biophys. Acta.*, **1603**, 113–28.

Olanow, C. W. and Tatton, W. G. (1999). Etiology and pathogenesis of Parkinson's disease. *Ann. Rev. Neurosci.*, **22**, 123–44.

Olkowski, Z. L. (1998). Mutant AP endonuclease in patients with amyotrophic lateral sclerosis. *NeuroReport*, **9**, 239–42.

Olney, J. W. (1971). Glutamate-induced neuronal necrosis in the infant mouse hypothalamus. An electron microscopic study. *J. Neuropathol. Exp. Neurol.*, **30**, 75–90.

Olshansky, S. J., Carnes, B. A., and Cassel, C. K. (1993). The aging of the human species. *Sci. Am.*, **268**, 46–52.

Orrenius, S., Zhivotovsky, B., and Nicotera, P. (2003). Regulation of cell death: the calcium-apoptosis link. *Nat. Rev.*, **4**, 552–65.

Paradis, E., Douillard, H., Koutroumanis, M., Goodyer, C., and LeBlanc, A. (1997). Amyloid β peptide of Alzheimer's disease downregulates Bcl-2 and upregulates Bax expression in human neurons. *J. Neurosci.*, **16**, 7533–9.

Passer, B. J., Pellegrini, L., Vito, P., Ganjei, J. K., and D'Adamio, L. (1999). Interaction of Alzheimer's presenilin-1 and presenilin-2 with Bcl-X(L). A potential role in modulating the threshold of cell death. *J. Biol. Chem.*, **274**, 24007–13.

Patrick, G. N., Zukerberg, L., Nikolic, M., de la Monte, S., Dikkes, P., and Tsai, L. H. (1999). Conversion of p35 to p25 deregulates Cdk5 activity and promotes neurodegeneration. *Nature*, **402**, 615–22.

Pelvig, D. P., Pakkenberg, H., Regeur, L., Oster, S., and Pakkenberg, B. (2003). Neocortical glial cell numbers in Alzheimer's disease. A stereological study. *Dement. Geriatr. Cogn. Disord.*, **16**, 212–19.

Perrelet, D., Ferri, A., Liston, P., Muzzini, P., Korneluk, R. G., and Kato, A. C. (2002). IAPs are essential for GDNF-mediated neuroprotective effects in injured motor neuron in vivo. *Nat. Cell Biol.*, **4**, 175–9.

Petersen, A., Larsen, K. E., Behr, G. G. *et al.* (2001). Expanded CAG repeats in exon 1 of the Huntington's disease gene stimulate dopamine-mediated striatal neuron autophagy and degeneration. *Hum. Mol. Genet.*, **10**, 1243–54.

Pilar, G. and Landmesser, L. (1976). Ultrastructural differences during embryonic cell death in normal and peripherally deprived ciliary ganglia. *J. Cell Biol.*, **68**, 339–56.

Pompl, P. N., Yemul, S., Xiang, Z. *et al.* (2003). Caspase gene expression in the brain as a function of the clinical progression of Alzheimer's disease. *Arch. Neurol.*, **60**, 369–76.

Poorkaj, P., Bird, T. D., Wijsman, E. *et al.* (1998). Tau is a candidate gene for chromosome 17 frontotemporal dementia. *Ann. Neurol.*, **43**, 815–25.

Portera-Cailliau, C., Price, D. L., and Martin, L. J. (1997a). Excitotoxic neuronal death in the immature brain is an apoptosis-necrosis morphological continuum. *J. Comp. Neurol.*, **378**, 70–87.

(1997b). Non-NMDA and NMDA receptor-mediated excitotoxic neuronal deaths in adult brain are morphologically distinct: further evidence for an apoptosis-necrosis continuum. *J. Comp. Neurol.*, **378**, 88–104.

Pozniak, C. D., Radinovic, S., Yang, A., McKeon, F., Kaplan, D. R., and Miller, F. D. (2000). An anti-apoptotic role for the p53 family member, p73, during developmental neuron death. *Science*, **289**, 304–6.

Prestige, M. (1970). Differentiation, degeneration, and the role of the periphery: quantitative considerations. In *The Neurosciences: Second Study Program*, ed. F. O. Schmitt. New York: Rockefeller University Press, pp. 73–82.

Proskuryakov, S. Y., Konoplyannikov, A. G., and Gabai, V. L. (2003). Necrosis: a specific form of programmed cell death. *Exp. Cell Res.*, **283**, 1–16.

Prudlo, J., Koenig, J., Graser, J. *et al.* (2000). Motor neuron cell death in a mouse model of FALS is not mediated by the p53 cell survival regulator. *Brain Res.*, **897**, 183–7.

Putcha, G. V., Deshmukh, M., and Johnson, E. M. Jr. (1999). Bax translocation is a critical event in neuronal apoptosis: regulation by neuroprotectants, Bcl-2, and caspases. *J. Neurosci.*, **19**, 7476–85.

Raffray, M. and Cohen, G. M. (1997). Apoptosis and necrosis in toxicology: a continuum or distinct modes of cell death? *Pharmacol. Ther.*, **75**, 153–77.

Raoul, C., Estvez, A. G., Nishimune, H. *et al.* (2002). Motoneuron death triggered by a specific pathway downstream of Fas: potentiation by ALS-linked SOD1 mutations. *Neuron*, **35**, 1067–83.

Rapoport, M., Dawson, H. N., Binder, L. I., Vitek, M. P., and Ferreira, A. (2002). Tau is essential to β-amyloid-induced neurotoxicity. *Proc. Natl. Acad. Sci. USA*, **99**, 6364–9.

Rathke-Hartlieb, S., Schlomann, U., Heimann, P., Meisler, M. H., Jockusch, H., and Bartsch, J. W. (2002). Progressive loss of striatal neurons causes motor dysfunction in MND2 mutant mice and is not prevented by Bcl-2. *Exp. Neurol.*, **175**, 87–97.

Reddy, P. H., Williams, M., Charles, V. *et al.* (1998). Behavioral abnormalities and selective neuronal loss in HD transgenic mice expressing mutated full-length HD cDNA. *Nat. Genet.*, **20**, 198–202.

Robertson, J. D., Enoksson, M., Suomela, M., Zhivotovsky, B., and Orrenius, S. (2002). Caspase-2 acts upstream of mitochondria to promote cytochrome c release during etoposide-induced apoptosis. *J. Biol. Chem.*, **277**, 29803–9.

Robison, S. H., Tandan, R., and Bradley, W. G. (1993). Repair of N-methylpurines in DNA from lymphocytes of patients with amyotrophic lateral sclerosis. *J. Neurol. Sci.*, **115**, 201–7.

Rohn, T. T., Rissman, R. A., Davis, M. C., Kim, Y. E., Cotman, C. W., and Head, E. (2002). Caspase-9 activation and caspase cleavage of tau in the Alzheimer's disease brain. *Neurobiol. Dis.*, **11**, 341–54.

Romanes, G. J. (1946). Motor localization and the effects of nerve injury on the ventral horn cells of the spinal cord. *J. Anat.*, **80**, 117–31.

Rosen, D. R., Siddique, T., Patterson, D. *et al.* (1993). Mutations in Cu/Zn superoxide dismutase gene are associated with familial amyotrophic lateral sclerosis. *Nature*, **362**, 59–62.

Rosenmann, H., Meiner, Z., Kahana, E. *et al.* (2003). An association study of the codon 72 polymorphism in the pro-apoptotic gene p53 and Alzheimer's disease. *Neurosci. Lett.*, **340**, 29–32.

Roses, A. D. (1996). Apolipoprotein E alleles as risk factors in Alzheimer's disease. *Annu. Rev. Med.*, **47**, 387–400.

Rothstein, J. D., Martin, L. J., and Kuncl, R. W. (1992). Decreased glutamate transport by the brain and spinal cord in amyotrophic lateral sclerosis. *N. Engl. J. Med.*, **326**, 1464–8.

Rothstein, J. D., Van Kammen, M., Levey, A. I., Martin, L. J., and Kuncl, R. W. (1995). Selective loss of glial glutamate transporter GLT-1 in amyotrophic lateral sclerosis. *Ann. Neurol.*, **38**, 73–84.

Rowland, L. P. and Shneider, N. A. (2001). Amyotrophic lateral sclerosis. *N. Engl. J. Med.*, **344**, 1688–700.

Roy, N., Mahadevan, M. S., McLean, M. *et al.* (1995). The gene for neuronal apoptosis inhibitory protein is partially deleted in individuals with spinal muscular atrophy. *Cell,* **80**, 167–78.

Sakahira, H., Enari, M., Ohsawa, Y., Uchiyama, Y., and Nagata, S. (1999). Apoptotic nuclear morphological change without DNA fragmentation. *Curr. Biol.,* **9**, 543–6.

Sano, T., Lin, H., Chen, X. *et al.* (1999). Differential expression of MMAC/PTEN in glioblastoma multiforme: relationship to localization and prognosis. *Cancer Res.,* **59**, 1820–4.

Sathasivam, S., Ince, P. G., and Shaw, P. J. (2001). Apoptosis in amyotrophic lateral sclerosis: a review of the evidence. *Neuropathol. Appl. Neurobiol.,* **27**, 257–74.

Satoh, T., Nakatsuka, D., Watanabe, Y., Nagata, I., Kikuchi, H., and Namura, S. (2000). Neuroprotection by MAPK/ERK kinase inhibition with U0126 against oxidative stress in a mouse neuronal cell line and rat primary cultured cortical neurons. *Neurosci. Lett.,* **288**, 163–6.

Saudou, F., Finkbeiner, S., Devys, D., and Greenberg, M. E. (1998). Huntingtin acts in the nucleus to induce apoptosis but death does not correlate with the formation of intranuclear inclusions. *Cell,* **95**, 55–66.

Saunders, J. W. (1966). Death in embryonic systems. *Science,* **154**, 604–12.

Schilling, G., Becher, M. W., Sharp, A. H. *et al.* (1999). Intranuclear inclusions and neuritic aggregates in transgenic mice expressing a mutant N-terminal fragment of huntingtin. *Hum. Mol. Genet.,* **8**, 397–407.

Schreiber, S. S., Tocco, G., Najm, I., Thompson, R. F., and Baudry, M. (1993). Cycloheximide prevents kainate-induced neuronal death and c-fos expression in adult rat brain. *J. Mol. Neurosci.,* **4**, 149–59.

Schwartz, L. M. and Milligan, C. E. (1996). Cold thoughts of death: the role of ICE proteases in neuronal cell death. *Trends Neurosci.,* **19**, 555–62.

Schwartz, L. M., Smith, S. W., Jones. M. E. E. M, and Osborne, B. A. (1993). Do all programmed cell deaths occur via apoptosis? *Proc. Natl. Acad. Sci. USA,* **90**, 980–4.

Schweichel, J. U. and Merker, H. J. (1973). The morphology of various types of cell death in prenatal tissues. *Teratology,* **7**, 253–66.

Scorrano, L., Oakes, S. A., Opferman, T. J. *et al.* (2003). Bax and Bak regulation of endoplasmic reticulum Ca^{2+}: a control point for apoptosis. *Science,* **300**, 135–9.

Selkoe, D. J. (2002). Alzheimer's disease is a synaptic failure. *Science,* **298**, 789–91.

Selznick, L. A., Holtzman, D. M., Han, B. H. *et al.* (1999). In situ immunodetection of neuronal caspase-3 activation in Alzheimer's disease. *J. Neuropath. Exp. Neurol.,* **58**, 1020–6.

Shaikh, A. Y. and Martin, L. J. (2002). DNA base-excision repair enzyme apurinic/apyrimidinic endonuclease/redox factor-1 is increased and competent in brain and spinal cord of individuals with amyotrophic lateral sclerosis. *NeuroMolecular Med.,* **2**, 47–60.

Sherrington, R., Rogaev, E. I., Liang, Y. *et al.* (1995). Cloning of a gene bearing missense mutations in early-onset familial Alzheimer's disease. *Nature,* **375**, 754–60.

Shieh, S.-Y., Ikeda, M., Taya, Y., and Prives, C. (1997). DNA damage-induced phosphorylation of p53 alleviates inhibition by MDM2. *Cell,* **91**, 325–34.

Shimizu, S., Ide, T., Yanagida, T., and Tsujimoto, Y. (2000). Electrophysiological study of a novel large pore formed by Bax and the voltage-dependent anion channel that is permeable to cytochrome c. *J. Biol. Chem.,* **275**, 12321–5.

Shimohama, S., Tanino, H., and Fujimoto, S. (2001). Differential subcellular localization of caspase family protein in the adult rat brain. *Neurosci. Lett.*, **315**, 125–8.

Simonian, N. A., Getz, R. L., Leveque, J. C., Konradi, C., and Coyle, J. T. (1996). Kainate induces apoptosis in neurons. *Neuroscience*, **74**, 675–83.

Simons, M., Beinroth, S., Gleichmann, M. *et al.* (1999). Adenovirus-mediated gene transfer of inhibitors of apoptosis proteins delays apoptosis in cerebellar granule neurons. *J. Neurochem.*, **72**, 292–301.

Smale, G., Nichols, N. R., Brady, D. R., Finch, C. E., and Horten, W. E. Jr. (1995). Evidence for apoptotic cell death in Alzheimer's disease. *Exp. Neurol.*, **133**, 225–30.

Song, Q., Kuang, Y., Dixit, V. M., and Vincenz, C. (1999). Boo, a negative regulator of cell death, interacts with Apaf-1. *EMBO J.*, **18**, 167–78.

Srivastava, R. K., Sollott, S. J., Khan, L., Hansford, R., Lakatta, E. G., and Longo, D. L. (1999). Bcl-2 and Bcl-X(L) block thapsigargin-induced nitric oxide generation, c-Jun NH(2)-terminal kinase activity, and apoptosis. *Mol. Cell Biol.*, **19**, 5659–74.

Stadelmann, C., Deckwerth, T. L., Srinivasan, A., Bancher, C., Brock, W., and Lassmann, H. (1999). Activation of caspase-3 in single neurons and autophagic granules of granulovacuolar degeneration in Alzheimer's disease. Evidence for apoptotic cell death. *Am. J. Pathol.*, **155**, 1459–66.

Stanciu, M., Wang, Y., Kentor, R. *et al.* (2000). Persistent activation of ERK contributes to glutamate-induced oxidative toxicity in a neuronal cell line and primary cortical neuron cultures. *J. Biol. Chem.*, **275**, 12200–6.

Steffan, J. S., Kazantsev, A., Spasic-Boskovic, O. *et al.* (2000). The Huntington's disease protein interacts with p53 and CREB-binding protein and represses transcription. *Proc. Natl. Acad. Sci. USA*, **97**, 6763–8.

Stennicke, H. R., Deveraux, Q. L., Humke, E. W., Reed, J. C., Dixit, V. M., and Salvesen, G. S. (1999). Caspase-9 can be activated without proteolytic processing. *J. Biol. Chem.*, **274**, 8359–62.

Su, J. H., Anderson, A. J., Cribbs, D. H. *et al.* (2003). Fas and Fas ligand are associated with neuritic degeneration in the AD brain and participate in β-amyloid-induced neuronal death. *Neurobiol. Dis.*, **12**, 182–93.

Su, J. H., Zhao, M., Anderson, A. J., Srinivasan, A., and Cotman, C. W. (2001). Activated caspase-3 expression in Alzheimer's and aged control brain: correlation with Alzheimer pathology. *Brain Res.*, **898**, 350–7.

Susin, S. A., Lorenzo, H. K., Zamzami, N. *et al.* (1999). Molecular characterization of mitochondrial apoptosis-inducing factor. *Nature*, **397**, 441–6.

Sze, C.-I., Bi, H., Kleinschmidt-DeMasters, B. K., Filley, C. M., and Martin, L. J. (2001). N-Methyl-D-aspartate receptor subunit proteins and their phosphorylation status are altered selectively in Alzheimer's disease. *J. Neurol. Sci.*, **182**, 151–9.

Sze, C.-I., Troncoso, J. C., Kawas, C., Mouton, P., Price, D. L., and Martin, L. J. (1997). Loss of the presynaptic vesicle protein synaptophysin in hippocampus correlates with cognitive decline in Alzheimer's disease. *J. Neuropathol. Exp. Neurol.*, **56**, 933–94.

Takashima, A., Noguchi, K., Sato, K., Hoshino, T., and Imahori, K. (1993). Tau protein kinase I is essential for amyloid beta-protein-induced neurotoxicity. *Proc. Natl. Acad. Sci. USA*, **90**, 7789–93.

Tandan, R., Robison, S. H., Munzer, J. S. and Bradley, W. G. (1987). Deficient DNA repair in amyotrophic lateral sclerosis cells. *J. Neurol. Sci.*, **79**, 189–203.

Tata, J. R. (1966). Requirement for RNA and protein synthesis for induced regression of tadpole tail in organ culture. *Dev. Biol.*, **13**, 77–94.

Tatton, N. A., MacLean-Fraser, A., Tatton, W. G., Perl, D. P., and Olnanow, C. W. (1998). A fluorescent double-labeling method to detect and confirm apoptotic nuclei in Parkinson's disease. *Ann. Neurol.*, **44**, S142–8.

Tatton, W. G., Chalmers-Redman, R., Brown, D., and Tatton, N. (2003). Apoptosis in Parkinson's disease: signals for neuronal degeneration. *Ann. Neurol.*, **53**, S61–72.

Tenneti, L. and Lipton, S. A. (2000). Involvement of activated caspase-3-like proteases in *N*-methyl-*D*-aspartate-induced apoptosis in cerebrocortical neurons. *Ann. Neurol.*, **74**, 134–42.

Terry, R. D., Masliah, E., Salmon, D. P. *et al.* (1991). Physical basis of cognitive alterations in Alzheimer's disease: synapse loss is the major correlate of cognitive impairment. *Ann. Neurol.*, **30**, 572–80.

Tomkins, J., Dempster, S., Banner, S. J., Cookson, M. R., and Shaw, P. J. (2000). Screening of AP endonuclease as a candidate gene for amyotrophic lateral sclerosis. *NeuroReport*, **11**, 1695–7.

Tompkins, M. M., Basgall, E. J., Zamrini, E., and Hill, W. D. (1997). Apoptotic-like changes in Lewy-body-associated disorders and normal aging in substantia nigra neurons. *Am. J. Pathol.*, **150**, 119–31.

Torvik, A. (1976). Central chromatolysis and the axon reaction. A reappraisal. *Neuropathol. Appl. Neurobiol.*, **2**, 423–32.

Trimmer, P. A., Smith, T. S., Jung, A. B., and Bennett, J. P. Jr. (1996). Dopamine neurons from transgenic mice with a knockout of the p53 gene resist MPTP neurotoxicity. *Neurodegeneration*, **5**, 233–9.

Troy, C. M., Friedman, J. E., and Friedman, W. J. (2002). Mechanisms of p75-mediated death of hippocampal neurons. Role of caspases. *J. Biol. Chem.*, **277**, 34295–302.

Trump, B. F. and Berezesky, I. K. (1996). The role of altered $[Ca^{2+}]_i$ regulation in apoptosis, oncosis, and necrosis. *Biochim. Biophys. Acta.*, **1313**, 173–8.

Trump, B. J., Goldblatt, P. J., and Stowell, R. E. (1964). Studies on necrosis of mouse liver in vitro. Ultrastructural alterations in the mitochondria of hepatic parenchymal cells. *Lab. Invest.*, **14**, 343–71.

Turmaine, M., Raza, A., Mahal, A., Mangiarini, L., Bates, G. P., and Davies, S. W. (2000). Non-apoptotic neurodegeneration in a transgenic mouse model of Huntington's disease. *Proc. Natl. Acad. Sci. USA*, **97**, 8093–7.

Uberti, D., Carsana, T., Bernardi, E. *et al.* (2002). Selective impairment of p53-mediated cell death in fibroblasts from sporadic Alzheimer's disease patients. *J. Cell Sci.*, **115**, 3131–8.

Uetsuki, T., Takemoto, K., Nishimura, I. *et al.* (1999). Activation of neuronal caspase-3 by intracellular accumulation of wild-type Alzheimer amyloid precursor protein. *J. Neurosci.*, **19**, 6955–64.

Uppuluri, S., Knipling, L., Sackett, D. L., and Wolff, J. (1993). Localization of the colchicine-binding site of tubulin. *Proc. Natl. Acad. Sci. USA*, **90**, 11598–602.

van Bekkum, D. W. (1957). The effect of x-rays on phosphorylations in vivo. *Biochim. Biophys. Acta.*, **25**, 487–92.

Vander Heiden, M. G., Chandel, N. S., Williamson, E. K., Schumacker, P. T., and Thompson, C. B. (1997). Bcl-x$_L$ regulates the membrane potential and volume homeostasis of mitochondria. *Cell*, **91**, 627–37.

van Gurp, M., Festjens, N., van Loo, G., Saelens, X., and Vandenabeele, P. (2003). Mitochondrial intermembrane proteins in cell death. *Biochem. Biophys. Res. Comm.*, **304**, 487–97.

van Lookeren Campagne, M., Lucassen, P. J., Vermeulen, J. P., and Balázs, R. (1995). NMDA and kainate induced internucleosomal DNA cleavage associated with both apoptotic and necrotic cell death in the neonatal rat brain. *Eur. J. Neurosci.*, **7**, 1627–40.

Verhagen, A. M., Ekert, P. G., Pakusch, M. *et al.* (2000). Identification of DIABLO, a mammalian protein that promotes apoptosis by binding to and antagonizing IAP proteins. *Cell*, **102**, 43–53.

Verhagen, A. M., Silke, J., Ekert, P. G. *et al.* (2002). HtrA2 promotes cell death through its serine protease activity and its ability to antagonize inhibitor of apoptosis proteins. *J. Biol. Chem.*, **277**, 445–54.

Vila, M., Jackson-Lewis, V., Vukosavic, S. *et al.* (2001). Bax ablation prevents dopaminergic neurodegeneration in the 1-methyl-4-phenyl-1,2,3,6-tetrahydropyridine mouse model of Parkinson's disease. *Proc. Natl. Acad. Sci.*, **98**, 2837–42.

Villunger, A., Michalak, E. M., Coultas, L. *et al.* (2003). p53- and drug-induced apoptotic responses mediated by BH3-only proteins Puma and Noxa. *Science*, **302**, 1036–8.

Virchow, R. (1858). *Cellular Pathology as Based Upon Physiological and Pathological Histology* (Translation). London: Churchill.

Vonsattel, J. P. and DiFiglia, M. (1998). Huntington's disease. *J. Neuropathol. Exp. Neurol.*, **57**, 369–84.

Vonsattel, J. P., Myers, R. H., Stevens, T. J., Ferrante, R. J., Bird, E. D., and Richardson, E. P. Jr. (1985). Neuropathological classification of Huntington's disease. *J. Neuropathol. Exp. Neurol.*, **44**, 559–77.

Vukosavic, S., Dubois-Dauphin, N., Romero, N., and Przedborski, S. (1999). Bax and Bcl-2 interaction in a transgenic mouse model of amyotrophic lateral sclerosis. *J. Neurochem.*, **73**, 2460–8.

Wang, H.-G., Pathan, N., Ethell, I. M. *et al.* (1999). Ca^{2+}-induced apoptosis through calcineurin dephosphorylation of Bad. *Science*, **284**, 339–43.

Wang, H.-G., Rapp, U. R., and Reed, J. C. (1996). Bcl-2 targets the protein kinase raf-1 to mitochondria. *Cell*, **87**, 629–38.

Wei, M. C., Zong, W-X., Cheng, E. H.-Y. *et al.* (2001). Proapoptotic Bax and Bak: a requisite gateway to mitochondrial dysfunction and death. *Science*, **292**, 727–30.

Weidemann, A., Paliga, K., Drrwang, U. *et al.* (1999). Proteolytic processing of the Alzheimer's disease amyloid precursor protein within its cytoplasmic domain by caspase-like proteases. *J. Biol. Chem.*, **274**, 5823–9.

West, M. J., Kawas, C. H., Martin L. J., and Troncoso, J. C. (2000). The CA1 region of the human hippocampus is a hot spot in Alzheimer's disease. *Ann. NY Acad. Sci.*, **908**, 255–9.

White, E. and Prives, C. (1999). DNA damage enables p73. *Nature*, **399**, 734–7.

Whitehouse, P. J., Price, D. L., Struble, R. G., Clark, A. W., Coyle, J. T., and DeLong, M. R. (1982). Alzheimer's disease and senile dementia: loss of neurons in the basal forebrain. *Science*, **215**, 1237–9.

Widmann, C., Gerwins, P., Johnson, N., Jarpe, M. B., and Johnson, G. L. (1998). MEK kinase 1, a substrate for DEVD-directed caspases, is involved in genotoxin-induced apoptosis. *Mol. Cell Biol.*, **18**, 2416–29.

Wolf, B. B. and Green, D. R. (1999). Suicidal tendencies: apoptotic cell death by caspase family proteinases. *J. Biol. Chem.*, **274**, 20049–52.

Wolter, K. G., Hsu, Y.-T., Smith, C. L., Nechushtan, A., Xi, X.-G., and Youle, R. L. (1997). Movement of Bax from the cytosol to mitochondria during apoptosis. *J. Cell Biol.*, **139**, 1281–92.

Wong, P. C., Cai, H., Borchelt, D. R., and Price, D. L. (2002). Genetically engineered mouse models of neurodegenerative diseases. *Nat. Neurosci.*, **5**, 633–9.

Wong, P. C., Pardo, C. A., Borchelt, D. R. *et al.* (1995). An adverse property of a familial ALS-linked SOD1 mutation causes motor neuron disease characterized by vacuolar degeneration of mitochondria. *Neuron*, **14**, 1105–16.

Wüllner, U., Kornhuber, J., Weller, M. *et al.* (1999). Cell death and apoptosis regulating proteins in Parkinson's disease – a cautionary note. *Acta. Neuropathol.*, **97**, 408–12.

Wyllie, A. H. (1980). Glucocorticoid-induced thymocyte apoptosis is associated with endogenous endonuclease activation. *Nature*, **284**, 555–6.

Wyllie, A. H., Kerr, J. F. R., and Currie, A. R. (1980). Cell death: the significance of apoptosis. *Int. Rev. Cytol.*, **68**, 251–306.

Xanthoudakis, S. and Curran, T. (1992). Identification and characterization of Ref-1, a nuclear protein that facilitates AP-1 DNA-binding activity. *EMBO J.*, **11**, 653–65.

Xiang, H., Kinoshita, Y., Knudson, C. M., Korsmeyer, S. J., Schwartzkroin, P. A., and Morrison, R. S. (1998). Bax involvement in p53-mediated neuronal cell death. *J. Neurosci.*, **18**, 1363–73.

Xu, D. G., Korneluk, R. G., Tamai, K. *et al.* (1997). Distribution of neuronal apoptosis inhibitory protein-like immunoreactivity in the rat central nervous system. *J. Comp. Neurol.*, **382**, 247–59.

Xu, J., Kao, S. Y., Lee, F. J., Song, W., Jin, L. W., and Yankner, B. A. (2002). Dopamine-dependent neurotoxicity of alpha-synuclein: a mechanism for selective neurodegeneration in Parkinson disease. *Nat. Med.*, **8**, 600–6.

Xue, L. Z., Fletcher, G. C., and Tolkovsky, A. M. (1999). Autophagy is activated by apoptotic signalling in sympathetic neurons: an alternative mechanism of death execution. *Mol. Cell Neurosci.*, **14**, 180–98.

Yaar, M., Zhai, S., Fine, R. E. *et al.* (2002). Amyloid β binds trimers as well as monomers of the 75-kDa neurotrophin receptor and activates receptor signaling. *J. Biol. Chem.*, **277**, 7720–5.

Yan, X. X., Najbauer, J., Woo, C. C., Dashtipour, K., Ribak, C. E., and Leon, M. (2001). Expression of active caspase-3 in mitotic and postmitotic cells or rat forebrain. *J. Comp. Neurol.*, **433**, 4–22.

Yang, A., Walker, N., Bronson, R. *et al.* (2000). p73-deficient mice have neurological, pheromonal and inflammatory defects but lack spontaneous tumors. *Nature*, **404**, 99–103.

Yang, E., Zha, J., Jockel, J., Boise, L. H., Thompson, C. B., and Korsmeyer, S. J. (1995). Bad, a heterodimeric partner for Bcl-xL and Bcl-2, displaces Bax and promotes cell death. *Cell*, **80**, 285–91.

Yang, J., Liu, X., Bhalla, K. *et al.* (1997). Prevention of apoptosis by bcl-2: release of cytochrome c from mitochondria blocked. *Science*, **275**, 1129–32.

Yang, L., Matthews, R. T., Schulz, J. B. *et al.* (1998). 1-methyl-4-phenyl-1,2,3,6-tetrahydropyride neurotoxicity is attenuated in mice overexpressing Bcl-2. *J. Neurosci.*, **18**, 8145–52.

Yankner, B. A., Dawes, L. R., Fisher, S. *et al.* (1989). Neurotoxicity of a fragment of the amyloid precursor associated with Alzheimer's disease. *Science*, **245**, 417–20.

Yoshio, H., Kong, Y.-Y., Yoshida, R. *et al.* (1998). Apaf1 is required for mitochondrial pathways of apoptosis and brain development. *Cell*, **94**, 739–50.

Younkin, S. G. (1995). Evidence that A beta 42 is the real culprit in Alzheimer's disease. *Ann. Neurol.*, **37**, 287–8.

Yu, Z. X., Li, S. H., Evans, J., Pillarsetti, A., Li, H., and Li, X. J. (2003). Mutant huntingtin causes context – dependent neurodegeneration in mice with Huntington's disease. *J. Neurosci.*, **23**, 2193–202.

Yuan, J., Lipinski, M., and Degterev, A. (2003). Diversity in the mechanisms of neuronal cell death. *Neuron*, **40**, 401–13.

Yue, Z., Horton, A., Bravin, M., DeJager, P. L., Selimi, F., and Heintz, N. (2002). A novel protein complex linking the δ2 glutamate receptor and autophagy: implications for neurodegeneration in Lurcher mice. *Neuron*, **35**, 921–33.

Zha, J., Harada, H., Yang, E., Jockel, J., and Korsmeyer, S. J. (1996). Serine phosphorylation of death agonist Bad in response to survival factor results in binding to 14-3-3 not Bcl-x$_L$. *Cell*, **87**, 619–28.

Zhang, Y., McLaughlin, R., Goodyer, C., and LeBlanc, A. (2002). Selective cytotoxicity of intracellular amyloid β peptide$_{1-42}$ through p53 and Bax in cultured primary human neurons. *J. Cell Biol.*, **156**, 519–29.

Zong, W.-X., Li, C., Hatzivassiliou, G. *et al.* (2003). Bax and Bak can localize to the endoplasmic reticulum to initiate apoptosis. *J. Cell Biol.*, **162**, 59–69.

Zou, H., Li, Y., Liu, X., and Wang, X. (1999). An Apaf-1-cytochrome c multimeric complex is a functional apoptosome that activates procaspase-9. *J. Biol. Chem.*, **274**, 11549–56.

Zuch, C. L., Nordstroem, V. K., Briedrick, L. A., Hoering, G. R., Granholm, A.-C., and Bickford, P. C. (2000). Time course of degenerative alterations in nigral dopaminergic neurons following a 6-hydroxydopamine lesion. *J. Comp. Neurol.*, **427**, 440–54.

Apoptosis in the cardiovascular system: incidence, regulation, and therapeutic options

Martin R. Bennett

Unit of Cardiovascular Medicine, Addenbrooke's Centre for Clinical Investigation, Addenbrooke's Hospital, Cambridge, UK

5.1 Incidence of apoptosis

5.1.1 Apoptosis in the heart

The adult cardiomyocyte is a terminally differentiated cell that cannot divide. Therefore, by definition, apoptosis within the adult heart cannot be physiologic, as no turnover of myocytes is possible. Indeed, apoptosis is observed very infrequently in adult hearts (Gottlieb *et al.*, 1994; Cheng *et al.*, 1995; Liu *et al.*, 1995; Kajstura *et al.*, 1996; Sharov *et al.*, 1996; Bialik *et al.*, 1997). In contrast, cardiomyocyte apoptosis in development plays a critical role in formation of the heart, and is an increasingly important feature of many diseases of the cardiovascular system (Table 5.1).

5.1.1.1 Apoptosis in cardiac development and aging

During development, organs and tissues are remodeled using the processes of cell division, cell migration, and cell death. Most, if not all, of this cell death occurs through apoptosis. Thus, apoptosis is seen when the notochord fuses in the developing spinal cord and apoptosis causes the breakdown of interdigital webs to sculpt the fingers (see Chapter 2). Within the heart, cell death may be involved in the formation of septa between cardiac chambers or of valves, suggesting that defects in apoptosis can result in congenital heart disease (Krstic and Pexieder, 1973; Pexieder, 1975). Major foci of apoptosis include the atrioventricular cushions and their zones of fusion, the bulbar cushions and their zones of fusion, and the aortic and pulmonary valves and arteries (Pexieder, 1975; Cheng *et al.*, 2002). Indeed, apoptosis has been found in the developing rat bulbus cordis within mesenchymal cells (Takeda *et al.*, 1996). Most of the apoptosis seen is in non-myocytes, although

Apoptosis in Health and Disease: Clinical and Therapeutic Aspects, ed. Martin Holcik, Alex E. MacKenzie, Robert G. Korneluk, and Eric C. LaCasse. Published by Cambridge University Press. © Cambridge University Press 2004.

Table 5.1 *Cardiovascular diseases in which apoptosis has been implicated*

Cardiac (myocyte)
Idiopathic dilated cardiomyopathy
Ischemic cardiomyopathy
Acute myocardial infarction
Arrhythmogenic right ventricular dysplasia
Myocarditis

Cardiac (conducting tissues)
Pre-excitation syndromes
Heart block, congenital complete atrioventricular heart block,
    long QT syndromes

Vascular
Atherosclerosis
Restenosis after angioplasty/stenting
Vascular graft rejection
Arterial aneurysm formation

---

in mid-gestation there is focal myocyte apoptosis in the interventricular septum close to the atrioventricular valves (Takeda *et al.*, 1996). Apoptosis of right ventricular and interventricular septal myocytes is also seen immediately after birth, where it may contribute to right ventricular (RV) remodeling as part of the transition from fetal to adult circulations. The importance of apoptosis in determining correct ventriculo-arterial connections has been demonstrated by inhibiting apoptosis with synthetic caspase inhibitors or an adenovirus-encoding inhibitor of apoptosis, XIAP; both maneuvres resulted in failure of outflow tract rotation and shortening, in addition to effects on the cardiac valves and trabeculae (Watanabe *et al.*, 2001). The conducting tissue also undergoes apoptosis (Cheng *et al.*, 2002), and both increased or decreased apoptosis have been implicated in the pathogenesis of congenital heart block and long QT syndrome (James *et al.*, 1993; James, 1994; James *et al.*, 1996) or the persistence of accessory pathways, respectively (James, 1994).

5.1.1.2 Apoptosis in ischemia/reperfusion

Ischemia is now a well-characterized inducer of myocyte apoptosis in vitro. Deprivation of oxygen alone, or in addition to serum withdrawal and deprivation of glucose, induces neonatal myocyte apoptosis that does not require protein synthesis, i.e. the machinery for executing apoptosis is already functional (Tanaka *et al.*, 1994; Umansky *et al.*, 1995).

Reperfusion is the mainstay of treatment for acute myocardial infarction, as demonstrated by numerous thrombolysis trials demonstrating improved left ventricular function and reduced mortality following successful reperfusion. However, reperfusion is also associated with myocyte death (Braunwald and Kloner, 1985), and neutrophil accumulation, inflammation, calcium overload, and oxidative stress, for example, have all been implicated in the pathogenesis of reperfusion injury (Zweier *et al.*, 1987; Nayler *et al.*, 1988; Smith *et al.*, 1988; Litt *et al.*, 1989; Entman and Smith, 1994). Reperfusion has been shown to increase apoptosis, and apoptosis therefore may represent the underlying mechanism of reperfusion injury. Ischemia alone can induce apoptosis in the ischemic territory in vivo, and this may be reduced by reperfusion. However, although reperfusion may limit ischemia-induced apoptosis, it may accelerate the appearance of apoptosis in the reperfused regions (Gottlieb *et al.*, 1994; Buerke *et al.*, 1995; Fliss and Gattinger, 1996). In vitro, hypoxia is mostly associated with death through non-apoptotic mechanisms, although re-oxygenation potently induces apoptosis (Kang *et al.*, 2000).

5.1.1.3   Apoptosis in myocardial infarction

Myocardial infarction has been considered to be a prima facie example of necrotic cell death, due to the breakdown of cellular energy metabolism. However, there is increasing evidence that apoptosis of cardiomyocytes may occur in a temporally and spatially specific manner. Thus, the acute stage of myocardial infarction may be associated with both forms of cell death (Itoh *et al.*, 1995; Bardales *et al.*, 1996; Saraste *et al.*, 1997; Takemura *et al.*, 1998). In particular, apoptotic myocytes are apparent at the hypoperfused "border" zones, between a central area of necrosis and the viable myocardium (Olivetti *et al.*, 1996; Saraste *et al.*, 1997; Takemura *et al.*, 1998). In contrast, apoptosis may occur in the central, unperfused region of an infarct (Fliss and Gattinger, 1996; Kajstura *et al.*, 1996; Bialik *et al.*, 1997), particularly within the first 6 hours after infarction. Although apoptosis increases in a time-dependent manner, reaching maximal levels 18–48 hours after the onset of ischemia in mouse models (Bialik *et al.*, 1997), between 6 and 24 hours necrosis may be more common than apoptosis, and may be due to secondary necrosis of apoptotic myocytes (Kajstura *et al.*, 1996). Apoptosis also occurs in the remote non-infarcted myocardium, where it may be partly responsible for myocardial remodeling and dilatation after myocardial infarction (Cheng *et al.*, 1996; Olivetti *et al.*, 1996; Palojoki *et al.*, 2001), and may be amenable to treatment (Li, Q. *et al.*, 1997). In human autopsy specimens, apoptosis of cardiomyocytes has been observed 12 hours to several days following the onset of myocardial infarction (Itoh *et al.*, 1995).

5.1.1.4   Apoptosis in heart failure

One of the most exciting advances in recent years has been the finding that cardiomyocyte apoptosis occurs in the end-stage human heart and may contribute to heart failure in a variety of situations (Narula *et al.*, 1996; Olivetti *et al.*, 1997). Thus, in models of heart failure induced by either pacing or coronary microemboli, cardiomyocyte death is associated with TUNEL-positive cells (Liu *et al.*, 1995; Cesselli *et al.*, 2001), which is particularly marked at the border zones of infarcts (Sharov *et al.*, 1996). Aging is also associated with myocardial cell loss in both human and animal hearts, and cardiomyocyte apoptosis has been suggested to be the mechanism responsible for the gradual deterioration in cardiac function in aging (Anversa *et al.*, 1986; Olivetti *et al.*, 1995). However, although significant differences in apoptotic indices are observed between men and women, no definite correlation of apoptosis with aging has been found (Mallat *et al.*, 2001).

Apoptosis can be observed in humans undergoing transplantation (Olivetti *et al.*, 1997; Guerra *et al.*, 1999), with some studies suggesting higher levels in ischemic versus idiopathic dilated cardiomyopathy (Narula *et al.*, 1996). The transition from compensated to decompensated hypertrophy is also associated with myocyte apoptosis in animals (Li, Z. *et al.*, 1997), and cardiomyocyte and fibroblast apoptosis is seen in hypertension in humans (Gonzalez *et al.*, 2002). In addition, high levels of apoptosis are seen in arrhythmogenic right ventricular dysplasia, a condition characterized by myocardial replacement with fibrofatty material (Mallat *et al.*, 1996). Finally, there is increasing evidence that toxic cardiomyopathies, such as those induced by doxorubicin (adriamycin) or endotoxin, are associated with cardiomyocyte apoptosis (Nakamura *et al.*, 2000; Dowd *et al.*, 2001; Fauvel *et al.*, 2001).

Although the evidence that apoptosis directly contributes to heart failure is persuasive, the present problem is defining by how much. Vastly different rates of apoptosis have been reported in both human and animal heart failure, with reported rates of up to 35.5% (Narula *et al.*, 1996). While these rates of death may be seen only in very localized areas, given that apoptosis takes <24 hours to complete, such rates would result in rapid involution of the whole heart. The variable and prolonged period of time over which heart failure develops and the gradual cell drop-out in failing hearts would suggest that very few myocytes would be undergoing apoptosis at any one time. More recently, rates of <0.5% have been consistently reported in end-stage heart failure (Liu *et al.*, 1995; Olivetti *et al.*, 1997), and between 0% and 0.0437% in normal hearts (Mallat *et al.*, 2001), which make far more physiologic sense. Unfortunately, while such low rates implicate apoptosis in disease, they make detecting any differences after therapeutic maneuvres very difficult.

Thus, although the evidence for apoptosis in heart failure is convincing, the role of apoptosis *per se* is unknown. In end-stage heart failure, necrotic cells are still (up to 7×) more likely to be found than apoptotic cells (Guerra *et al.*, 1999).

5.1.2 Apoptosis in the vessel wall

Vascular smooth muscle cells (VSMCs) within the vessel wall are able to proliferate, migrate, and synthesize and degrade extracellular matrix upon receiving appropriate stimuli. The normal adult artery shows very low levels of VSMC turnover, and apoptotic and mitotic indices are low in this tissue (Gordon *et al.*, 1990). In diseased tissue, additional factors are present both locally – such as inflammatory cytokines, inflammatory cells, and the presence of modified cholesterol – and systemically – such as blood pressure and flow. These factors can substantially alter the normal balance of cell proliferation and apoptosis, although the degree to which they are altered is dependent on the vascular disease under study.

5.1.2.1 Remodeling

Vessel wall remodeling defines a condition in which alterations in luminal size can occur through processes that do not necessarily require large changes in overall cell number or tissue mass. Thus, redistribution of cells, either toward or away from the lumen, through processes such as selective cell proliferation/apoptosis or matrix synthesis/degradation can significantly alter lumen dimensions. Physiologic remodeling by cell proliferation/apoptosis occurs in closure of the ductus arteriosus (Slomp *et al.*, 1997), and reduction in lumen size of infra-umbilical arteries after birth (Cho and Langille, 1993; Cho *et al.*, 1995). Surgical reduction in flow also results in compensatory VSMC apoptosis (Cho *et al.*, 1997; Kumar and Lindner, 1997), but remodeling in both low flow (shrinkage) or high flow (enlargement) of vessels is accompanied by medial VSMC apoptosis (Buus *et al.*, 2001). Remodeling also occurs in primary atherosclerosis, after angioplasty, and in angioplasty restenosis. Although apoptosis undoubtedly occurs in all of these conditions (see below), the role of VSMC apoptosis in determining the outcome of remodeling is unclear.

5.1.2.2 Arterial injury

Acute arterial injury, such as that occurring at angioplasty, is followed by rapid induction of medial cell apoptosis, at least in animal models. Thus, in rat or rabbit vessels, balloon overstretch injury results in medial cell apoptosis from 30 minutes to 4 hours after injury (Perlman *et al.*, 1997; Pollman *et al.*, 1998; Pollman *et al.*, 1999). In pigs, apoptotic cells occur within the media at 6 hours, with peaks in the media, adventitia, and neointima at 18 hours, 6 hours, and 7 days after percutaneous transluminal coronary angioplasty (PTCA), respectively (Malik *et al.*, 1998). Although we have no direct evidence, the consistency of this response in animal models suggests that human vessels may behave similarly. Repair of the vessel after injury is also associated with VSMC apoptosis, both in the media and in the intima, and in the rat occurs 8–21 days after injury (Bochaton-Piallat *et al.*, 1995). In humans, restenosis after angioplasty has been reported to be associated with either an increase (Isner *et al.*, 1995) or decrease (Bauriedel *et al.*, 1998) in VSMC apoptosis. Thus, the role of VSMC apoptosis in either the initial injury or the remodeling process in restenosis is again unclear in human vessels.

5.1.2.3   Aneurysm formation

The commonest form of arterial aneurysm in humans is associated with advanced atherosclerosis, and is characterized by a loss of VSMCs from the vessel media, with fragmentation of elastin and matrix degradation, leading to progressive dilatation and, eventually, rupture. Apoptosis of VSMCs is increased in aortic aneurysms (LopezCandales *et al.*, 1997; Thompson *et al.*, 1997; Henderson *et al.*, 1999) compared with normal aorta, and associated with an increase in expression of a number of pro-apoptotic molecules, such as death receptors, caspases, Bax, and p53 (Lopez-Candales *et al.*, 1997; Henderson *et al.*, 1999; Jacob *et al.*, 2001). Macrophages and T lymphocytes are found in aneurysmal lesions, suggesting that inflammatory mediators released by these cells may promote VSMC apoptosis. Moreover, the production of tissue metalloproteinases by macrophages may accelerate cell death by degrading the extracellular matrix from which VSMCs derive survival signals (see below) (Jacob *et al.*, 2001). This evidence suggests that VSMC apoptosis is a major cause of the development of arterial aneurysms.

5.1.2.4   Atherosclerosis

Rupture of atherosclerotic plaques is associated with a thinning of the VSMC-rich fibrous cap overlying the core. Rupture occurs particularly at the shoulder regions of plaques, which are noted for their lack of VSMCs and the presence of macrophages and other inflammatory cells. Apoptotic VSMC are evident in advanced human plaques (Geng and Libby, 1995; Han *et al.*, 1995; Isner *et al.*, 1995), including the shoulder regions, prompting the suggestion that VSMC apoptosis may hasten plaque rupture. Indeed, increased VSMC apoptosis occurs in unstable versus stable angina lesions (Bauriedel *et al.*, 1998).

Although loss of VSMCs would be expected to promote plaque rupture, there is no direct evidence of the effect of apoptosis *per se* in advanced human atherosclerosis. Most apoptotic cells in histologic sections are found in advanced lesions next to the lipid core (Kockx, 1998), and most of these apoptotic cells are macrophages, not VSMCs. Loss of macrophages from atherosclerotic lesions is likely to promote plaque stability rather than rupture, since macrophages can promote VSMC apoptosis by both direct interactions (Boyle *et al.*, 2001) and by the release of cytokines (Geng *et al.*, 1996). Of interest, apoptosis also occurs in the early stages of atherosclerosis induced by cholesterol feeding in animals, at the fatty streak stage before morphologic evidence of lesion formation (Hasdai *et al.*, 1999). Again, the effect of apoptosis at this early stage of lesion development is unknown.

5.1.2.5 Effect of VSMC apoptosis

The effect of VSMC apoptosis is clearly context dependent. Thus, VSMC apoptosis in advanced atherosclerotic plaques would be expected to promote plaque rupture as well as medial atrophy in aneurysm formation (Figure 5.1). In neointima formation post-injury, VSMC apoptosis of both the intima and media can limit neointimal formation (Pollman *et al.*, 1998; Sata *et al.*, 1998; Pollman *et al.*, 1999) at a defined time point, although long-term studies have not been performed to ensure that the neointima is not simply delayed. It is not yet known whether such inhibition of neointimal formation in an animal model can translate into suppression of restenosis following angioplasty or stenting.

Therapeutic induction of apoptosis in the vessel wall may also be limited by important sequelae. In contrast to the dogma that apoptosis is silent (that is, it does not elicit an immune response), a number of deleterious effects of apoptotic cells have emerged within the vasculature. In particular, exposure of phosphatidylserine on the surface of apoptotic cells provides a potent substrate for the generation of thrombin and activation of the coagulation cascade (Bombeli *et al.*, 1997; Flynn *et al.*, 1997). Apoptotic cells can also release membrane-bound microparticles into the circulation, which remain pro-coagulant and are increased in patients with unstable versus stable coronary syndromes (Mallat *et al.*, 1998, 1999). Although apoptotic cells are not the only source of circulating microparticles, such microparticles may contribute to the increased pro-coagulant state in these syndromes. Finally, there is increasing evidence that induction of VSMC apoptosis may be directly pro-inflammatory, with release of chemoattractants and cytokines from inflammatory cells (Schaub *et al.*, 2000), and may act as a focus for calcification in the vessel wall (Proudfoot, 2000).

5.2   Regulation of apoptosis

5.2.1   Apoptosis via death receptors

Many stimuli can trigger apoptosis in cells, but in vascular disease it is likely that specific alterations within the cell itself elicit sensitivity to a particular stimulus that is disease associated. Thus, remodeling may trigger apoptosis following reduction in blood flow, and the major stimulus may therefore be flow-dependent stimuli such as nitric oxide or shear stress. In contrast, apoptosis in atherosclerosis or aneurysm formation may be due to the surrounding influences of inflammatory cells that express death ligands on their surface or secrete pro-apoptotic cytokines. Whatever

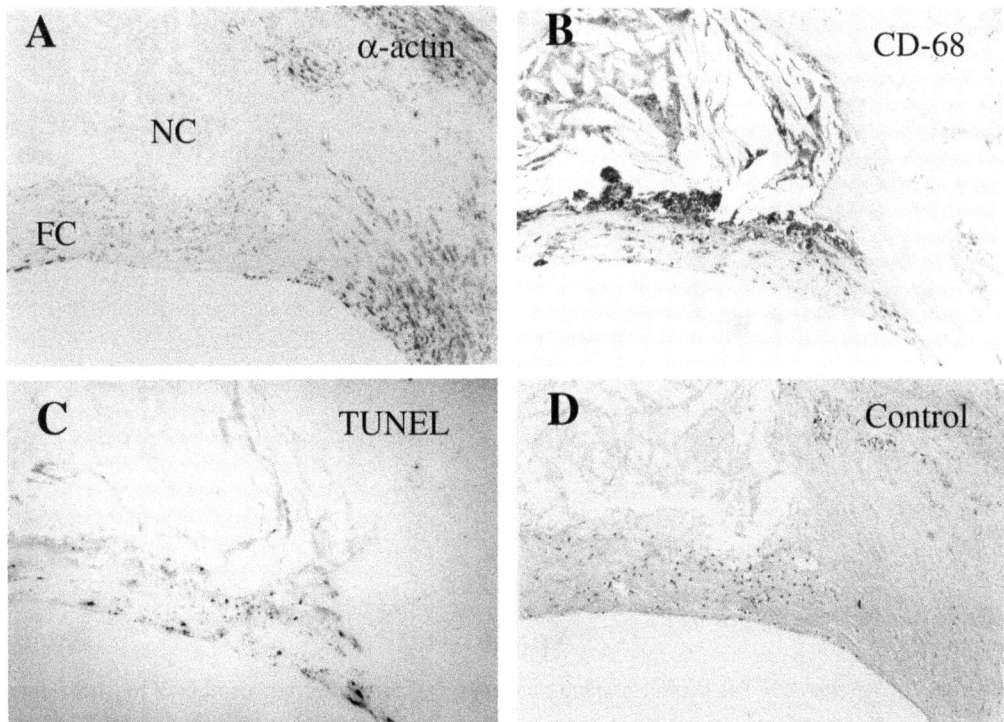

Figure 5.1   *Typical vulnerable atherosclerotic plaque, prone to rupture.* The plaque consists of a lipid and debris-filled "necrotic" core (NC), surrounded by macrophages, covered by a thin fibrous cap (FC) consisting of smooth muscle cells and collagen. (A) VSMC marker, α-sm-actin; (B) macrophage marker CD68; (C) apoptosis detected by TUNEL localized to the fibrous cap; (D) negative control antibody for (A) and (B).
(For a colour version of this figure, please see www.cambridge.org/9780521159449.)

the stimulus, many of the downstream pathways by which the apoptotic stimuli are transmitted are similar.

The regulation of apoptosis within the cell can be simplified into two major pathways (see Figures 5.2 and 5.3). First, membrane-bound death receptors of the tumor necrosis receptor family (TNFR), such as Fas (CD95), TNFR1, death receptor (DR)3, DR4, and DR5, bind their trimerized ligands causing receptor aggregation and subsequent recruitment of a number of adapter proteins through protein–protein interactions (Ashkenazi and Dixit, 1998) (see Figure 5.2). The major active caspases in Fas-mediated apoptosis are caspases-8, -3, -6, and -7 (Hirata *et al.*, 1998), with stepwise appearance of active caspases suggesting a caspase cascade. Cells in which caspase-8 is expressed in abundance, recently termed type I cells, use this pathway of direct caspase-3 activation. Moreover, in these cells, Fas-mediated cell death cannot be inhibited by anti-apoptotic factors such as Bcl-2 and Bcl-X$_L$

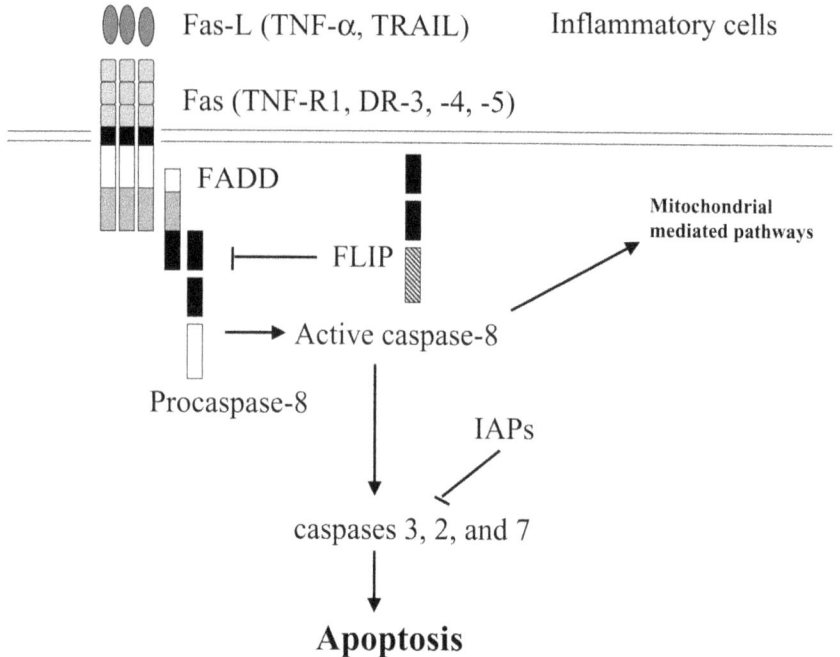

Figure 5.2   *Schematic of Fas death signaling pathways.* Fas, the prototypic member of the TNF death
receptor family, binds to its cognate ligand. Recruitment of the adapter molecule FADD
and procaspase-8 results in activation of the latter. Caspase-8 activation directly activates
downstream caspases (-3, -6, and -7) that results in DNA fragmentation and cleavage of
cellular proteins. This pathway is thought to occur in type I cells and does not involve mitochondrial pathways. Alternatively, caspase-8 activation also results in cleavage of Bid, which
translocates and interacts with other Bcl-2 family members (see Figure 5.3). Within the
cardiovascular system, death ligands are expressed predominantly on inflammatory cells in
myocardial infarction or inflammatory vascular disease (atherosclerosis, aneurysm formation).

since the pathway does not require amplification by pro-apoptotic factors released
by mitochondria (Scaffidi *et al.*, 1998).

5.2.2   Apoptosis via mitochondrial amplification

In contrast, cells in which caspase-8 is not abundantly expressed cannot activate
caspase-3 and other downstream caspases directly. Instead, in these cells, termed
type II cells (Scaffidi *et al.*, 1998), caspase-8 activation causes cleavage of proteins
of the bcl-2 family such as bid (Li *et al.*, 1998) (see Figure 5.3). Thus, death induced
by Fas signaling may or may not be inhibitable by Bcl-2 family members, suggesting that high levels of expression of anti-apoptotic Bcl-2 family members do not
automatically correlate with suppression of cell death. The classification of human

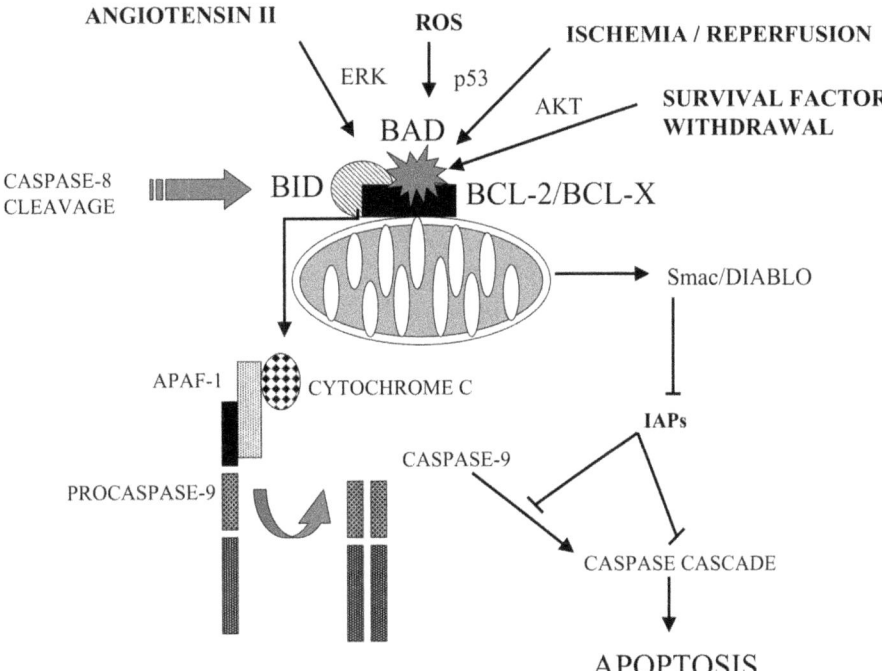

Figure 5.3   *Schematic of mitochondrial death signaling pathways.* Anti-apoptotic members of the Bcl-2 family, such as Bcl-2 and Bcl-X$_L$, are located on the mitochondrial outer membrane. Here, they act to prevent the release of apoptogenic factors from the inner mitochondrial space. Binding of the pro-apoptotic proteins Bid (after cleavage by caspase-8) or Bad (after dephosphorylation) to Bcl-2 mitigates the protective effect of Bcl-2 and triggers release of cytochrome c. Cytochrome c, in concert with the adapter protein apaf-1 and caspase-9, activates caspase-3 and the downstream caspase cascade. Stimuli acting through type II pathways include angiotensin II, reactive oxygen species (ROS), ischemia/reperfusion or DNA damage, and survival factor withdrawal. In addition, death receptor activation in vascular cells or cardiomyocytes often requires mitochondrial amplification for apoptosis.

VSMCs into type I or type II cells has yet to be made; however, the kinetics of cell death in response to anti-Fas antibodies suggests that they are type II cells.

5.2.3   Regulation of cardiomyocyte apoptosis

The stimulus for cardiomyocyte apoptosis clearly depends on the clinical or experimental setting. Ischemia is associated with many changes in the intracellular and extracellular milieu of cardiomyocytes, many of which are potent stimuli to apoptosis. Hypoxia is associated with cardiomyocyte apoptosis, both in vitro and in vivo (Tanaka *et al.*, 1994; Long *et al.*, 1997), although the effectors of this pathway are unknown. In addition, hypoxia *per se* seems to be less effective at inducing apoptosis than hypoxia/re-oxygenation both in vitro and in vivo (Bitar *et al.*, 2002). p53 may promote both hypoxia-induced apoptosis (Long *et al.*, 1997), or apoptosis related to inhibition of the vacuolar ATPase (see below) (Long *et al.*, 1998) but is not an absolute requirement for apoptosis post-myocardial infarction (Bialik *et al.*, 1997). Ischemia/reperfusion and hypoxia/re-oxygenation are associated with increased expression of Fas (Kajstura *et al.*, 1996; Yue *et al.*, 1998) and Fas-L (Stephanou *et al.*, 2001), although the evidence that Fas/Fas-L induces apoptosis in cardiomyocytes after ischemia/reperfusion is controversial. Decreased serum and glucose concentrations do trigger cytochrome c release from mitochondria in cardiomyocytes (Bialik *et al.*, 1999), suggesting that the effects of ischemia may be mediated by mitochondrial amplification of apoptosis signals.

In ischemia/reperfusion models, reactive oxygen species may promote apoptosis (Jeroudi *et al.*, 1994), again by triggering pathways involving mitochondria with release of cytochrome c and the activation of caspases (Kang *et al.*, 2000). In addition, prolonged hypoxia and depletion of high-energy phosphates may inhibit the ATP-dependent vacuolar proton pump ATPase, resulting in intracellular acidification and apoptosis (Karwatowska-Prokopczuk *et al.*, 1998; Long *et al.*, 1998). In contrast to the chronic effects of ischemia/reperfusion, a period of sub-lethal ischemia can protect myocytes from cell death due to a subsequent stimulus, a phenomenon known as ischemic pre-conditioning. Pre-conditioning is associated with a reduction in both apoptosis and necrosis in myocytes, and signaling through specific protein kinase C isoforms and adenosine receptors has been implicated (Liu *et al.*, 2001).

In heart failure, a huge variety of initial stimuli have been propounded. In vitro, mechanical stretch can induce apoptosis, indicating a possible role for volume overload and raised ventricular end diastolic pressure (Cheng *et al.*, 1995). This study also directly implicated reactive oxygen species in the induction of stretch-induced apoptosis. Pressure overload following aortic banding can also induce early myocyte apoptosis, before significant hypertrophy, emphasizing the role of mechanical stimuli (Teiger *et al.*, 1996), and there is increased myocyte apoptosis in the hearts of spontaneously hypertensive rats (Hamet *et al.*, 1995). In models of chronic heart failure, such as following microembolization of coronary arteries in dogs, apoptotic myocytes tend to cluster in peri-infarct zones, also suggesting that ongoing ischemia or mechanical stress is responsible (Sharov *et al.*, 1996). Four weeks of rapid ventricular pacing can induce myocyte apoptosis in dogs associated with heart failure, suggesting that catecholamine responses may be directly toxic to myocytes (Liu *et al.*, 1995). Indeed, in vitro data suggest that catecholamines are potent inducers of cardiomyocyte apoptosis, and sustained or excessive activation of either

Gq- or Gs-signaling pathways results in apoptotic loss of cardiomyocytes both in vitro and in vivo (reviewed in Adams and Brown, 2001). In vivo, over-expression of β1 adrenergic receptors promotes apoptosis, while engagement of β2 receptors promotes survival through Akt (Zhu *et al.*, 2001). Thus, within a clinical context, there may be a multitude of both pro- and anti-apoptotic signaling pathways functioning.

A large number of circulating factors have also been implicated in cardiomyocyte apoptosis in heart failure, including both angiotensin II (Ang II) (Kajstura *et al.*, 1997) and atrial natriuretic factor (ANF). Both of these agents increase apoptosis in cultured cardiomyocytes, although the evidence that Ang II induces apoptosis in the heart in vivo is controversial, as is the role of any specific Ang II receptor subtype (Leri *et al.*, 2000; Sugino *et al.*, 2001). In contrast, intraperitoneally administered insulin like growth factor 1 (IGF-1) in vivo or insulin in vitro may limit ischemia/reperfusion-induced apoptosis (Buerke *et al.*, 1995; Jonassen *et al.*, 2000). Many of the downstream pathways induced by these agents converge on common secondary messenger pathways, including activation of Akt, p38, ERK, or SAPK. Inhibition of Akt, for example, by over-expression of the tumor suppressor gene PTEN promotes apoptosis in cardiomyocytes (Schwartzbauer and Robbins, 2001) while SAPK induces apoptosis in cardiomyocytes and inhibition of SAPK protects (Au-Yeung *et al.*, 2001). Similarly, p42/44 MAPK (ERK 1 and 2) are activated during ischemia reperfusion and protect against apoptosis in cardiomyocytes (Baxter *et al.*, 2001), and over-expression of p42/44 MAPK is protective alone (Bueno *et al.*, 2000). Indeed, activation of p42/44 MAPK may be the mechanism of the protection afforded by TGF-β (Baxter *et al.*, 2001) and cardiotrophin-1 (Brar *et al.*, 2001).

Whatever the upstream stimulus, activation of caspase-3 in the heart induces myocyte death, and increases myocardial infarction size in ischemia/reperfusion (Condorelli *et al.*, 2001). This study, utilizing heart-specific caspase-3 expression, also provided direct evidence that apoptosis *per se* promotes expansion of myocardial infarctions, in addition to chronic effects on myocardial function.

5.2.4 Regulation of vascular smooth muscle cell apoptosis

Human VSMCs express both Fas and TNFR1 (Geng *et al.*, 1997), and given the widespread occurrence of tumor necrosis factor-related apoptosis-inducing ligand receptors (TRAIL-Rs) are also likely to express these receptors. T lymphocytes and macrophages within the atherosclerotic plaque express Fas ligand and tumor necrosis factor (TNF)α, and interaction between membrane-bound ligands on T cells and macrophages and receptors on VSMC may induce the death of VSMCs. VSMCs can also express Fas-L and TNFα, which exist as membrane-bound or soluble forms. The soluble forms are cleaved from the membrane-bound forms by a metalloproteinase of the ADAM family. Interestingly, a recent study of the tissue inhibitors of metalloproteinases (TIMPs) indicates that TIMP-3 can induce VSMC apoptosis in vitro and in vivo (Baker *et al.*, 1998); TIMP-3 may be acting by stabilizing expression of the more potent membrane forms of death ligands on the VSMC surface.

Although over-expression of FADD can induce VSMC apoptosis in vitro and in vivo (Schaub *et al.*, 2000), soluble ligand binding to death receptors is a very weak inducer of apoptosis in VSMCs (Geng *et al.*, 1996; Bennett *et al.*, 1998; Belanger *et al.*, 2001), and does not induce apoptosis in the absence of "priming" of the cell, usually with cycloheximide or a cytokine cocktail. Some of this resistance can be explained by the observation that death receptors are sequestered intracellularly in VSMCs (Bennett *et al.*, 1998), and priming may be associated with increased surface expression. Physiologically, increased death receptor expression can be achieved via the combination of cytokines such as interleukin (IL)-1-β, interferon (IFN)-γ, and TNFα (Fukuo *et al.*, 1997; Geng *et al.*, 1997; Hoffmann *et al.*, 1998), possibly via nitric oxide and p53 stabilization (Fukuo *et al.*, 1996; Geng *et al.*, 1996; Hoffmann *et al.*, 1998; Boyle *et al.*, 1999; Hayden *et al.*, 2001), or via direct activation of p53 (Bennett *et al.*, 1998). Thus, DNA damage induced by nitric oxide, anoxia, or free radical formation, for instance, may stabilize p53, effectively priming VSMCs to death receptor-mediated apoptosis. Oxidized lipids, free radicals, and nitric oxide can also induce apoptosis which may be independent of p53 but associated with caspase-3 activation (Pollman *et al.*, 1996; Li, P. F. *et al.*, 1997; Nishio and Watanabe, 1997; Zhao *et al.*, 1997; Iwashina *et al.*, 1998; Wang and Keiser, 1998; Hsieh *et al.*, 2001), but not associated with death receptor signaling. Oxidized lipids also exert their effect in part by the upregulation of Fas-L and Fas (Lee and Chau, 2001).

In contrast, efficient death receptor-mediated apoptosis of VSMCs can be achieved by high level expression of death ligands such as Fas-L (Sata *et al.*, 1998), possibly by increasing the expression of membrane-bound forms of these ligands, resulting in reduction in neointima formation after injury. Proteosomal inhibitors may sensitize VSMCs to apoptosis, in part by stabilizing membrane-bound death receptors or ligands (Kim, 2001), and they have been found to reduce neointima formation after injury, in part by inducing apoptosis (Meiners *et al.*, 2002). Membrane-bound rather than soluble ligands are also responsible for monocyte-macrophage-induced apoptosis of VSMCs, in combination with nitric oxide generated by monocytes (Boyle *et al.*, 2001; Seshiah *et al.*, 2002). Both differentiation and activation of monocytes are required for this effect (Boyle *et al.*, 2001; Seshiah *et al.*, 2002). This may explain the observation that most VSMC apoptosis occurs in regions of plaque with high inflammatory cell content (Kockx *et al.*, 1998a). In addition, membrane-bound Fas-L on endothelial cells may promote VSMC apoptosis in human atherosclerosis (Imanishi *et al.*, 2001). Finally, chymase from mast cells within lesions can induce VSMC apoptosis (Leskinen *et al.*, 2001).

Irrespective of the local environment, VSMCs derived from atherosclerotic plaques show increased rates of apoptosis in culture compared with cells from normal vessels, reflecting an intrinsic sensitivity to apoptosis (Bennett *et al.*, 1995). This appears to be a stable property, and is part of the phenotype of plaque VSMCs that includes slow proliferation and early senescence. Rat neo-intimal VSMCs also show increased spontaneous apoptosis compared with normal medial VSMCs (Niemann-Jonsson *et al.*, 2001). Heterogeneity of sensitivity to apoptosis within VSMCs in the vessel wall has also been demonstrated in animal vessels after injury (Bochaton-Piallat *et al.*, 1996) and in medial VSMCs from normal human arteries (Chan *et al.*, 1998). This is likely to reflect differences in expression of pro- and anti-apoptotic molecules, specifically those regulating signaling from survival cytokines, cell–cell and cell–matrix interactions, and members of the bcl-2 family.

VSMC apoptosis in the vessel wall is suppressed by signaling through survival factors, extracellular matrix contacts, and cyclic strain (Chen, A. H. *et al.*, 2001; Ikari *et al.*, 2001), many acting through the serine threonine kinase Akt. Akt inhibits many downstream pro-apoptotic proteins including Fas-L (Suhara *et al.*, 2001), Bad, and caspase-9. One of the potent survival factors for normal VSMCs is IGF-1, although IGF-1 cannot completely inhibit plaque VSMC apoptosis in vitro after serum withdrawal (Bennett *et al.*, 1995). This lack of survival signaling in plaque VSMCs is due to reduced IGF-1R expression and Akt phosphorylation, and increased expression of IGF-1-binding proteins (IGF-1bp) in plaque VSMCs (Patel *et al.*, 2001) – both the reduced IGF-1R and increased IGF-1bp expression may be the effects of oxidized low-density lipoprotein (LDL) (Scheidegger *et al.*, 2000). Both IGF-1 and IGF-1R have been shown to be reduced in VSMCs in advanced lesions compared with normal medial VSMCs (Okura *et al.*, 2001; Patel *et al.*, 2001). Similarly, inhibition of bFGF binding induces VSMC apoptosis in vitro (Fox and Shanley, 1996; Miyamoto *et al.*, 1998) and in vivo (Neschis *et al.*, 1998), possibly by induction of the oncogene *mdm2*, which inactivates p53 (Shaulian *et al.*, 1997).

Evidence suggesting the critical role of bcl-2 family members in regulating VSMC apoptosis has come from both in vitro and in vivo studies. Human VSMCs express low levels of Bcl-2 (Bennett *et al.*, 1995; Konstadoulakis *et al.*, 1998), but Bax is seen particularly in atherosclerotic plaques, both in human and animal models of atherosclerosis and injury (Kockx *et al.*, 1998b; Konstadoulakis *et al.*, 1998; Igase *et al.*, 1999). In addition, spontaneous and growth factor withdrawal-induced apoptosis can be inhibited by over-expression of bcl-2 (Bennett *et al.*, 1995). In vivo, rat VSMCs express minimal Bcl-2, but high levels of Bcl-X can be found after injury (Igase *et al.*, 1999). Indeed, inhibition of Bcl-X can dramatically induce apoptosis of VSMCs after balloon injury (Pollman *et al.*, 1998) and differences in expression of Bcl-X may account for differences in sensitivity to apoptosis after injury of intimal versus medial VSMCs (Pollman *et al.*, 1999). The reduced levels of VSMCs apoptosis seen after cholesterol lowering in rabbit models of atherosclerosis is also associated with a loss of Bax immunoreactivity (Kockx *et al.*, 1998b), arguing for a pro-apoptotic role of this protein in VSMC apoptosis. However, it should be noted that excessive reliance on immunocytochemistry of one member of the bcl-2 family to ascertain a role for that protein in vivo should be avoided. Although Bcl-X is upregulated after injury, in rats it is the short Bcl-x_S or pro-apoptotic form that appears to predominate (Igase *et al.*, 1999).

Regulation of sensitivity to apoptosis in VSMCs is also mediated by the expression of inhibitor of apoptosis (IAP) proteins (Erl *et al.*, 1999) and individual caspases (Krajewska *et al.*, 1997; Chan *et al.*, 1998), and it is likely that there are marked differences in the expression of multiple proteins that regulate VSMC apoptosis of individual VSMCs in response to specific stimuli. This heterogeneity may underlie observations that, despite (apparently) the same stimulus for apoptosis, VSMC apoptosis in either the normal or diseased vessel wall is highly localized. Heterogeneity of response is also seen in different cell populations from the normal media in response to both death receptors (Chan *et al.*, 2000; Belanger *et al.*, 2001) and angiotensin II (Bascands *et al.*, 2001). Some of this effect may be due to differing angiotensin II receptor subtype densities, as apoptosis is selectively induced via the AT1 subtype in VSMCs (Cui *et al.*, 2001), although both subtypes may be pro-apoptotic in cardiomyocytes (Goldenberg *et al.*, 2001). Heterogeneity of response is also seen in subtypes of VSMCs that express abundant Fas-L and undergo autocrine/paracrine Fas/Fas-L-induced apoptosis (Kavurma *et al.*, 2001), or undergo apoptosis in response to retinoic acid (Orlandi *et al.*, 2001).

5.3 Therapeutic options for apoptosis treatment

From the above discussion, prevention of cardiomyocyte apoptosis would appear to be a very important therapeutic aim. Although the mechanisms of induction of apoptosis and the precise significance of apoptosis in the development of each disease entity are unclear, it would seem obvious that a live myocyte is better than a dead one. However, critical to determining therapeutic benefit of any treatment is not only inhibition of markers of apoptosis at a single defined time point but improvement of cardiac function. Many agents may prevent the development of the morphologic appearance of an apoptotic cell or a defined biochemical marker (for example, DNA fragmentation) without inhibiting cell death. The ability to delay death may serve no useful purpose and may be deleterious if that cell is destined to undergo necrosis at a later time point, with concomitant inflammation.

In contrast, some studies have indicated that inhibition of apoptosis improves ventricular remodeling and contractility after infarction (Li, Q. *et al.*, 1997; Goussev *et al.*, 1998). For example, in a rat model of myocardial infarction (ischemia/reperfusion), treatment with the caspase inhibitor zVAD.fmk reduced infarct size, with an improvement in acute myocardial function (Yaoita *et al.*, 1998). The long-term effects of this inhibitor are, however, unknown, and prevention of apoptosis may simply delay death, but not prevent it. In addition, improvement in function can be seen after ischemia reperfusion with caspase inhibitors irrespective of any effect on apoptosis (Ruetten *et al.*, 2001). In contrast, neovascularization of infarct zones using bone marrow-derived angioblasts can reduce infarct size long term and improve function, in part by suppression of cardiomyocyte apoptosis (Kocher *et al.*, 2001). In addition, transgenic over-expression of p35 can improve contractility in an animal model of heart failure over extended periods of time (Laugwitz *et al.*, 2001).

Apoptosis can be interrupted at many points in the signaling pathway. Prevention of apoptotic myocyte death may be directed at: (1) inhibiting/preventing the apoptotic stimulus; (2) inhibiting the regulatory mechanisms determining the decision to die; or (3) inhibiting the effector pathways executing apoptosis (Table 5.2). The cascade of events leading to cardiomyocyte apoptosis, and also the point at which a cell is irreversibly committed to die, crucially determine the approach to inhibiting apoptosis. Clearly, a multitude of signaling pathways are activated in complex pathophysiologic states such as ischemia and heart failure. Interruption of a single pathway would, therefore, be predicted not to inhibit apoptosis if there are multiple, redundant pathways inducing apoptosis. Such arguments would indicate that mediators that act beyond the convergence of multiple signaling pathways would be better targets to inhibit apoptosis. However, many of the identified downstream mediators are enzymes required for effective disintegration and packaging of the cell, and, as such, are beyond the point at which the cell is committed to die. Inhibition at these points would prevent the cellular appearances and markers of apoptosis, but the cell would still die. In addition, these molecules are critical to apoptosis in many tissues and such non-cardiac specificity may be unwelcome. From this argument, inhibiting the stimulus to apoptosis, particularly if specific to the heart at one point in time, would be more effective. The timing and delivery of therapy is also dependent on the clinical situation. Clearly, it is easier to inhibit apoptosis transiently in an acute situation such as myocardial infarction than with chronic therapy in diseases such as heart failure.

Table 5.2 *Potential inhibitors and signaling pathways of cardiomyocyte apoptosis*

Stimulus	Signaling pathway	Potential inhibitor
Ischemia/reperfusion	ERK/SAPK	Activation of ERK, inhibition of SAPK signaling
Pressure overload	ERK/SAPK	
Neurohormonal factors (e.g. catecholamines)	G-protein coupling	β-blockers
Ischemia	Lack of growth factor signaling	Activation of Akt/ERK pathways (e.g. by IGF-1)
Death receptor ligands	Adapter molecules/caspases	Decoy receptors/receptor antagonists IAPs/caspase inhibitors

5.3.1   Inhibiting/preventing the pro-apoptotic stimulus

Ischemia/reperfusion, hypertrophy due to increased afterload, and myocardial remodeling following infarction are all associated with myocyte apoptosis (Tanaka *et al.*, 1994; Itoh *et al.*, 1995; Bardales *et al.*, 1996; Olivetti *et al.*, 1996; Teiger *et al.*, 1996; Saraste *et al.*, 1997). This suggests that current therapy of proven benefit in these diseases may already act by inhibition of apoptosis. The beneficial effects of β-blockers in chronic heart failure and ischemic heart disease may counteract the pro-apoptotic effect of excess catecholamines. Indeed, carvedilol can inhibit ischemia/reperfusion-induced apoptosis of myocytes (Yue *et al.*, 1998), and ACE inhibitors would be predicted to protect against angiotensin II-induced apoptosis (Kajstura *et al.*, 1997). Similarly, calcium antagonists may reduce the calcium overload that produces reperfusion injury (Gao *et al.*, 2001). Clearly, approaches aimed at reducing myocardial stretch, or oxidative stress, or improving myocardial perfusion would be predicted to have the same effect. Finally, many pathways leading to apoptosis are triggered by specific death ligands, with either apoptosis or the disease entity being associated with upregulation of death receptors or transgenic expression of the death ligand directly inducing the disease (Tanaka *et al.*, 1994; Torre-Amione *et al.*, 1995; Kubota *et al.*, 1997; Yue *et al.*, 1998). Inhibition of the delivery of death ligands – e.g. by scavenging ligands through soluble receptors or receptor antagonists – would be predicted to reduce apoptosis mediated through these pathways.

5.3.2   Protection against apoptosis

The expression of many molecules, including anti-apoptotic members of the Bcl-2 family, IAPs, decoys for death receptors etc., protects cells from apoptosis. Although many of these agents are promiscuous enough to inhibit apoptosis mediated by a variety of stimuli, and may therefore be clinically useful, at present they cannot be selectively expressed without gene transfer into the heart, with all its inherent problems. More promising is the potential administration of soluble survival factors following the apoptotic stimulus. Many growth factors, including IGF-1, cardiotrophin-1, and the neuregulins, inhibit apoptosis following ischemia, serum withdrawal, myocyte stretch, and cytotoxic drugs (Buerke *et al.*, 1995; Li, Q. *et al.*, 1997; Sheng *et al.*, 1997; Wang, L. *et al.*, 1998; Zhao *et al.*, 1998; Leri *et al.*, 1999; Mehrhof *et al.*, 2001). Over-expression of IGF-1 can reduce apoptosis in non-infarcted remote zones and promote favorable remodeling after myocardial infarction (Li, Q. *et al.*, 1997), and a single intraperitoneal injection of IGF-1 can inhibit ischemia/reperfusion-induced apoptosis in the rat by 50% (Buerke *et al.*, 1995). Activation of the cardiotrophin-1 receptor also inhibits cardiac dilatation following aortic banding, suggesting that reduced cardiomyocyte apoptosis can be translated into improved function (Hirota *et al.*, 1999). These agents signal through the Akt and ERK pathways, respectively, which are known to be anti-apoptotic in many cell types. Indeed, over-expression of a constitutively active Akt or IGF-1 reduces infarct size and improves function after ischemia/reperfusion (Fujio *et al.*, 2000; Miao *et al.*, 2000; Matsui *et al.*, 2001; Yamashita *et al.*, 2001). Similarly, inhibition of the ERK signaling pathway increases ischemia/reperfusion-induced apoptosis and worsens reperfusion injury, so those agents that increase signaling through this pathway may be beneficial (Yue *et al.*, 1998, 2000). These agents could be given either locally or systemically at the time of infarction. Stimulation of the MKK/p38α pathway also causes cardiomyocyte apoptosis (Wang, Y. *et al.*, 1998; Zechner *et al.*, 1998), as does over-expression of Gsα or Gqα. Antagonists to these more "classical" myocyte signaling pathways may, therefore, be of benefit.

In contrast, a number of agents have been suggested as potential therapeutics for long-term administration. Heart failure is characterized by increased plasma levels of catecholamines and TNFα. The beneficial effects of β-blockers in heart failure may, therefore, be by the prevention of myocyte apoptosis. In contrast, long-term inhibition of TNFα has been found not to be beneficial in human heart failure trials. Evidence identifying the type 2 AT-II receptor as inducing apoptosis in many cell types, including the myocardium, especially in models of heart failure, has suggested that inhibition of this receptor may also be beneficial.

5.3.3   Preventing execution of apoptosis

Execution of apoptosis and cellular disintegration and packaging requires the activation of downstream signaling pathways, including mitochondrial amplification and activation of caspases. Augmentation of endogenous inhibitors of caspases, such as the IAPs, could therefore inhibit apoptosis induced by a variety of stimuli. Pharmacologic inhibition of caspases using cell permeable analogs of caspase cleavage sites can inhibit myocyte apoptosis over the short term, if given close to the time of the stimulus (Mocanu *et al.*, 2000; Fauvel *et al.*, 2001). However, the long-term benefits of such drugs are unknown, as it may be that cells which are destined to die do so anyway, and delaying apoptosis may not provide long-term benefit. In addition, some inducers of apoptosis may absolutely require caspases for the execution of cell death (Susin *et al.*, 1999).

5.3.4   Induction of apoptosis

In contrast to cardiomyocyte apoptosis, therapeutic induction of apoptosis has been suggested as a mechanism of prevention of restenosis after angioplasty/intravascular stenting, or in bypass grafts. A number of studies have identified that induction of apoptosis after arterial injury reduces neointima formation at a later time point. These studies have utilized either agents that specifically induce apoptosis, such as Fas-L, p53, or antisense oligonucleotides/ribozymes to protective genes (Pollman *et al.*, 1998; Sata *et al.*, 1998; Mano *et al.*, 2000; Perlman *et al.*, 2000; George *et al.*, 2001; Luo *et al.*, 2001) or that induce apoptosis in addition to the prevention of cell proliferation or migration, such as brachytherapy or photodynamic therapy (Waksman *et al.*, 1997; Chen, Z. *et al.*, 2001; Granville *et al.*, 2001).

A number of caveats should be considered, however, before advocating therapeutic induction of apoptosis in restenosis. First, most positive studies have been performed in normal (rodent) vessels after injury, and induction of apoptosis in diseased vessels may accelerate atherosclerosis (Schneider *et al.*, 2000). In addition, aggressive induction of apoptosis would be predicted to promote local thrombosis through plaque rupture and inefficient vessel healing, as has been seen with studies utilizing radioactive stents. All of the sequelae of VSMC apoptosis outlined above are potential limitations of this strategy.

5.4   Outstanding questions

There are many unresolved questions regarding anti-apoptotic therapeutics in cardiovascular disease, particularly in those relating to myocardial apoptosis.
1. There is considerable debate as to the role that apoptosis plays in, and amount that apoptosis contributes to, cardiovascular disease, particularly in disease entities that are characterized by complex pathophysiology. For example, heart failure is characterized by impairments of excitation–contraction coupling, alterations in intracellular filaments, interstitial fibrosis, side–side slippage of myocytes, and neurohormonal deregulation. How much myocyte apoptosis contributes to this

complex situation is unknown. Clearly, anti-apoptotic therapy may inhibit only a very small part of such a complex pathophysiology.

2. It is presently unknown whether inhibition of apoptosis prevents or simply delays disease progression. Although many studies have shown reduced apoptosis after therapy, few have shown that this translates into improved left ventricular function, and fewer still have demonstrated that this is maintained long term and associated with increased survival. A number of studies have shown that agents can significantly inhibit both apoptosis and expression of pro-apoptotic genes, yet have no effect on cardiac function (e.g. Moudgil *et al.*, 2001).

3. The specificity of action of any therapy is unknown. Many of the apoptotic/anti-apoptotic signaling pathways have many other functions, and, given the conserved nature of apoptosis signaling, no pathway is likely to be cardiac specific. While this may not be important for myocardial salvage after infarction where short-term therapy is planned, it is a critical consideration for long-term therapy in heart failure. It must be remembered that apoptosis is a vital element in the prevention and destruction of tumors, and any long-term therapy has to be examined for tumorigenicity. Apoptosis is also crucial to such physiologic functions as normal fertility, shedding of the skin and gut lining, and deletion of leucocytes. Inhibition of normal apoptosis may therefore cause sterility and autoimmune disease such as hemolytic anemia, and also thrombocytopenia. Genetic knock-out of many components of the apoptotic machinery is also lethal in the embryonic period. It is, therefore, possible that targeting apoptosis is directly teratogenic.

4. Furthermore, although apoptosis is signaled by death receptors, other signals also emanate from these receptors. For example, Fas activation reduces the membrane potential and induces after-depolarizations in cardiac myocytes. Thus, therapy inhibiting Fas-induced apoptosis may allow the escape of other Fas-signaling, -promoting arrhythmias (Felzen *et al.*, 1998). Similarly, interrupting TNFα-induced apoptosis may still allow the negatively inotropic effect of this cytokine.

5.5 Summary

VSMC apoptosis occurs in the vasculature in both physiologic and pathologic contexts. These deaths are regulated by specific proteins that serve to either induce or protect against apoptosis. We are now beginning to understand the complex biology observed in lesions such as atherosclerosis and to identify potential pro-apoptotic factors that may lead to the loss of cells from the vasculature. Sensitivity to apoptosis is determined by expression of cell death receptors and ligands, and by multiple protein species below receptor level. In addition, sensitivity to apoptosis is determined by the presence and response to survival cytokines, mito-gens, and local cell and matrix interactions, and by the growth status of the cell. Although much of this research has been carried out in vitro, future studies in animal models should help to identify which of the pro- and anti-apoptotic factors that are effective in vitro are also relevant in vivo. Moreover, a closer examination of the population dynamics of vascular cells within the vessel wall will aid in the understanding of the timing and triggers of VSMC apoptosis in disease.

In contrast, in the adult cardiomyocyte, apoptosis is almost always pathologic. Apoptosis of cardiac myocytes is part of many disease states including myocardial infarction and heart failure. At present, the precise role of cardiac myocyte apoptosis in the pathogenesis of these diseases is unknown, and therefore the likely benefit from anti-apoptotic therapy is unproven. Prevention of cardiomyocyte apoptosis may involve inhibiting both the pro-apoptotic stimulus and apoptosis signaling within the cell. Given the lack of cardiac specificity of apoptosis signaling, such strategies are far more likely to be beneficial in short-lived insults, such as myocardial infarction or unstable angina, rather than chronic administration in heart failure. However, it is also highly likely that proven conventional therapy for heart failure works at least in part by inhibiting apoptosis.

REFERENCES

Adams, J. W. and Brown, J. H. (2001). G-proteins in growth and apoptosis: lessons from the heart. *Oncogene*, **20**, 1626–34.

Anversa, P., Hiler, B., Ricci, R., Guideri, G., and Olivetti, G. (1986). Myocyte cell loss and myocyte hypertrophy in the aging rat heart. *J. Am. Coll. Cardiol.*, **8**, 1441–8.

Ashkenazi, A. and Dixit, V. (1998). Death receptors: signaling and modulation. *Science*, **281**, 1305–8.

Au-Yeung, K. K., Zhu, D. Y., O, K. and Siow, Y. L. (2001). Inhibition of stress-activated protein kinase in the ischemic/reperfused heart: role of magnesium tanshinoate B in preventing apoptosis. *Biochem. Pharmacol.*, **62**, 483–93.

Baker, A. H., Zaltsman, A. B., George, S. J., and Newby, A. C. (1998). Divergent effects of tissue inhibitor of metalloproteinase-1, -2, or -3 overexpression on rat vascular smooth muscle cell invasion, proliferation, and death in vitro – TIMP-3 promotes apoptosis. *J. Clin. Invest.*, **101**, 1478–87.

Bardales, R. H., Hailey, L. S., Xie, S. S., Schaefer, R. F., and Hsu, S. M. (1996). In situ apoptosis assay for the detection of early acute myocardial infarction. *Am. J. Pathol.*, **149**, 821–9.

Bascands, J. L., Girolami, J. P., Troly, M. *et al.* (2001). Angiotensin II induces phenotype-dependent apoptosis in vascular smooth muscle cells. *Hypertension*, **38**, 1294–9.

Bauriedel, G., Schluckebier, S., Hutter, R. *et al.* (1998). Apoptosis in restenosis versus stable-angina atherosclerosis. *Arterioscler. Thromb. Vasc. Biol.*, **18**, 1132–9.

Baxter, G. F., Mocanu, M. M., Brar, B. K., Latchman, D. S., and Yellon, D. M. (2001). Cardioprotective effects of transforming growth factor-beta1 during early reoxygenation or reperfusion are mediated by p42/p44 MAPK. *J. Cardiovasc. Pharmacol.*, **38**, 930–9.

Belanger, A. J., Scaria, A., Lu, H. *et al.* (2001). Fas ligand/Fas-mediated apoptosis in human coronary artery smooth muscle cells: therapeutic implications of fratricidal mode of action. *Cardiovasc. Res.*, **51**, 749–61.

Bennett, M., Macdonald, K., Chan, S.-W., Simari, R., Luzio, J., and Weissberg, P. (1998). Cell surface trafficking of Fas: a rapid mechanism of p53-mediated apoptosis. *Science*, **282**, 290–3.

Bennett, M. R., Evan, G. I., and Schwartz, S. M. (1995). Apoptosis of human vascular smooth muscle cells derived from normal vessels and coronary atherosclerotic plaques. *J. Clin. Invest.*, **95**, 2266–74.

Bialik, S., Cryns, V. L., Drincic, A. *et al.* (1999). The mitochondrial apoptotic pathway is activated by serum and glucose deprivation in cardiac myocytes. *Circ. Res.*, **85**, 403–14.

Bialik, S., Geenen, D. L., Sasson, I. E. *et al.* (1997). Myocyte apoptosis during acute myocardial infarction in the mouse localizes to hypoxic regions but occurs independently of p53. *J. Clin. Invest.*, **100**, 1363–72.

Bitar, F. F., Bitar, H., El Sabban, M. *et al.* (2002). Modulation of ceramide content and lack of apoptosis in the chronically hypoxic neonatal rat heart. *Pediatr. Res.*, **51**, 144–9.

Bochaton-Piallat, M., Gabbiani, F., Redard, M., Desmouliere, A., and Gabbiani, G. (1995). Apoptosis participates in cellularity regulation during rat aortic intimal thickening. *Am. J. Path.*, **146**, 1059–64.

Bochaton-Piallat, M.-L., Ropraz, P., Gabbiani, F., and Gabbiani, G. (1996). Phenotypic heterogeneity of rat aortic smooth muscle cell clones. *Arterioscler. Thromb. Vasc. Biol.*, **16**, 815–20.

Bombeli, T., Karsan, A., Tait, J. F., and Harlan, J. M. (1997). Apoptotic vascular endothelial cells become procoagulant. *Blood*, **89**, 2429–42.

Boyle, J., Bowyer, D., Weissberg, P., and Bennett, M. (2001). Human blood-derived macrophages induce apoptosis in human plaque-derived vascular smooth muscle cells by Fas ligand/Fas interactions. *Arterioscler. Thromb. Vasc. Biol.*, **21**, 1402–7.

Boyle, J. J., Bowyer, D. E., Weissberg, P. L., and Bennett, M. R. (1999). Interactions between TNF alpha and nitric oxide in human macrophage-induced vascular smooth muscle cell apoptosis. *J. Pathol.*, **187**, A12.

Brar, B. K., Stephanou, A., Liao, Z. *et al.* (2001). Cardiotrophin-1 can protect cardiac myocytes from injury when added both prior to simulated ischaemia and at reoxygenation. *Cardiovasc. Res.*, **51**, 265–74.

Braunwald, E. and Kloner, R. (1985). Myocardial reperfusion: a double-edged sword? *J. Clin. Invest.*, **76**, 1713.

Bueno, O. F., De Windt, L. J., Tymitz, K. M. *et al.* (2000). The MEK1-ERK1/2 signaling pathway promotes compensated cardiac hypertrophy in transgenic mice. *EMBO J.*, **19**, 6341–50.

Buerke, M., Murohara, T., Skurk, C., Nuss, C., Tomaselli, K., and Lefer, A. (1995). Cardioprotective effect of insulin-like growth factor I in myocardial ischaemia followed by reperfusion. *Proc. Natl. Acad. Sci. USA*, **92**, 8031–5.

Buus, C. L., Pourageaud, F., Fazzi, G. E., Janssen, G., Mulvany, M. J., and De Mey, J. G. (2001). Smooth muscle cell changes during flow-related remodeling of rat mesenteric resistance arteries. *Circ. Res.*, **89**, 180–6.

Cesselli, D., Jakoniuk, I., Barlucchi, L. *et al.* (2001). Oxidative stress-mediated cardiac cell death is a major determinant of ventricular dysfunction and failure in dog dilated cardiomyopathy. *Circ. Res.*, **89**, 279–86.

Chan, S., Weissberg, P., and Bennett, M. (1998). Heterogeneity of caspase regulation of human vascular smooth muscle cell apoptosis. *Heart*, **71**, 12 (Abstract).

Chan, S.-W., Hegyi, L., Scott, S., Cary, N., Weissberg, P., and Bennett, M. (2000). Sensitivity to Fas-mediated apoptosis is determined below receptor level in vascular smooth muscle cells. *Circ. Res.*, **86**, 1038–46.

Chen, A. H., Gortler, D. S., Kilaru, S., Araim, O., Frangos, S. G., and Sumpio, B. E. (2001). Cyclic strain activates the pro-survival Akt protein kinase in bovine aortic smooth muscle cells. *Surgery*, **130**, 378–81.

Chen, Z., Woodburn, K. W., Shi, C., Adelman, D. C., Rogers, C., and Simon, D. I. (2001). Photodynamic therapy with motexafin lutetium induces redox-sensitive apoptosis of vascular cells. *Arterioscler. Thromb. Vasc. Biol.*, **21**, 759–64.

Cheng, G., Wessels, A., Gourdie, R. G., and Thompson, R. P. (2002). Spatiotemporal and tissue specific distribution of apoptosis in the developing chick heart. *Dev. Dyn.*, **223**, 119–33.

Cheng, W., Kajstura, J., Nitahara, J. A. *et al.* (1996). Programmed myocyte cell death affects the viable myocardium after infarction in rats. *Exp. Cell. Res.*, **226**, 316–27.

Cheng, W., Li, B., Kajstura, J., Li, P. *et al.* (1995). Stretch-induced programmed myocyte cell death. *J. Clin. Invest.*, **96**, 2247–59.

Cho, A. and Langille, B. L. (1993). Arterial smooth-muscle cell turnover during the postnatal-period in lambs. *FASEB J.*, **7**, A756–A6.

Cho, A., Courtman, D., and Langille, L. (1995). Apoptosis (programmed cell death) in arteries of the neonatal lamb. *Circ. Res.*, **76**, 168–75.

Cho, A., Mitchell, L., Koopmans, D., and Langille, B. L. (1997). Effects of changes in blood flow rate on cell death and cell proliferation in carotid arteries of immature rabbits. *Circ. Res.*, **81**, 328–37.

Condorelli, G., Roncarati, R., Ross, J., Jr. *et al.* (2001). Heart-targeted overexpression of caspase3 in mice increases infarct size and depresses cardiac function. *Proc. Natl. Acad. Sci. USA*, **98**, 9977–82.

Cui, T., Nakagami, H., Iwai, M. *et al.* (2001). Pivotal role of tyrosine phosphatase SHP-1 in AT2 receptor-mediated apoptosis in rat fetal vascular smooth muscle cell. *Cardiovasc. Res.*, **49**, 863–71.

Dowd, N. P., Scully, M., Adderley, S. R., Cunningham, A. J., and Fitzgerald, D. J. (2001). Inhibition of cyclooxygenase-2 aggravates doxorubicin-mediated cardiac injury in vivo. *J. Clin. Invest.*, **108**, 585–90.

Entman, M. and Smith, C. (1994). Postreperfusion inflammation: a model for reaction to injury in cardiovascular disease. *Cardiovasc. Res.*, **28**, 1301.

Erl, W., Hansson, G., de Martin, R., Draude, G., Weber, K., and Weber, C. (1999). Nuclear factor-κβ regulates induction of apoptosis and inhibitor of apoptosis protein-1 expression in vascular smooth muscle cells. *Circ. Res.*, **84**, 668–77.

Fauvel, H., Marchetti, P., Chopin, C., Formstecher, P., and Neviere, R. (2001). Differential effects of caspase inhibitors on endotoxin-induced myocardial dysfunction and heart apoptosis. *Am. J. Physiol. Heart Circ. Physiol.*, **280**, H1608–14.

Felzen, B., Shilkrut, M., Less, H. *et al.* (1998). Fas (CD95/Apo-1)-mediated damage to ventricular myocytes induced by cytotoxic T lymphocytes from perforin-deficient mice: a major role for inositol 1,4,5-trisphosphate. *Circ. Res.*, **82**, 438–50.

Fliss, H. and Gattinger, D. (1996). Apoptosis in ischemic and reperfused rat myocardium. *Circ. Res.*, **79**, 949–56.

Flynn, P., Byrne, C., Baglin, T., Weissberg, P., and Bennett, M. (1997). Thrombin generation by apoptotic vascular smooth muscle cells. *Blood*, **89**, 4373–84.

Fox, J. and Shanley, J. (1996). Antisense inhibition of basic fibroblast growth factor induces apoptosis in vascular smooth muscle cells. *J. Biol. Chem.*, **271**, 12578–84.

Fujio, Y., Nguyen, T., Wencker, D., Kitsis, R. N., and Walsh, K. (2000). Akt promotes survival of cardiomyocytes in vitro and protects against ischemia-reperfusion injury in mouse heart. *Circulation*, **101**, 660–7.

Fukuo, K., Hata, S., Suhara, T. *et al.*(1996). Nitric oxide induces upregulation of fas and apoptosis in vascular smooth muscle. *Hypertension*, **27**, 823–6.

Fukuo, K., Nakahashi, T., Nomura, S. *et al.* (1997). Possible participation of Fas-mediated apoptosis in the mechanism of atherosclerosis. *Gerontology*, **43**, 35–42.

Gao, F., Gong, B., Christopher, T. A., Lopez, B. L., Karasawa, A., and Ma, X. L. (2001). Anti-apoptotic effect of benidipine, a long-lasting vasodilating calcium antagonist, in ischaemic/reperfused myocardial cells. *Br. J. Pharmacol.*, **132**, 869–78.

Geng, Y. and Libby, P. (1995). Evidence for apoptosis in advanced human atheroma: colocalization with interleukin-1β converting enzyme. *Am. J. Path.*, **147**, 251–66.

Geng, Y., Wu, Q., Muszynski, M., Hansson, G., and Libby, P. (1996). Apoptosis of vascular smooth-muscle cells induced by in vitro stimulation with interferon-gamma, tumor necrosis factor-alpha, and interleukin-1-beta. *Arterioscler. Thromb. Vasc. Biol.*, **16**, 19–27.

Geng, Y. J., Henderson, L. E., Levesque, E. B., Muszynski, M., and Libby, P. (1997). Fas is expressed in human atherosclerotic intima and promotes apoptosis of cytokine-primed human vascular smooth muscle cells. *Arterioscler. Thromb. Vasc. Biol.*, **17**, 2200–08.

George, S. J., Angelini, G. D., Capogrossi, M. C., and Baker, A. H. (2001). Wild-type p53 gene transfer inhibits neointima formation in human saphenous vein by modulation of smooth muscle cell migration and induction of apoptosis. *Gene. Ther.*, **8**, 668–76.

Goldenberg, I., Grossman, E., Jacobson, K. A., Shneyvays, V., and Shainberg, A. (2001). Angiotensin II-induced apoptosis in rat cardiomyocyte culture: a possible role of AT1 and AT2 receptors. *J. Hypertens.*, **19**, 1681–9.

Gonzalez, A., Lopez, B., Ravassa, S. *et al.* (2002). Stimulation of cardiac apoptosis in essential hypertension: potential role of angiotensin II. *Hypertension*, **39**, 75–80.

Gordon, D., Reidy, M. A., Benditt, E. P., and Schwartz, S. M. (1990). Cell proliferation in human coronary arteries. *Proc. Natl. Acad. Sci. USA*, **87**, 4600–4.

Gottlieb, R. A., Burleson, K. O., Kloner, R. A., Babior, B. M., and Engler, R. L. (1994). Reperfusion injury induces apoptosis in rabbit cardiomyocytes. *J. Clin. Invest.*, **94**, 1621–8.

Goussev, A., Sharov, V. G., Shimoyama, H. *et al.* (1998). Effects of ACE inhibition on cardiomyocyte apoptosis in dogs with heart failure. *Am. J. Physiol.*, **275**, H626–31.

Granville, D. J., Cassidy, B. A., Ruehlmann, D. O. *et al.* (2001). Mitochondrial release of apoptosis-inducing factor and cytochrome c during smooth muscle cell apoptosis. *Am. J. Pathol.*, **159**, 305–11.

Guerra, S., Leri, A., Wang, X. *et al.* (1999). Myocyte death in the failing human heart is gender dependent. *Circ. Res.*, **85**, 856–66.

Hamet, P., Richard, L., Dam, T. *et al.* (1995). Apoptosis in target organs of hypertension. *Hypertension*, **26**, 642–8.

Han, D., Haudenschild, C., Hong, M., Tinkle, B., Leon, M., and Liau, G. (1995). Evidence for apoptosis in human atherosclerosis and in a rat vascular injury model. *Am. J. Path.*, **147**, 267–77.

Hasdai, D., Sangiorgi, G., Spagnoli, L. G. *et al.* (1999). Coronary artery apoptosis in experimental hypercholesterolemia. *Atherosclerosis*, **142**, 317–25.

Hayden, M. A., Lange, P. A., and Nakayama, D. K. (2001). Nitric oxide and cyclic guanosine monophosphate stimulate apoptosis via activation of the Fas-FasL pathway. *J. Surg. Res.*, **101**, 183–9.

Henderson, E. L., Gang, Y. J., Sukhova, G. K., Whittemore, A. D., Knox, J., and Libby, P. (1999). Death of smooth muscle cells and expression of mediators of apoptosis by T lymphocytes in human abdominal aortic aneurysms. *Circulation*, **99**, 96–104.

Hirata, H., Takahashi, A., Kobayashi, S. *et al.* (1998). Caspases are activated in a branched protease cascade and control distinct downstream processes in Fas-induced apoptosis. *J. Exp. Med.*, **187**, 587–600.

Hirota, H., Chen, J., Betz, U. A. *et al.* (1999). Loss of a gp130 cardiac muscle cell survival pathway is a critical event in the onset of heart failure during biomechanical stress. *Cell.*, **97**, 189–98.

Hoffmann, G., Kenn, S., Wirleitner, B. *et al.* (1998). Neopterin induces nitric oxide-dependent apoptosis in rat vascular smooth muscle cells. *Immunobiology*, **199**, 63–73.

Hsieh, C. C., Yen, M. H., Yen, C. H., and Lau, Y. T. (2001). Oxidized low density lipoprotein induces apoptosis via generation of reactive oxygen species in vascular smooth muscle cells. *Cardiovasc. Res.*, **49**, 135–45.

Igase, M., Okura, T., Kitami, Y., and Hiwada, K. (1999). Apoptosis and Bcl-xs in the intimal thickening of balloon-injured carotid arteries. *Clin. Sci.*, **96**, 605–12.

Ikari, Y., Mulvihill, E., and Schwartz, S. M. (2001). α1-Proteinase inhibitor, alpha 1-antichymotrypsin, and alpha 2-macroglobulin are the antiapoptotic factors of vascular smooth muscle cells. *J. Biol. Chem.*, **276**, 11798–803.

Imanishi, T., Hano, T., Nishio, I., Han, D. K., and Schwartz, S. M. (2001). Apoptosis of vascular smooth muscle cells is induced by Fas ligand derived from endothelial cells. *Jpn. Circ. J.*, **65**, 556–60.

Isner, J., Kearney, M., Bortman, S., and Passeri, J. (1995). Apoptosis in human atherosclerosis and restenosis. *Circulation*, **91**, 2703–11.

Itoh, G., Tamura, J., Suzuki, M. *et al.* (1995). DNA fragmentation of human infarcted myocardial cells demonstrated by the nick end labeling method and DNA agarose gel electrophoresis. *Am. J. Pathol.*, **146**, 1325–31.

Iwashina, M., Shichiri, M., Marumo, F., and Hirata, Y. (1998). Transfection of inducible nitric oxide synthase gene causes apoptosis in vascular smooth muscle cells. *Circulation*, **98**, 1212–18.

Jacob, T., Ascher, E., Hingorani, A., Gunduz, Y., and Kallakuri, S. (2001). Initial steps in the unifying theory of the pathogenesis of artery aneurysms. *J. Surg. Res.*, **101**, 37–43.

James, T. N. (1994). Normal and abnormal consequences of apoptosis in the human heart. From postnatal morphogenesis to paroxysmal arrhythmias. *Circulation*, **90**, 556–73.

James, T. N., St. Martin, E., Willis, P. W., 3rd, and Lohr, T. O. (1996). Apoptosis as a possible cause of gradual development of complete heart block and fatal arrhythmias associated with absence of the AV node, sinus node, and internodal pathways. *Circulation*, **93**, 1424–38.

James, T. N., Terasaki, F., Pavlovich, E. R., and Vikhert, A. M. (1993). Apoptosis and pleomorphic micromitochondriosis in the sinus nodes surgically excised from five patients with the long QT syndrome. *J. Lab. Clin. Med.*, **122**, 309–23.

Jeroudi, M. O., Hartley, C. J., and Bolli, R. (1994). Myocardial reperfusion injury: role of oxygen radicals and potential therapy with antioxidants. *Am. J. Cardiol.*, **73**, 2B–7B.

Jonassen, A. K., Brar, B. K., Mjos, O. D., Sack, M. N., Latchman, D. S., and Yellon, D. M. (2000). Insulin administered at reoxygenation exerts a cardioprotective effect in myocytes by a possible anti-apoptotic mechanism. *J. Mol. Cell. Cardiol.*, **32**, 757–64.

Kajstura, J., Cheng, W., Reiss, K. *et al.* (1996). Apoptotic and necrotic myocyte cell deaths are independent contributing variables of infarct size in rats. *Lab. Invest.*, **74**, 86–107.

Kajstura, J., Cigola, E., Malhotra, A. *et al.* (1997). Angiotensin II induces apoptosis of adult ventricular myocytes in vitro. *J. Mol. Cell. Cardiol.*, **29**, 859–70.

Kang, P. M., Haunstetter, A., Aoki, H., Usheva, A., and Izumo, S. (2000). Morphological and molecular characterization of adult cardiomyocyte apoptosis during hypoxia and reoxygenation. *Circ. Res.*, **87**, 118–25.

Karwatowska-Prokopczuk, E., Nordberg, J. A., Li, H. L., Engler, R. L., and Gottlieb, R. A. (1998). Effect of vacuolar proton ATPase on pHi, Ca2+, and apoptosis in neonatal cardiomyocytes during metabolic inhibition/recovery. *Circ. Res.*, **82**, 1139–44.

Kavurma, M. M., Santiago, F. S., Bonfoco, E., and Khachigian, L. M. (2001). Sp1 phosphorylation regulates apoptosis via extracellular FasL-Fas engagement. *J. Biol. Chem.*, **276**, 4964–71.

Kim, K. (2001). Proteasome inhibitors sensitize human vascular smooth muscle cells to Fas (CD95)-mediated death. *Biochem. Biophys. Res. Commun.*, **281**, 305–10.

Kocher, A. A., Schuster, M. D., Szabolcs, M. J. *et al.* (2001). Neovascularization of ischemic myocardium by human bone-marrow-derived angioblasts prevents cardiomyocyte apoptosis, reduces remodeling and improves cardiac function. *Nat. Med.*, **7**, 430–6.

Kockx, M. M. (1998). Apoptosis in the atherosclerotic plaque – quantitative and qualitative aspects. *Arterioscler. Thromb. Vasc. Biol.*, **18**, 1519–22.

Kockx, M. M., DeMeyer, G., Muhring, J., Jacob, W., Bult, H., and Herman, A. G. (1998a). Apoptosis and related proteins in different stages of human atherosclerotic plaques. *Circulation*, **97**, 2307–15.

Kockx, M. M., DeMeyer, G. Y., Buyssens, N., Knaapen, M. W. M., Bult, H., and Herman, A. G. (1998b). Cell composition, replication, and apoptosis in atherosclerotic plaques after 6 months of cholesterol withdrawal. *Circ. Res.*, **83**, 378–87.

Konstadoulakis, M. M., Kymionis, G. D., Karagiani, M. *et al.* (1998). Evidence of apoptosis in human carotid atheroma. *J. Vasc. Surg.*, **27**, 733–9.

Krajewska, M., Wang, H. G., Krajewski, S. *et al.* (1997). Immunohistochemical analysis of in vivo patterns of expression of CPP32 (caspase-3), a cell death protease. *Cancer Res.*, **57**, 1605–13.

Krstic, R. and Pexieder, T. (1973). Ultrastructure of cell death in bulbar cushions of chick embryo heart. *Z. Anat. Entwicklungsgesch.*, **140**, 337–50.

Kubota, T., McTiernan, C., Frye, C. *et al.* (1997). Dilated cardiomyopathy in transgenic mice with cardiac-specific overexpression of tumor necrosis factor-α. *Circ. Res.*, **81**, 627–35.

Kumar, A. and Lindner, V. (1997). Remodeling with neointima formation in the mouse carotid artery after cessation of blood flow. *Arterioscler. Thromb. Vasc. Biol.*, **17**, 2238–44.

Laugwitz, K. L., Moretti, A., Weig, H. J. *et al.* (2001). Blocking caspase-activated apoptosis improves contractility in failing myocardium. *Hum. Gene. Ther.*, **12**, 2051–63.

Lee, T. and Chau, L. (2001). Fas/Fas ligand-mediated death pathway is involved in oxLDL-induced apoptosis in vascular smooth muscle cells. *Am. J. Physiol. Cell. Physiol.*, **280**, C709–18.

Leri, A., Fiordaliso, F., Setoguchi, M. *et al.* (2000). Inhibition of p53 function prevents renin–angiotensin system activation and stretch-mediated myocyte apoptosis. *Am. J. Pathol.*, **157**, 843–57.

Leri, A., Liu, Y., Claudio, P. P. *et al.* (1999). Insulin-like growth factor-1 induces Mdm2 and down-regulates p53, attenuating the myocyte renin–angiotensin system and stretch-mediated apoptosis. *Am. J. Pathol.*, **154**, 567–80.

Leskinen, M., Wang, Y., Leszczynski, D., Lindstedt, K. A., and Kovanen, P. T. (2001). Mast cell chymase induces apoptosis of vascular smooth muscle cells. *Arterioscler. Thromb. Vasc. Biol.*, **21**, 516–22.

Li, H. L., Zhu, H., Xu, C. J., and Yuan, J. Y. (1998). Cleavage of BID by caspase 8 mediates the mitochondrial damage in the Fas pathway of apoptosis. *Cell*, **94**, 491–501.

Li, P. F., Dietz, R., and vonHarsdorf, R. (1997). Differential effect of hydrogen peroxide and superoxide anion on apoptosis and proliferation of vascular smooth muscle cells. *Circulation*, **96**, 3602–9.

Li, Q., Li, B., Wang, X. *et al.* (1997). Overexpression of insulin-like growth factor-1 in mice protects from myocyte death after infarction, attenuating ventricular dilation, wall stress, and cardiac hypertrophy. *J. Clin. Invest.*, **100**, 1991–9.

Li, Z., Bing, O. H., Long, X., Robinson, K. G., and Lakatta, E. G. (1997). Increased cardiomyocyte apoptosis during the transition to heart failure in the spontaneously hypertensive rat. *Am. J. Physiol.*, **272**, H2313–19.

Litt, M. R., Jeremy, R. W., Weisman, H. F., Winkelstein, J. A., and Becker, L. C. (1989). Neutrophil depletion limited to reperfusion reduced myocardial infarct size after 90 minutes of ischemia: evidence for neutrophil-mediated reperfusion injury. *Circulation*, **80**, 1816–27.

Liu, H., McPherson, B. C., and Yao, Z. (2001). Preconditioning attenuates apoptosis and necrosis: role of protein kinase C-epsilon and -delta isoforms. *Am. J. Physiol. Heart Circ. Physiol.*, **281**, H404–10.

Liu, Y., Cigola, E., Cheng, W. *et al.* (1995). Myocyte nuclear mitotic division and programmed myocyte cell death characterize the cardiac myopathy induced by rapid ventricular pacing in dogs. *Lab. Invest.*, **73**, 771–87.

Long, X., Crow, M. T., Sollott, S. J. *et al.* (1998). Enhanced expression of p53 and apoptosis induced by blockade of the vacuolar proton ATPase in cardiomyocytes. *J. Clin. Invest.*, **101**, 1453–61.

Long, X. L., Boluyt, M. O., Hipolito, M. D. *et al.* (1997). p53 and the hypoxia-induced apoptosis of cultured neonatal rat cardiac myocytes. *J. Clin. Invest.*, **99**, 2635–43.

LopezCandales, A., Holmes, D. R., Liao, S. X., Scott, M. J., Wickline, S. A., and Thompson, R. W. (1997). Decreased vascular smooth muscle cell density in medial degeneration of human abdominal aortic aneurysms. *Am. J. Pathol.*, **150**, 993–1007.

Luo, Z., Garron, T., Palasis, M. *et al.* (2001). Enhancement of fas ligand-induced inhibition of neointimal formation in rabbit femoral and iliac arteries by coexpression of p35. *Hum. Gene. Ther.*, **12**, 2191–202.

Malik, N., Francis, S. E., Holt, C. M. *et al.* (1998). Apoptosis and cell proliferation after porcine coronary angioplasty. *Circulation*, **98**, 1657–65.

Mallat, Z., Benamer, H., Hugel, B., Steg, P., Freyssinet, J., and Tedgui, A. (1998). Elevated plasma levels of shed membrane microparticles in patients with acute coronary syndromes. *Circulation*, **98**, I-172 (Abstract).

Mallat, Z., Fornes, P., Costagliola, R. *et al.* (2001). Age and gender effects on cardiomyocyte apoptosis in the normal human heart. *J. Gerontol. A. Biol. Sci. Med. Sci.*, **56**, M719–23.

Mallat, Z., Hugel, B., Ohan, J., Leseche, G., Freyssinet, J. M., and Tedgui, A. (1999). Shed membrane microparticles with procoagulant potential in human atherosclerotic plaques – a role for apoptosis in plaque thrombogenicity. *Circulation*, **99**, 348–53.

Mallat, Z., Tedgui, A., Fontaliran, F., Frank, R., Durigon, M., and Fontaine, G. (1996). Evidence of apoptosis in arrhythmogenic right ventricular dysplasia. *New. Eng. J. Med.*, **335**, 1190–6.

Mano, T., Luo, Z., Suhara, T., Smith, R. C., Esser, S., and Walsh, K. (2000). Expression of wild-type and noncleavable Fas ligand by tetracycline-regulated adenoviral vectors to limit intimal hyperplasia in vascular lesions. *Hum. Gene. Ther.*, **11**, 1625–35.

Matsui, T., Tao, J., del Monte, F. *et al.* (2001). Akt activation preserves cardiac function and prevents injury after transient cardiac ischemia in vivo. *Circulation*, **104**, 330–5.

Mehrhof, F. B., Muller, F. U., Bergmann, M. W. *et al.* (2001). In cardiomyocyte hypoxia, insulin-like growth factor-I-induced antiapoptotic signaling requires phosphatidylinositol-3-OH-kinase-dependent and mitogen-activated protein kinase-dependent activation of the transcription factor cAMP response element-binding protein. *Circulation*, **104**, 2088–94.

Meiners, S., Laule, M., Rother, W. *et al.* (2002). Ubiquitin-proteasome pathway as a new target for the prevention of restenosis. *Circulation*, **105**, 483–9.

Miao, W., Luo, Z., Kitsis, R. N., and Walsh, K. (2000). Intracoronary, adenovirus-mediated Akt gene transfer in heart limits infarct size following ischemia–reperfusion injury in vivo. *J. Mol. Cell. Cardiol.*, **32**, 2397–402.

Miyamoto, T., Leconte, I., Swain, J. L. *et al.* (1998). Autocrine FGF signaling is required for vascular smooth muscle cell survival in vitro. *J. Cell. Physiol.*, **177**, 58–67.

Mocanu, M. M., Baxter, G. F., and Yellon, D. M. (2000). Caspase inhibition and limitation of myocardial infarct size: protection against lethal reperfusion injury. *Br. J. Pharmacol.*, **130**, 197–200.

Moudgil, R., Menon, V., Xu, Y., Musat-Marcu, S., Kumar, D., and Jugdutt, B. I. (2001). Post-ischemic apoptosis and functional recovery after angiotensin II type 1 receptor blockade in isolated working rat hearts. *J. Hypertens.*, **19**, 1121–9.

Nakamura, T., Ueda, Y., Juan, Y., Katsuda, S., Takahashi, H., and Koh, E. (2000). Fas-mediated apoptosis in adriamycin-induced cardiomyopathy in rats: in vivo study. *Circulation*, **102**, 572–8.

Narula, J., Haider, N., Virmani, R. *et al.* (1996). Apoptosis in myocytes in end-stage heart failure. *N. Engl. J. Med.*, **335**, 1182–9.

Nayler, W., Panagiotopoulos, S., Elz, J., and Daly, M. (1988). Calcium mediated damage during post ischemic reperfusion. *J. Molec. Cell. Cardiol.*, **20**, 41.

Neschis, D. G., Safford, S. D., Hanna, A. K., Fox, J. C., and Golden, M. A. (1998). Antisense basic fibroblast growth factor gene transfer reduces early intimal thickening in a rabbit femoral artery balloon injury model. *J. Vasc. Surg.*, **27**, 126–34.

Niemann-Jonsson, A., Ares, M. P., Yan, Z. Q. *et al.* (2001). Increased rate of apoptosis in intimal arterial smooth muscle cells through endogenous activation of TNF receptors. *Arterioscler. Thromb. Vasc. Biol.*, **21**, 1909–14.

Nishio, E. and Watanabe, Y. (1997). NO induced apoptosis accompanying the change of onco-protein expression and the activation of CPP32 protease. *Life Sci.*, **62**, 239–45.

Okura, Y., Brink, M., Zahid, A. A., Anwar, A., and Delafontaine, P. (2001). Decreased expression of insulin-like growth factor-1 and apoptosis of vascular smooth muscle cells in human atherosclerotic plaque. *J. Mol. Cell. Cardiol.*, **33**, 1777–89.

Olivetti, G., Abbi, R., Quaini, F. *et al.* (1997). Apoptosis in the failing human heart. *N. Engl. J. Med.*, **336**, 1131–41.

Olivetti, G., Giordano, G., Corradi, D. *et al.* (1995). Gender differences and aging: effects on the human heart. *J. Am. Coll. Cardiol.*, **26**, 1068–79.

Olivetti, G., Quaini, F., Sala, R. *et al.* (1996). Acute myocardial infarction in humans is associated with activation of programmed myocyte cell death in the surviving portion of the heart. *J. Mol. Cell. Cardiol.*, **28**, 2005–16.

Orlandi, A., Francesconi, A., Cocchia, D., Corsini, A., and Spagnoli, L. G. (2001). Phenotypic heterogeneity influences apoptotic susceptibility to retinoic acid and cis-platinum of rat arterial smooth muscle cells in vitro: implications for the evolution of experimental intimal thickening. *Arterioscler. Thromb. Vasc. Biol.*, **21**, 1118–23.

Palojoki, E., Saraste, A., Eriksson, A. *et al.* (2001). Cardiomyocyte apoptosis and ventricular remodeling after myocardial infarction in rats. *Am. J. Physiol. Heart Circ. Physiol.*, **280**, H2726–31.

Patel, V., Zhang, Q.-J., Soos, M., Siddle, K., Weissberg, P. L., and Bennett, M. R. (2001). Defect in insulin-like growth factor 1 signaling underlies increased apoptosis of human atherosclerotic plaque-derived vascular smooth muscle cells. *Circ. Res.*, **88**, 895–902.

Perlman, H., Maillard, L., Krasinski, K., and Walsh, K. (1997). Evidence for the rapid onset of apoptosis in medial smooth muscle cells after balloon injury. *Circulation*, **95**, 981–7.

Perlman, H., Sata, M., Krasinski, K., Dorai, T., Buttyan, R., and Walsh, K. (2000). Adenovirus-encoded hammerhead ribozyme to Bcl-2 inhibits neointimal hyperplasia and induces vascular smooth muscle cell apoptosis. *Cardiovasc. Res.*, **45**, 570–8.

Pexieder, T. (1975). Cell death in the morphogenesis and teratogenesis of the heart. *Adv. Anat. Embryol. Cell. Biol.*, **51**, 3–99.

Pollman, M. J., Hall, J. L., and Gibbons, G. H. (1999). Determinants of vascular smooth muscle cell apoptosis after balloon angioplasty injury – influence of redox state and cell phenotype. *Circ. Res.*, **84**, 113–21.

Pollman, M. J., Hall, J. L., Mann, M. J., Zhang, L. N., and Gibbons, G. H. (1998). Inhibition of neointimal cell bcl-x expression induces apoptosis and regression of vascular disease. *Nat. Med.*, **4**, 222–7.

Pollman, M. J., Yamada, T., Horiuchi, M., and Gibbons, G. H. (1996). Vasoactive substances regulate vascular smooth-muscle cell apoptosis – countervailing influences of nitric-oxide and angiotensin-II. *Circ. Res.*, **79**, 748–56.

Proudfrot, D., Skepper, J. N., Hegyi, C. *et al.* (2000). Apoptosis regulates human vascular calcification in vitro: evidence for initiation of vascular calcification by apoptotic bodies. *Circ. Res.* **87**, 1055–62.

Ruetten, H., Badorff, C., Ihling, C., Zeiher, A. M., and Dimmeler, S. (2001). Inhibition of caspase-3 improves contractile recovery of stunned myocardium, independent of apoptosis-inhibitory effects. *J. Am. Coll. Cardiol.*, **38**, 2063–70.

Saraste, A., Pulkki, K., Kallajoki, M., Henriksen, K., Parvinen, M., and Voipio-Pulkki, L. (1997). Apoptosis in human acute myocardial infarction. *Circulation*, **95**, 320–3.

Sata, M., Perlman, H. R., Muruve, D. A. *et al.* (1998). Fas ligand gene transfer to the vessel wall inhibits neointima formation and overrides the adenovirus-mediated T cell response. *Proc. Natl. Acad. Sci. USA*, **95**, 1213–17.

Scaffidi, C., Fulda, S., Srinivasan, A. *et al.* (1998). Two CD95 (APO-1/Fas) signaling pathways. *EMBO J.*, **17**, 1675–87.

Schaub, F. J., Han, D. K., Conrad Liles, W. *et al.* (2000). Fas/FADD-mediated activation of a specific program of inflammatory gene expression in vascular smooth muscle cells. *Nat. Med.*, **6**, 790–6.

Scheidegger, K. J., James, R. W., and Delafontaine, P. (2000). Differential effects of low density lipoproteins on IGF-1 and IGF-1R expression in vascular smooth muscle cells. *J. Biol. Chem.*, **275**, 26864–9.

Schneider, D. B., Vassalli, G., Wen, S. *et al.* (2000). Expression of Fas ligand in arteries of hypercholesterolemic rabbits accelerates atherosclerotic lesion formation. *Arterioscler. Thromb. Vasc. Biol.*, **20**, 298–308.

Schwartzbauer, G. and Robbins, J. (2001). The tumor suppressor gene PTEN can regulate cardiac hypertrophy and survival. *J. Biol. Chem.*, **276**, 35786–93.

Seshiah, P. N., Kereiakes, D. J., Vasudevan, S. S. *et al.* (2002). Activated monocytes induce smooth muscle cell death: role of macrophage colony-stimulating factor and cell contact. *Circulation*, **105**, 174–80.

Sharov, V. G., Sabbah, H. N., Shimoyama, H., Goussev, A. V., Lesch, M., and Goldstein, S. (1996). Evidence of cardiocyte apoptosis in myocardium of dogs with chronic heart failure. *Am. J. Pathol.*, **148**, 141–9.

Shaulian, E., Resnitzky, D., Shifman, O. *et al.* (1997). Induction of Mdm2 and enhancement of cell survival by bFGF. *Oncogene*, **15**, 2717–25.

Sheng, Z., Knowlton, K., Chen, J., Hoshijima, M., Brown, J., and Chien, K. (1997). Cardiotrophin 1 (CT-1) inhibition of cardiac myocyte apoptosis via a mitogen-activated protein kinase-dependent pathway: divergence from downstream CT-1 signals for myocardial cell hypertrophy. *J. Biol. Chem.*, **272**, 5783–91.

Slomp, J., GittenbergerdeGroot, A. C., Glukhova, M. A. *et al.*(1997). Differentiation, dedifferentiation, and apoptosis of smooth muscle cells during the development of the human ductus arteriosus. *Arterioscler. Thromb. Vasc. Biol.*, **17**, 1003–9.

Smith, E. F. D., Egan, J. W., Bugelski, P. J., Hillegass, L. M., Hill, D. E., and Griswold, D. E. (1988). Temporal relation between neutrophil accumulation and myocardial reperfusion injury. *Am. J. Physiol.*, **255**, H1060–8.

Stephanou, A., Scarabelli, T. M., Brar, B. K. *et al.* (2001). Induction of apoptosis and Fas receptor/Fas ligand expression by ischemia/reperfusion in cardiac myocytes requires serine 727 of the STAT-1 transcription factor but not tyrosine 701. *J. Biol. Chem.*, **276**, 28340–7.

Sugino, H., Ozono, R., Kurisu, S. *et al.* (2001). Apoptosis is not increased in myocardium over-expressing type 2 angiotensin II receptor in transgenic mice. *Hypertension*, **37**, 1394–8.

Suhara, T., Mano, T., Oliveira, B. E., and Walsh, K. (2001). Phosphatidylinositol 3-kinase/Akt signaling controls endothelial cell sensitivity to Fas-mediated apoptosis via regulation of FLICE-inhibitory protein (FLIP). *Circ. Res.*, **89**, 13–19.

Susin, S. A., Lorenzo, H., Zamzami, N. *et al.* (1999). Molecular characterization of mitochondrial apoptosis-inducing factor. *Nature*, **397**, 441–6.

Takeda, K., Yu, Z. X., Nishikawa, T. *et al.* (1996). Apoptosis and DNA fragmentation in the bulbus cordis of the developing rat heart. *J. Mol. Cell. Cardiol.*, **28**, 209–15.

Takemura, G., Ohno, M., Hayakawa, Y. *et al.* (1998). Role of apoptosis in the disappearance of infiltrated and proliferated interstitial cells after myocardial infarction. *Circ. Res.*, **82**, 1130–8.

Tanaka, M., Ito, H., Adachi, S. *et al.* (1994). Hypoxia induces apoptosis with enhanced expression of Fas antigen messenger RNA in cultured neonatal rat cardiomyocytes. *Circ. Res.*, **75**, 426–33.

Teiger, E., Than, V. D., Richard, L. *et al.* (1996). Apoptosis in pressure overload-induced heart hypertrophy in the rat. *J. Clin. Invest.*, **97**, 2891–7.

Thompson, R. W., Liao, S. X., and Curci, J. A. (1997). Vascular smooth muscle cell apoptosis in abdominal aortic aneurysms. *Coron. Artery Dis.*, **8**, 623–31.

Torre-Amione, G., Kapadia, S., Lee, J., Bies, R., Lebovitz, R., and Mann, D. (1995). Expression and functional significance of tumor necrosis factor receptors in human myocardium. *Circulation*, **92**, 1487–93.

Umansky, S., Cuenco, G., Khutzlan, S. *et al.* (1995). Post ischemic apoptotic death of rat neonatal cardiomyocytes. *Cell Death Differ.*, **2**, 235.

Waksman, R., Rodriguez, J. C., Robinson, K. A. *et al.* (1997). Effect of intravascular irradiation on cell proliferation, apoptosis, and vascular remodeling after balloon overstretch injury of porcine coronary arteries. *Circulation*, **96**, 1944–52.

Wang, H. and Keiser, J. A. (1998). Molecular characterization of rabbit CPP32 and its function in vascular smooth muscle cell apoptosis. *Am. J. Physiol.*, **43**, H1132–H40.

Wang, L., Ma, W. Q., Markovich, R., Chen, J. W., and Wang, P. H. (1998). Regulation of cardiomyocyte apoptotic signaling by insulin-like growth factor I. *Circ. Res.*, **83**, 516–22.

Wang, Y., Huang, S., Sah, V. P. *et al.* (1998). Cardiac muscle cell hypertrophy and apoptosis induced by distinct members of the p38 mitogen-activated protein kinase family. *J. Biol. Chem.*, **273**, 2161–8.

Watanabe, M., Jafri, A., and Fisher, S. A. (2001). Apoptosis is required for the proper formation of the ventriculo-arterial connections. *Dev. Biol.*, **240**, 274–88.

Yamashita, K., Kajstura, J., Discher, D. J. *et al.* (2001). Reperfusion-activated Akt kinase prevents apoptosis in transgenic mouse hearts overexpressing insulin-like growth factor-1. *Circ. Res.*, **88**, 609–14.

Yaoita, H., Ogawa, K., Maehara, K., and Maruyama, Y. (1998). Attenuation of ischemia/ reperfusion injury in rats by a caspase inhibitor. *Circulation*, **97**, 276–81.

Yue, T. L., Ma, X. L., Wang, X. *et al.* (1998). Possible involvement of stress-activated protein kinase signaling pathway and Fas receptor expression in prevention of ischemia/reperfusion-induced cardiomyocyte apoptosis by carvedilol. *Circ. Res.*, **82**, 166–74.

Yue, T. L., Wang, C., Gu, J. L. *et al.* (2000). Inhibition of extracellular signal-regulated kinase enhances ischemia/reoxygenation-induced apoptosis in cultured cardiac myocytes and exaggerates reperfusion injury in isolated perfused heart. *Circ. Res.*, **86**, 692–9.

Zechner, D., Craig, R., Hanford, D. S., McDonough, P. M., Sabbadini, R. A., and Glembotski, C. C. (1998). MKK6 activates myocardial cell NF-kappaB and inhibits apoptosis in a p38 mitogen-activated protein kinase-dependent manner. *J. Biol. Chem.*, **273**, 8232–9.

Zhao, Y. Y., Sawyer, D. R., Baliga, R. R. *et al.* (1998). Neuregulins promote survival and growth of cardiac myocytes. Persistence of ErbB2 and ErbB4 expression in neonatal and adult ventricular myocytes. *J. Biol. Chem.*, **273**, 10261–9.

Zhao, Z. H., Francis, C. E., Welch, G., Loscalzo, J., and Ravid, K. (1997). Reduced glutathione prevents nitric oxide-induced apoptosis in vascular smooth muscle cells. *Biochim. Biophys. Acta.*, **1359**, 143–52.

Zhu, W. Z., Zheng, M., Koch, W. J., Lefkowitz, R. J., Kobilka, B. K., and Xiao, R. P. (2001). Dual modulation of cell survival and cell death by beta(2)-adrenergic signaling in adult mouse cardiac myocytes. *Proc. Natl. Acad. Sci. USA*, **98**, 1607–12.

Zweier, J., Flaherty, J., and Weisfeldt, M. (1987). Direct measurement of radical generation following reperfusion of ischemic myocardium. *Proc. Natl. Acad. Sci. USA*, **84**, 1404.

Cytotoxic lymphocytes, apoptosis, and autoimmunity

Pere Santamaria[1] and R. Chris Bleackley[2]

[1]Department of Microbiology and Infectious Diseases and Julia McFarlane Diabetes Research Centre, The University of Calgary, Calgary, Canada
[2]Department of Biochemistry, University of Alberta, Edmonton, Alberta, Canada

6.1 Introduction

Cytotoxic lymphocytes – you cannot live without them, but sometimes you have trouble living with them. The Jeckyl and Hyde character of these cells relates to their ability to induce death in target cells. On the one hand, they can recognize and destroy pathogenic cells, such as those infected with viruses, but, on the other hand, they can also mistakenly turn their attention to normal cells, resulting in autoimmunity. Lymphocytes can kill, either through direct cell contact or via the secretion of cytokines, and, in the case of B lymphocytes, antibodies. These secreted proteins are important in killing and autoimmune disorders. They can act directly but often function via the activation and/or recruitment of lytic and inflammatory effector cells. Most of the discussion in this chapter, however, will focus on the pathways that involve close apposition of effector and target cells. A knowledge of the molecular killing mechanisms used by cytotoxic lymphocytes may allow us to develop novel strategies to either curb or amplify target cell destruction. The current models for apoptosis induced by cytotoxic T lymphocytes (CTLs) and natural killer (NK) cells will be outlined, and some insights into which pathways are used in autoimmune disorders will be provided.

Over the last few years, it has become clear that CTLs and NK cells can kill via two distinct pathways. The first to be described involves the exocytosis of lytic proteins from dense granules in the cytoplasm of the effector cells toward the targets (Henkart, 1985). One of the characteristic features of this pathway is a dependence on calcium; however, it was later realized that some cells could kill independently of calcium (Ostergaard *et al.*, 1987) and in the absence of granules (Helgason *et al.*, 1992). We know now that this second route to death involves the expression of Fas

Apoptosis in Health and Disease: Clinical and Therapeutic Aspects, ed. Martin Holcik, Alex E. MacKenzie, Robert G. Korneluk, and Eric C. LaCasse. Published by Cambridge University Press. © Cambridge University Press 2004.

on the surface of the target and interaction with its ligand on the effector (Rouvier *et al.*, 1993). Many of the molecular details of the Fas-death pathway are now known as it has been one of the most studied apoptotic mechanisms (see Krammer, 2000).

6.2   Cell-mediated cytotoxicity: kissing the offender

6.2.1   Death via Fas–Fas ligand interaction

The binding of Fas ligand to Fas brings together Fas molecules in the membrane of the target cell. Once Fas is clustered, a number of proteins are recruited to the cytoplasmic tail to form a death inducing signaling complex (DISC) (Muzio *et al.*, 1996). Within the complex, caspase-8 is activated after recruitment to the complex via an adapter protein FADD (Boldin *et al.*, 1996; Chinnaiyan *et al.*, 1996). The activated caspase then initiates the apoptotic program by the cleavage of substrates such as caspase-3. In some cell types (Scaffidi *et al.*, 1998), caspase-8 does not act directly on -3 but, rather, goes via the mitochondria. This is achieved through the cleavage and translocation of bid which, in association with other pro-apoptotic proteins, translocates to the mitochondrial membrane (Li *et al.*, 1998; Luo *et al.*, 1998). Through a mechanism that is not yet understood, cytochrome c is then released from the mitochondria into the cytoplasm. As a result of this translocation, the apoptosome is formed where caspase-9 is activated and ultimately this protease acts on caspase-3. In addition, SMAC/Diablo (Du *et al.*, 2000; Verhagen *et al.*, 2000) and Omi (Hegde *et al.*, 2002) are released from mitochondria. Both of these proteins can relieve inhibition of caspase-3 and -9 by inhibitors of apoptosis (IAPs) and thus release the safety break on apoptosis. The activation of effector caspases such as caspase-3 has profound implications for the cell. A wide variety of substrates are then proteolyzed, resulting in the various morphologic and biochemical features of apoptosis. One of the hallmarks is DNA fragmentation and this is initiated by the cleavage and inactivation of an inhibitor (ICAD) of the nuclease (CAD, caspase-activated DNase) that degrades DNA (Enari *et al.*, 1998).

6.2.2   Granule exocytosis pathway

CTLs and NK cells contain electron-dense granules in their cytoplasms. Upon interaction with a target cell, these organelles move to the point of contact between the two cells. Thus, it is believed that killing is achieved through the directed exocytosis of lytic effectors into the synapse. Perforin and granzymes are contained within the granules and are believed to be key components in the killing pathway. Initially, it was believed that perforin was the sole effector until it was realized that CTL-mediated death was accompanied by DNA fragmentation, and perforin only induced membrane damage. The fragmenting activity was isolated and identified as a serine protease from the granules, granzyme B (Shi *et al.*, 1992). This is an interesting enzyme with an unusual substrate specificity for aspartate residues. The importance of perforin and granzyme B in the killing process has been underscored by results from gene deletion studies. CTLs generated from perforin deficient mice display defects in membrane damage and apoptosis (Kagi *et al.*, 1994; Lowin *et al.*, 1994). Mice deficient in granzyme B are impaired in CTL-induced DNA fragmentation in target cells (Heusel *et al.*, 1994). In addition, numerous studies have demonstrated that perforin and granzyme B in combination can mediate efficient cell death in vitro.

6.2.3   Delivery of granzyme B

In the initial model of granule exocytosis, perforin was believed to be the single lytic effector protein. The model evolved to account for a role for granzyme in the initiation of apoptosis where perforin created a channel through which the protease passed. Recently, a number of investigators have shown that granzyme B can enter a cell independently of perforin (Froelich *et al.*, 1996; Shi *et al.*, 1997), but the latter is required for death (Pinkoski *et al.*, 1998). In addition, it has been shown that perforin channels, if they exist, are not large enough to allow the passage of a macromolecule as large as granzyme (Browne *et al.*, 1999). These studies led to the hypothesis that a receptor for granzyme B existed on the surface of susceptible cells. A cell surface-binding protein for granzyme B has now been identified as the mannose-6-phosphate receptor (Motyka *et al.*, 2000). This protein binds and internalizes the protease via receptor-mediated endocytosis. Granzyme B builds up in an endosomal compartment until released into the cytoplasm by perforin. Interference with binding and uptake of granzyme B by the receptor may represent a novel strategy for immunosuppression.

6.2.4   Granule-mediated death

Perforin and granzymes act together to bring about apoptotic cell death. In the case of granzyme B, this is in part due to the cleavage and activation of caspase-3 by the aspase activity of the protease (Darmon *et al.*, 1995). As described for Fas, the activated caspase-3 proceeds to cleave a series of key substrates that bring about the membrane, nuclear, and genomic events involved in apoptosis. Other members of the granzyme family also act on substrates within the target cell to contribute to target cell demise. Notably, granzyme A induces a novel form of DNA damage through the cleavage of substrates such as histones and nuclear lamins (Zhang *et al.*, 2001a, b). This is likely to be why granzyme B knock-out CTLs are defective in early DNA fragmentation while granzyme A,B double knock-outs induce no DNA damage (Simon *et al.*, 1997; Shresta *et al.*, 1999). There are other granule proteins that contribute to cell death more indirectly, but another molecule with lytic activity, granulysin, has recently been described (Krensky, 2000). This protein is likely to be most important in protection from microbial pathogens, but it can also induce apoptosis in some target cells (Kaspar *et al.*, 2001).

In addition to its direct action on caspase-3, granzyme B has also been shown to cleave and activate the pro-apoptotic, BH3 only, protein bid (Barry *et al.*, 2000; Pinkoski *et al.*, 2001). Even though the cleavage site is different from that recognized by caspase-8 in the Fas pathway, the granzyme truncated (gt) bid is activated. In association with other members of the family, gt bid translocates to the mitochondria and induces cytochrome c release (Heibein *et al.*, 2000; Sutton *et al.*, 2000). This can contribute to apoptosis by the activation of caspase-9 and caspase-3 via the apoptosome as was described earlier. In addition, the loss of cytochrome c impairs ATP generation and this will lead to necrotic death. Thus, if the apoptotic pathway is blocked by, say, a virus-encoded protein, the target cell will still be destroyed. Apoptotic, or non-inflammatory, death is preferred but when the survival of the whole organism is at stake necrotic death is still effective.

Although bid is required for cytochrome c release, loss of the mitochondrial membrane potential still occurs in cells under attack by CTLs or treated with granzyme B and perforin. This has been reported to be due to a direct effect of the protease on mitochondria (Alimonti *et al.*, 2001), but others have suggested that an intermediate cytosolic protein is involved (Thomas *et al.*, 2001). Either way, this again represents a useful back-up system that would lead to necrotic death and also the release of other death effector proteins from the mitochondria such as apoptosis-inducing factor (AIF) and Endo G. Finally, it has also been reported that granzyme B can cleave iCAD directly and thus short-circuit the need for caspase activation (Sharif-Askari *et al.*, 2001).

The granzyme B-mediated pathway to death is not just a simple linear progression. Rather, it hits the cell at multiple targets simultaneously because the protease can cleave numerous substrates. There is direct cleavage and activation of caspase-3 and this is amplified by caspase-9 through the mitochondria/apoptosome route after bid cleavage. The disruption of mitochondrial integrity also leads to necrotic death, and further caspase-independent lethal events can occur through the release of toxic proteins from the mitochondria or through direct activation by granzyme. However, the system is not just dependent on granzyme B, as the granules contain other lytic effector proteins. Thus, the granule-mediated pathway has considerable functional redundancy built in with various effectors, multiple granzymes, and alternate pathways that ultimately all lead to death. Such a complex system has likely evolved in order to deal with the various immune evasion strategies employed by pathogens.

6.3   Tumor necrosis factor/tumor necrosis factor receptor family members and CTL killing without target cell engagement: looks can kill

6.3.1   Tumor necrosis factor and tumor necrosis factor receptors

In the previous sections, the two major pathways of cell-mediated cytotoxicity were discussed in some detail. Perforin-mediated lysis requires a cognate interaction between the T-cell receptor for antigen (TCR) on a T lymphocyte and specific antigen (usually a peptide)/major histocompatibility complex (MHC) molecules on the target cell's plasma membrane. Likewise, Fas-mediated cytotoxicity involves the ligation of Fas on the target cell by Fas ligand on T cells, but does not require a cognate interaction between the effector lymphocyte and its target, and thus has the potential to damage innocent bystanders. Although these two pathways of cell-mediated cytotoxicity are clearly important in autoimmunity, T lymphocytes can also kill their targets without engaging them. They can do so by secreting cytokines capable of ligating pro-apoptotic receptors on the target cell's surface or of rendering target cells susceptible to lymphocyte-independent death effector mechanisms, or by promoting the recruitment of different types of inflammatory cells into the target tissue. Members of the Tumor necrosis factor (TNF) and TNF receptor (TNFR) families play an important role in this regard.

Cytokines of the TNF superfamily are trimeric type II membrane proteins that can be cleaved from the membrane by metalloproteinases. Two members of this family, Fas ligand and TNFα, are major effectors of T cell-mediated cytotoxicity, but also play a critical role in the propriocidal death of T cells following repeated stimulation with antigen. TNFα binds to TNFR1 and/or TNFR2, which are type I membrane proteins that have characteristic cysteine-rich extracellular domains. Like Fas, TNFR1 has a death domain (DD) in its intracytoplasmic tail, and DD dock adaptor proteins with DDs. TNFR1 recruits TRADD (TNFR-associated DD) and TRADD recruits FADD (Fas-associated DD). FADD contains two death effector domains (DEDs) that trigger apoptosis by recruiting caspase-8 or caspase-10. cFLIP (FLICE-inhibitory protein) inhibits FADD-induced activation of caspase-8 (Irmler et al., 1997). Caspase-8 activates several caspases, including caspase-3, -6, and -7, which degrade nuclear lamins, and cytoskeletal proteins such as fodrin and gelsolin. Caspase-3 inactivates ICAD (inhibitor of CAD), which results in the activation of CAD and DNA fragmentation (Enari et al., 1998). Caspase-8 can directly cleave Bid to generate tBid, which binds to and inhibits Bcl-2 (Li et al., 1998; Kim, T. et al., 2000). IAPs, like XIAP, interfere with caspase activation (Deveraux and Reed, 1999). Although primarily pro-apoptotic, TNFR1-bound TRADD can also recruit RIP (TNFR-associated kinase receptor-interacting protein) and TRAF2 (TNFR-associated factor 2), leading to NF-kB or mitogen-activated protein (MAP) kinase activation, respectively.

TRAIL (tumor necrosis factor-related apoptosis-inducing ligand) is another type II transmembrane protein belonging to the TNF family produced by professional antigen-presenting cells (APCs) such as dendritic cells and by T cells (Wang *et al.*, 1999), and has been shown to be an additional pathway of cell-mediated cytotoxicity, especially of tumor cells (Kayagaki *et al.*, 1999). TRAIL receptors 1 and 2 (DR4 and DR5) have a DD. There are TRAIL decoy receptors that lack DDs and inhibit TRAILR-mediated cytotoxicity, particularly in normal cell types (Gura, 1997). TNFR1, Fas, and TRAILR1 form pre-assembled complexes that are silenced by superoxide dismutase (SODD) and signal rapidly upon ligand binding (Jiang *et al.*, 1999; Siegel *et al.*, 2000a). It is worth noting, however, that not all the cellular death induced by death receptor triggering is exclusively mediated by apoptosis, but is mediated also by necrosis (Kawahara *et al.*, 1998; Vercammen *et al.*, 1998; Li and Berg, 2000; Villunger *et al.*, 2000).

6.3.2   The immune system in balance: killing the killers

The killing of target cells is not the only place where apoptosis is important in lymphocytes. Self-reactive lymphocytes must be destroyed and, after immune stimulation, activated cells must be removed. A discussion of tolerance induction and memory generation is, however, beyond the scope of this chapter. Suffice it to say that a fine balance between lymphocyte proliferation and death is absolutely essential to maintain an effective yet not overly active immune system. In the periphery, when lymphocytes are activated, there is a massive increase in the population of effector T cells, followed by a precipitous contraction in number, achieved via apoptosis. In order to generate memory cells, some cells must survive but too little death will lead to autoimmunity.

In stimulated cloned T cells, death is driven mainly via Fas and TNF, and this can be negatively regulated through the expression of a FADD antagonist, FLIP. TNFα can cause T-cell death via TNFR1 or TNFR2, primarily of CD8+ T cells. TNFR2 does not have a death domain, but can recruit TRAF (TNFR-associated factors) and IAPs, limiting the ability of the TRAFF–cIAP complex to compete with FADD for TNFR1 binding (Chan and Lenardo, 2000). As a result, TNFR2 can markedly enhance TNFR1-induced cell death. In addition, it has been shown that ligation of TNFR2 and other TNFR family members lacking a DD can lead to apoptosis by inducing the production of TNF and autotropic or paratropic activation of TNFR1 (Grell *et al.*, 1999). The levels of the TNFR-associated RIP significantly influence the ability of TNFR2-induced apoptosis or survival. In the presence of low levels of RIP, for example, TNFα promotes T-cell growth. Upregulation of RIP following cell cycle progression increases susceptibility to apoptosis (Pimentel-Muinos and Seed, 1999).

However, the situation in vivo is likely to be more complex and is controlled through the integration of signals from antigen-presenting cells (APCs), co-stimulation, cytokine levels, and O-glycan expression, etc. (Hildeman *et al.*, 2002). Transcriptional regulation through NFκB appears to be very important, and it is likely this operates to control the levels of key apoptotic regulators such as members of the bcl-2 and IAP families. A decrease in anti-apoptotic bcl 2 proteins occurs in vivo, leading to increased sensitivity to death particularly in the presence of death-promoting family members. The pro-apoptotic member Bim appears to be a major player in lymphocyte cell death, both in the thymus and in the periphery (Bouillet *et al.*, 2002).

6.4 Apoptosis and autoimmunity

Notwithstanding the fact that dysregulation of lymphocyte homeostasis can cause lymphoproliferative disorders, most spontaneous autoimmune disorders arise when cells of specific tissues become the targets of destruction by self-reactive lymphocytes without the need to invoke any defects in lymphocyte homeostasis. Any discussion about effector mechanisms of target cell death in autoimmunity must, therefore, take into account the nature of the effector(s). The sections that follow discuss the relative contribution of different lymphocyte types and pathways of target cell death in autoimmunity.

6.4.1 Effector lymphocytes

Much of what is known about effector pathways of autoimmunity has been learned from studies of a handful of spontaneous and experimental autoimmune disorders. A prototypic example of organ-specific autoimmunity is type 1 diabetes (T1D) in non-obese diabetic (NOD) mice, which results from a complex T cell-dependent autoimmune process that is directed against the pancreatic beta cells. Since beta cells do not express MHC class II molecules, it is generally believed that naive auto-reactive CD4+ T cells differentiate into effector cells by engaging auto-antigens shed from the beta cells by a prior insult, in the context of MHC class II molecules on local APCs. Studies of β2 microglobulin-deficient and anti-CD8 mAb-treated NOD mice have suggested that the initial beta cell insult is effected by cytotoxic CD8+ T cells (Katz *et al.*, 1993; Serreze *et al.*, 1994; Wicker *et al.*, 1994; Wang *et al.*, 1996), which are consistently recruited to islets in NOD mice (Nagata *et al.*, 1994; Santamaria *et al.*, 1995; Verdaguer *et al.*, 1996; Verdaguer *et al.*, 1997; Anderson *et al.*, 1999; Amrani *et al.*, 2000a; DiLorenzo *et al.*, 1998). This hypothesis, however, is at odds with other observations suggesting that diabetogenesis is initiated by CD4+ T cells. For example, splenic CD4+ T cells from pre-diabetic NOD mice can transfer insulitis into NOD.*scid* mice without CD8+ T-cell co-transfer, and splenic CD8+ T cells from diabetic NOD mice cannot home into pancreatic islets of NOD.*scid* mice in the absence of CD4+ T cells (Christianson *et al.*, 1993). Furthermore, genetic susceptibility and resistance to both insulitis and diabetes are profoundly affected by polymorphisms of MHC class II genes (Tisch and McDevitt, 1996), which control the development and function of diabetogenic CD4+ T cells (Schmidt *et al.*, 1997, 1999).

A number of transgenic models of autoimmune diabetes in non-diabetes-prone genetic backgrounds have shown that CD8+ T cells can efficiently kill beta cells expressing transgenic neo-antigens (Ohashi *et al.*, 1991; Blanas *et al.*, 1996; Morgan *et al.*, 1996). Little is known, however, about the antigenic specificity(ies) of the CD8+ T cells that are putatively involved in the initiation and/or progression of spontaneous T1D in the NOD mouse, let alone humans. Wong and coworkers have reported the existence, in islets of 3–4-week-old NOD mice, of a CD8+ T-cell subpopulation that recognizes an insulin-derived peptide (Wong *et al.*, 1999). Although the size of this T-cell subpopulation in islets of young NOD mice is a matter of controversy (Amrani *et al.*, 2000a), it clearly vanishes with age (Amrani *et al.*, 2000a) and is replaced, in part, by another subpopulation of highly diabetogenic, CD8+ T cells that use highly homologous TCRα chains (Santamaria *et al.*, 1995; Verdaguer *et al.*, 1996, 1997; DiLorenzo *et al.*, 1998). This CD8+ T-cell subpopulation recognizes the peptide NOD-relevant peptide (NRP) (Anderson *et al.*, 1999) as well as a number of mimics (Amrani *et al.*, 2001), expands in size as the mice age, and undergoes a process of "avidity maturation" that overlaps with the exponential phase of diabetes penetrance through the colony (Amrani *et al.*, 2000a). It would be interesting to determine whether this phenomenon occurs with other (non-NRP reactive) auto-reactive CD8+ T-cell specificities (Graser *et al.*, 1999; Kanagawa *et al.*, 2000). Immunopathologic studies of pancreata from human diabetic individuals has suggested that destruction of beta cells in human T1D may also be effected by CD8+ T cells. CD8+ T lymphocytes are abundant among the mononuclear cells that infiltrate the pancreatic islets at the clinical onset of the disease (insulitis) (Bottazzo *et al.*, 1985; Hanninen *et al.*, 1992; Itoh *et al.*, 1993; Somoza *et al.*, 1994), and in diabetic patients given pancreas grafts from healthy identical twins or HLA-identical siblings (Sibley *et al.*, 1985; Santamaria *et al.*, 1992a, b). It should be noted, however, that beta cell destruction in T1D is not exclusively effected by CD8+ T cells, and that there is clear evidence indicating that CD4+ T-cells also play a role in this process (reviewed in Tisch and McDevitt, 1996; Delovitch and Singh, 1997).

Multiple sclerosis (MS) is a human autoimmune disease caused by T cell-induced demyelination in the white matter of the brain and spinal cord (Owens *et al.*, 2001).

Experimental autoimmune encephalomyelitis (EAE) is a model of MS that can be induced in susceptible strains of rodents (mice and rats) by immunization with myelin basic protein (MBP), proteolipid antigen (PLP), and myelin oligodendrocyte protein (MOG). Auto-reactive CD4+ T cells appear to be the major effectors of EAE, as determined by T cell-depletion studies, adoptive transfer experiments, and studies with T cell receptor (TCR) transgenic mice (Goverman, 1999). There is, however, evidence pointing to a role for CD8+ T cells in disease progression and severity (Koh *et al.*, 1992; Zhang *et al.*, 1997). In fact, MBP is processed in vivo by the endogenous MHC class I pathway (Huseby *et al.*, 1999), and a MOG-derived peptide has recently been shown to activate encephalitogenic CD8+ T cells in vivo (Sun *et al.*, 2001). There have also been reports of auto-reactive CD8+ T-cell responses to human MBP-derived peptides in MS (Tsuchida, 1994; Dressel *et al.*, 1997). Furthermore, there is evidence for an excess of CD8+ over CD4+ T cells in the lesions of some patient samples (Babbe *et al.*, 2000), and work by Babbe *et al.* (2000) has provided evidence for the clonal expansion of CD8+ T cells in active MS lesions. These data suggest that these T cells may trigger oligodendrocyte death (Babbe *et al.*, 2000). Although not relevant to the pathogenesis of EAE or MS, it is clear that CD8+ CTLs can also kill neurons in vitro (Medana *et al.*, 2000). This mechanism of neuronal loss might be important in certain viral infections or neurodegenerative disorders of poorly understood ethiopathogenesis.

Experimental autoimmune uveitis (uveoretinitis, EAU) is another experimental autoimmune disorder that is induced by active immunization with retinal antigens and that is primarily caused by CD4+ T cells. Hashimoto's thyroiditis results from the autoimmune destruction of thyroid follicular cells by CD8+ T cells (Arscott and Baker, 1998). CD8+ T cells also appear to play an important role in the initiation of iodine-induced thyroiditis in NOD and NOD-H2^{h4} mice and in the production of anti-thyroglobulin autoantibodies in Graves' disease (Verma *et al.*, 2000). It is clear, however, that development of these diseases also requires CD4+ T cells. Although not autoimmune in nature, some other diseases involve a cytolytic attack of T lymphocytes against self-targets "modified" by exogenous agents, such as viruses. Liver injury in viral hepatitis, for example, is effected by both CD4+ Th1 and CD8+ T cells (Ando *et al.*, 1993; Franco *et al.*, 1997).

There are a number of other autoimmune disorders in which tissue damage is effected by Th1- or Th2-dependent, autoantibody-secreting lymphocytes (Elson and Barker, 2000). In these disorders, T lymphocytes function as indirect effectors of autoimmunity, by driving the differentiation of B cells into autoantibody-secreting plasma cells. These diseases include, among others, autoimmune hemolytic anemia, autoimmune thrombocytopenic purpura, Graves' disease, Goodpasture's syndrome, myastenia gravis, pemphigus vulgaris, and systemic lupus erythematosus

(SLE). Autoantibodies binding the Rhesus complex proteins on red blood cells accelerate their disposal in the spleen (by macrophages), causing autoimmune hemolytic anemia. Other autoantibodies bind platelet membrane glycoproteins, triggering autoimmune thrombocytopenic purpura. In Graves' disease, circulating autoantibodies ligate the thyroid-stimulating hormone receptor, promoting thyroid hormone production. Goodpasture's syndrome is caused by anti-glomerular basement membrane autoantibodies against type IV collagen. Production of autoantibodies against acetylcholine receptors is responsible for myastenia gravis, and anti-desmoglein, desmocollin, and desmoplakin autoantibodies are responsible for pemphigus vulgaris.

Pathogenic autoantibodies are also produced in SLE, in a T helper-dependent manner. SLE-associated autoantibodies are directed against chromatin components, including double-stranded DNA (dsDNA), and cytoplasmic ribonucleoproteins. It is thought that SLE is the result of disregulated B-lymphocyte apoptosis; MRL/lpr (Fas-deficient) mice, mice expressing a Bcl-2 transgene in B cells, Bim-deficient mice, and BAFF-transgenic mice, for example, produce pathogenic autoantibodies and develop lupus-like syndromes. Bim inhibits anti-apoptotic genes such as Bcl-2 and Bcl-xl, and Bim deficiency leads to autoimmune kidney disease (Bouillet *et al.*, 1999). BAFF is a TNF family member that induces B-cell proliferation by engaging BCMA or TACI receptors (Mackay *et al.*, 1999; Schneider *et al.*, 1999; Gross *et al.*, 2000; Khare *et al.*, 2000; Xia *et al.*, 2000). The pathologic features of SLE, however, are not always caused by autoantibodies; T cells also contribute to the pathogenesis of SLE. It has been proposed that the excessive production of certain cytokines by T cells, such as interleukin-10 (IL-10) or interferon (IFN)-γ, promotes the production of anti-dsDNA and anti-histone antibodies in SLE (Peng *et al.*, 1997; Haas *et al.*, 1998). Production of anti-dsDNA autoantibodies might also be induced by the killing of macrophages (a source of antigenic DNA) by cytotoxic T cells (Kaplan *et al.*, 2000). T cells in skin lesions of patients with SLE or systemic sclerosis (a related disorder) produce high levels of IL-17 which, in turn, induces IL-1 and IL-6 secretion by fibroblasts and endothelial cells. Locally, this augments the secretion of matrix metalloproteinases and collagenase, leading to fibrosis. Systemically, this results in B-cell hyper-reactivity and autoantibody production (Kurasawa *et al.*, 2000).

IL-2Rβ chain-deficient mice have increased production of plasma cells and develop a form of autoimmune hemolytic anemia that requires CD4+ T cells (Suzuki *et al.*, 1995). T cells can also induce the production of pathogenic autoantibodies, by recognizing peptide/MHC complexes on macrophages, dendritic cells, and B cells. This was shown in studies of KRN TCR transgenic mice, which express a foreign antigen-specific TCR that cross-reacts with a ubiquitously expressed self-antigen. In these animals, KRN TCR-specific T cells undergo activation by recognizing I-A^{g7} on macrophages and/or dendritic cells. The activated T cells activate B cells, enhancing Ig production and promoting arthritis (Ji *et al.*, 1999). However, autoantibodies are not always pathogenic. Most type 1 diabetic patients, for example, bear high levels of circulating autoantibodies against a number of beta cell autoantigens. Although these autoantibodies have an excellent predictive value, they are not thought to play an important role in diabetogenesis (Abiru and Eisenbarth, 1999).

6.4.2 Fas versus perforin/granzymes

Studies in the lymphocytic choriomeningitis virus (LCMV)-induced model of autoimmune diabetes have suggested a major role for perforin as an effector mechanism of beta cell death in T1D (Kagi *et al.*, 1996). Using a similar model, Seewaldt *et al.* (2000) have argued for a major role for cytokines as (direct or indirect) effectors of beta cell death in virus-induced T1D. In the rat insulin promoter (RIP)–influenza virus hemagglutinin transgenic model of diabetes, the contribution of the perforin pathway is also more important than that of Fas (Kreuwel *et al.*, 1999). Paradoxically, adoptive transfer of a diabetogenic, insulin-reactive, CD8+ T-cell clone into Fas-deficient NOD.*lpr* mice suggested that CD8+ T cell-induced destruction of beta cells in NOD mice is exclusively mediated via Fas (Chervonsky *et al.*, 1997). Studies of perforin- or Fas-deficient NOD mice have also yielded apparently contradictory results: perforin-deficient NOD mice develop severe insulitis but rarely become diabetic (Kagi *et al.*, 1997), and NOD.*lpr* mice develop neither diabetes nor insulitis despite expressing perforin (Itoh *et al.*, 1997). Studies with NOD mice expressing a transgenic NRP-reactive TCR have shown that NRP-reactive CD8+ T cells kill via cells exclusively via Fas (Amrani *et al.*, 1999). This is also the case for CD4+ T cells bearing a highly diabetogenic, I-A^{g7}-restricted beta cell-reactive T-cell receptor (4.1-TCR) (Amrani *et al.*, 2000b). The beta cell cytotoxic activity of these CD4+ T cells not only requires the ability of beta cells to express Fas but can only be effected upon exposure of the beta cell targets to IL-1α, IL-1β, TNFα, and/or IFN-γ. Interestingly, IL-1 has been shown to induce selectively Fas expression on beta cells, but not on other endocrine cells (Heitmeier *et al.*, 1997; Stassi *et al.*, 1997). Moriwaki *et al.* (1999) have shown that >90% of the beta cells of human pancreatic tissue from acutely diabetic patients express Fas, albeit only in the presence of insulitis, and that IL-1 induces Fas expression in human pancreatic islet cells (Stassi *et al.*, 1997). Interestingly, IFN-γ-induced upregulation of Fas on at least some target cells appears to be mediated by the induction of the inducible form of nitric oxide synthase (iNOS) and by NO. NO is thought to operate by inactivating the binding activity of the transcriptional repressor yin-yang 1 to the silencer region of the Fas promoter (Garban and Bonavida,

2001). Whether this mechanism is relevant to autoimmune responses remains to be determined.

The molecular nature of the target auto-antigens may influence the choice of cytotoxicity pathway that is employed by a CTL in a given autoimmune disorder. Fas-mediated cytotoxicity by CD8+ CTL can be triggered independently of perforin-mediated cytotoxicity, depending on the structure of the triggering antigenic peptide/MHC complex (Cao *et al.*, 1995; Brossart and Bevan, 1996; Esser *et al.*, 1996). In this regard, it is worth noting that perforin-expressing 8.3-CTL could not lyse Fas-deficient beta cells but could efficiently kill Fas-deficient fibroblasts pulsed with a high affinity mimotope (Amrani *et al.*, 1999). Other autoreactive CTLs appear to be capable of eliciting the perforin pathway as well, although this has not been formally demonstrated (Kreuwel *et al.*, 1999; Mathews *et al.*, 2001). The avidity of the T cells involved at different stages of disease may be another contributing factor; Fas-mediated cytotoxicity may be elicited at a lower threshold of avidity (of antigen-specific target/CTL interactions) than perforin-mediated cytotoxicity. Alternatively, beta cells may change their susceptibility to Fas- versus perforin-mediated cytotoxicity as the disease progresses. Certain cell types, such as neurons (Medana *et al.*, 2000) and leukemic cells (Lehmann *et al.*, 2000), display greater susceptibility to Fas- than to perforin-mediated lysis. This may be related to the ability of target cells to bind perforin (Lehmann *et al.*, 2000) or granzyme B (Motyka *et al.*, 2000). Alternatively, expression of FasL on target cells might block degranulation of CTLs and perforin-dependent cytotoxicity, as shown recently for neurons (Medana *et al.*, 2001). When taken together, it would be reasonable to hypothesize that diabetogenesis in NOD mice is initiated by CD8+ and CD4+ CTL clonotypes capable of lysing beta cells exclusively via Fas, and later amplified by clonotypes that can kill via other death effector pathways, including perforin. This interpretation of the data would explain why NOD.*lpr* mice do not develop insulitis (Itoh *et al.*, 1997), why Fas-deficient islet grafts are readily destroyed in spontaneously diabetic NOD mice (Allison and Strasser, 1998) or following adoptive transfer of splenic T cells from NOD or NOD mice transgenic for the BDC-2.5 TCR (Pakala *et al.*, 1999), and why perforin-deficient NOD mice develop severe insulitis and yet remain diabetes resistant (Kagi *et al.*, 1997). Nevertheless, and despite all these studies, there is still no consensus on this issue; a number of recent papers continue to support diametrically opposed views (Thomas *et al.*, 1999; Kim, S. *et al.*, 2000; Su *et al.*, 2000).

Fas/FasL interactions also appear to play a critical role in the induction and progression of EAE (Sabelko *et al.*, 1997; Waldner *et al.*, 1997; Sabelko-Downes *et al.*, 1999) but not in its effector phase (Suvannavejh *et al.*, 2000a). A recent study employing naked DNA vaccines encoding Fas ligand have indicated that anti-FasL autoantibodies downregulate TNFα production by T cells and prevent the onset of EAE, suggesting a critical role for Fas–FasL interactions in the initiation of this disease (Wildbaum *et al.*, 2000). Although there is evidence suggesting that oligodendrocytes are susceptible to perforin-mediated cytotoxicity, it is not clear what the role of perforin is in the disease process (Zeine *et al.*, 1998). Studies in TCR transgenic mice have also provided support for the involvement of Fas/FasL interactions in the pathogenesis of EAE: however, they have indicated that disease occurrence does not require Fas (Dittel *et al.*, 1999). Interestingly, bone marrow chimeras expressing Fas and FasL on T cells but not radioresistant tissues (i.e. occular tissues) develop EAU with normal intensity and incidence, but expression of Fas and Fas ligand is indispensable (Wahlsten *et al.*, 2000). This observation has suggested that the role of Fas in EAU is dissociated from its pro-apoptotic activity, or that it somehow enhances presentation of auto-antigen. This interpretation is compatible with the observation that disruption of Fas–FasL interactions inhibits autoantibody production in pristane-induced lupus auto-antibodies which usually target auto-antigens that cluster in apoptotic blebs (Satoh *et al.*, 2000). Alternatively, Fas signaling co-stimulates auto-reactive T-cell responses (Siegel *et al.*, 2000b). In fact, ligation of Fas has been shown to induce the secretion of pro-inflammatory cytokines and chemokines, such as IL-1, IL-6, IL-8, and macrophage inflammatory protein (MIP)-1α (Abreu-Martin *et al.*, 1995; Saas *et al.*, 1999; Lee *et al.*, 2000), possibly by activating the ERK1/ERK2 and P38 MAPK signaling pathways (Choi *et al.*, 2001). This may have pathogenic significance in some autoimmune disorders. Thus, ligation of Fas on T cells from SLE patients induces proliferation, rather than cell death (Sakata *et al.*, 1998). Furthermore, dendritic cells are resistant to Fas-induced cell death, owing to increased expression levels of c-FLIP (Ashany *et al.*, 1999), and ligation of Fas on dendritic cells upregulates B7.1, B7.2, and class II MHC molecules (Rescigno *et al.*, 2000). It has recently been shown that FLIP can divert Fas signaling toward ERK and NFκB, possibly by virtue of its ability to associate with Raf-1 and, therefore, to activate MEK1. FLIP can also associate with TRAF1 and TRAF2, leading to NFκB activation (Kataoka *et al.*, 2000). Therefore, Fas (and TNFα, see below) may be involved at different stages of an autoimmune process via different mechanisms and different cell types.

Fas and its ligand have also been proposed to play a critical role in Hashimoto's thyroiditis. Thyroiditis results from autoimmune destruction of thyroid follicular cells by CD8+ CTLs (Arscott and Baker, 1998). FasL was found to be constitutively expressed in thyrocytes and it was proposed that IL-1β-induced upregulation of Fas elicits fraticidal lysis of thyrocytes, leading to hypothyroidism (Giordano *et al.*, 1997). However, it has also been reported that thyroid follicular cells constitutively express Fas and that Fas expression is not affected by IFN-γ and IL-1β (Arscott *et al.*, 1997). Bretz *et al.* (1999a) found that, although no single cytokine could enhance the susceptibility of thyroid follicular cells to Fas-mediated apoptosis, Fas-dependent cytotoxicity could be readily induced with a combination of cytokines such as IFN-γ and either TNFα or IL-1β. It has also been shown that normal thyrocytes do not express FasL despite expressing FasL messenger RNA (Hiromatsu *et al.*, 1999; Mitsiades *et al.*, 1999). Finally, it has been reported that eczematous dermatitis, including atopic dermatitis and allergic contact dermatitis, involves T cell-mediated, Fas-dependent apoptosis of IFN-γ-stimulated keratinocytes (Trautmann *et al.*, 2000).

Fas-induced apoptosis does not always have a pathogenic effect in autoimmunity. For example, synovial cells of rheumatoid arthritis patients express Fas in response to both cytokines and CD44 (the hyaluronan receptor) signaling. Ligation of CD44 contributes to the inhibition of synovial hyperplasia (Fujii *et al.*, 2001).

6.4.3 TNFα, TNFR, and IL-1

TNFα plays an important role in the development of T1D. Although TNFα is both cytostatic and cytocidal for beta cells in vitro, production of transgenic TNFα in situ does not cause diabetes in the absence of T cells and thus the role of TNFα as effector cytokine in vivo remains unclear. When expressed in the islets of pre-diabetic NOD mice, TNFα enhances the presentation of beta cell auto-antigens to auto-reactive T cells, by ligating TNFR1 on APCs (Green and Flavell, 2000) and, possibly, islet cells (Pakala *et al.*, 1999). Beta cells express low levels of TNFR1 constitutively, but can be induced to express both TNFR1 and TNFR2 during inflammation (Walter *et al.*, 2000). This pro-immunogenic effect of TNFR1 signaling in T1D has also been observed in experimental myasthenia gravis (EMG) following immunization with *Torpedo* acethylcholine receptor: TNFR1-deficient mice do not develop EMG unless antigen is co-injected with IL-12 (Wang *et al.*, 2000).

TNFα can damage the myelin sheath and can induce oligodendrocyte apoptosis, suggesting an effector role for TNFα/TNFR1 interactions in the pathogenesis of EAE and MS (Powell *et al.*, 1990; Chung *et al.*, 1991; Renno *et al.*, 1995). This possibility is consistent with the observation that, unlike TNFR2-deficient mice, TNFR1-deficient mice do not develop MOG-induced EAE (Eugster *et al.*, 1999; Suvannavejh *et al.*, 2000b; Kassiotis and Kollias, 2001). This view, however, is at odds with observations suggesting that TNFα plays an anti-inflammatory role in EAE (Liu *et al.*, 1998). Although lymphotoxin-α can be secreted by T cells, can trigger TNFR1, and has been implicated as an effector molecule in EAE, its role in EAE is less clear than that of TNFα (Riminton *et al.*, 1998; Owens *et al.*, 2001).

TNFα also plays an effector role in inflammatory bowel disease (IBD) and rheumatoid arthritis (RA). TNFα can induce apoptosis of intestinal epithelial cells (Piguet *et al.*, 1998). Interestingly, in a T-cell transfer model of colitis, CD4+

CD45Rb^{hi} cells induce colitis by stimulating the production of TNFα by non-T cells of the colonic mucosa (Corazza *et al.*, 1999). The contribution of TNFα to these disorders is further supported by the observation that mice expressing elevated levels of TNFα have an increased susceptibility to developing arthritis and IBD (Taylor *et al.*, 1996; Kontoyiannis *et al.*, 1999). Furthermore, TNFR2–Ig fusion proteins or humanized anti-TNFα mAbs can inhibit IBD and RA (Galon *et al.*, 2000; Lovell *et al.*, 2000; Choy and Panayi, 2001). TNFα, along with IFN-γ, has also been shown to be indispensable for the CD4+ and CD8+ T cell-mediated destruction of liver cells in viral hepatitis (Ohta *et al.*, 2000). While the mechanisms of destruction are not clear, they do not involve perforin or Fas (Nakamoto *et al.*, 1997; Ohta *et al.*, 2000). TNFα, along with IL-1, is also a key inflammatory cytokine in pemphigus (Feliciani *et al.*, 2000) and both are involved in the pathogenesis of experimental SLE in mice. Abrogation of the production of these two cytokines in the early stages of experimental EAE is clinically beneficial (Segal *et al.*, 2001).

The cytopathic effect of TNFα in autoimmunity need not be direct. TNFα can upregulate Fas on target cells (Yamada *et al.*, 1996; Bretz *et al.*, 1999a; Thomas *et al.*, 1999; Amrani *et al.*, 2000b), can promote the secretion of tissue-destroying matrix metalloproteinases, and can sensitize target cells to TRAIL-induced cytotoxicity (Bretz *et al.*, 1999b). TNF family members are also powerful paracrine inducers of other cytotoxic cytokines, such as IL-1α and IL-1β, which can be secreted by activated macrophages in response to TNFα. Whether or not TNFα can induce the production of IL-1 by non-immune, TNFR-expressing cells is not known; however, pancreatic beta cells can produce mature IL-1β in response to double-stranded RNA and IFN-γ (Heitmeier *et al.*, 2001) and thus this is a possibility.

IFN-γ and IL-1β enhance the activity of iNOS in beta cells and augment the endogenous production of NO, which impairs glucose-stimulated insulin secretion (Arnush *et al.*, 1998a, b). Over-expression of Bcl-2 in beta cells in vitro results in improved resistance to stress-induced apoptosis, but cannot overcome the effects of IL-1 and IFN-γ signaling (Dupraz *et al.*, 1999). Over-expression of a dominant-negative MyD88, which couples the IL-1R to downstream signaling, also impaired iNOS activation, NO production, and cytokine-induced beta cell death, without blocking cytokine-induced, glucose-induced insulin secretion (Dupraz *et al.*, 2000). Recent studies using beta cells from mice expressing a dominant–negative IFNγR exclusively in beta cells have shown that they are resistant to IL-1+IFN-γ-induced inhibition of insulin secretion and DNA damage, suggesting that NO must be produced by the beta cells themselves to induce beta cell damage (Thomas *et al.*, 2002). Over-expression of the suppressor of cytokine signaling-1, which inhibits the Janus kinase/STAT pathway, was also able to prevent IL-1+IFN-γ-induced

NO production, inhibition of insulin secretion, and DNA damage in a beta cell line, lending further support to the above studies (Cottet *et al.*, 2001). Although NO is directly responsible for inducing necrosis of beta cells in response to IL-1β (Hoorens *et al.*, 2001) or combinations of TNFα, IFN-γ, and IL-1β, it appears to be dispensable for cytokine-induced beta cell apoptosis (Liu *et al.*, 2000; Hoorens *et al.*, 2001).

The contribution of free-radical damage to beta cell loss in T1D is supported by the observation that beta cells from ALR/Lt mice, which express high systemic levels of molecules associated with the dissipation of free-radical stress, are unusually resistant to cytotoxic cytokines and diabetogenic CTLs (Mathews *et al.*, 2001). Furthermore, expression of transgene-encoded thioredoxin in beta cells prevents diabetes, indicating that superoxide-induced beta cell death does play a role as an effector mechanism of beta cell destruction (Hotta *et al.*, 1998). Thus, although death receptor triggering induces cell death primarily by apoptosis, TNF/TNFR interactions can indirectly lead to necrosis (Kawahara *et al.*, 1998; Vercammen *et al.*, 1998; Li and Berg, 2000; Villunger *et al.*, 2000). This phenomenon is not unique to beta cells. IL-1β can degrade joint cartilage in autoimmune arthritis (Choy and Panayi, 2001). Finally, as was the case for Fas, TNFR1 and TRAIL-R signaling is not invariably associated with cell death. They can both trigger ERK1/2 activation, which can override apoptotic signaling (Tran *et al.*, 2001).

6.4.4 TRAIL, TWEAK, and RANKL

TRAIL is another member of the TNF family that is produced by T cells, among other cell types (Wang *et al.*, 1999). TRAIL binds to the DD-bearing receptors TRAILR1 (DR4) and TRAILR2 (DR5). There are TRAIL decoy receptors that lack DDs and inhibit TRAILR-mediated cytotoxicity, particularly in normal cell types (Gura, 1997). Although TRAIL can induce cell death, particularly of tumor cells (Kayagaki *et al.*, 1999), its role in T lymphocyte-mediated autoimmunity is unclear. On the one hand, TRAIL and TRAILRs have been shown to be expressed in intrathyroidal lymphocytes and in cytokine-stimulated thyroid follicular cells (Bretz *et al.*, 1999b), suggesting a role in autoimmune thyroid disease. On the other hand, systemic administration of TRAIL can prevent autoimmune disorders such as collagen-induced arthritis (Song *et al.*, 2000) and EAE (Hilliard *et al.*, 2001). CD4+ CTLs can also kill certain targets (macrophages) via TWEAK (TNF-like weak inducer of apoptosis) (Kaplan *et al.*, 2000), but the role of TWEAK in organ-specific autoimmunity has not been explored. RANKL (TRANCE, osteo-protegerin ligand), a TNF family member that is secreted by activated T cells, can trigger bone erosion in the joints of arthritic patients (Kong *et al.*, 1999; Teng *et al.*, 2000).

6.4.5 Other pathways

IL-6, which can be produced by CD4+ T cells, is elevated in ex vivo organ cultures of inflamed colonic mucosa from both ulcerative colitis and Crohn's disease-affected patients. IL-6 is thought to contribute to disease pathogenesis by increasing the production of cytopathic cytokines, such as IL-1β and TNFα, and matrix metalloproteinases, as well as by inhibiting T-cell apoptosis (Atreya *et al.*, 2000). IL-6 may also play a role in EAE, as IL-6-deficient mice are resistant to EAE (Mendel *et al.*, 1998; Okuda *et al.*, 1998; Samoilova *et al.*, 1998). IL-6, however, can protect liver cells from Fas-induced apoptosis, by upregulating FLIP, Bcl-2, and Bcl-xL (Kovalovich *et al.*, 2001). Although usually associated with protection from autoimmunity, IL-4-producing Th2 cells may act as effectors of autoimmunity. These cells, for example, have a pathogenic role in oxazolone-induced colitis and contribute to the pathogenesis of IBD in TCRα-deficient mice, perhaps by increasing intestinal epithelium permeability and neutrophil adhesion to epithelial layers (Boirivant *et al.*, 1998; Iijima *et al.*, 1999). IL-17, produced by CD4+ T cells in affected joints of RA patients, is an effector of joint destruction in collagen-induced arthritis, independently of IL-1 (Lubberts *et al.*, 2001). By producing INF-γ, auto-reactive CD4+ T cells upregulate the expression of MHC class I and Fas on target tissues, including thyroid epithelial cells and beta cells, and mark them for lysis by CTLs, among other effects (Yamada *et al.*, 1996; Bretz *et al.*, 1999a; Thomas *et al.*, 1999; Amrani *et al.*, 2000b). IFN-γ signaling in beta cells, however, while playing a critical role in upregulating MHC class I, is dispensable in autoimmune diabetes (Serreze *et al.*, 2000).

6.5 Concluding remarks

The effector mechanisms of tissue destruction in autoimmune disorders are very complex and involve multiple immune cell types, antigenic specificities, and pathways of cell- and cytokine-mediated cytotoxicity. This chapter has attempted to summarize current concepts in our understanding of how death signaling molecules trigger apoptosis or necrosis of target cells in autoimmune disorders. The importance of a given pathway in a given autoimmune disease is influenced by a number of factors, including the nature of target auto-antigen, the type of effector lymphocyte involved, and the genetic background of the affected individual or animal model. This complexity is compounded by changes in the susceptibility of target cells to different cell death pathways during disease progression that are caused by the upregulation/downregulation of death receptors and signaling intermediates by inflammatory stimuli. This is further complicated by the fact that death signaling receptors can, in some instances, promote cell survival and growth rather than death. Breathtaking progress has been made in our understanding of how apoptosis contributes to the initiation and/or progression of autoimmunity; however, many controversies remain unresolved.

ACKNOWLEDGEMENTS

We thank the members of our laboratories for exciting discussions. This work was supported by grants from the Canadian Institutes of Health Research (CIHR), the National Cancer Institute of Canada, the Natural Sciences and Engineering Research Council of Canada (NSERC), the Juvenile Diabetes Research Foundation (JDRF), and the Canadian Diabetes Association (CDA). P.S. is a Scientist of the Alberta Heritage Foundation for Medical Research (AHFMR). R.C.B. is a Scientist of the AHFMR, a Distinguished Scientist of the CIHR, and an International Research Scholar of the Howard Hughes Medical Institute.

REFERENCES

Abiru, N. and Eisenbarth, G. (1999). Autoantibodies and autoantigens in type 1 diabetes: role in pathogenesis, prediction and prevention. *Can. J. Diab. Care*, **23**, 59–65.

Abreu-Martin, M., Vidrich, A., Lynch, D., and Targan, S. (1995). Divergent induction of apoptosis and IL-8 secretion in HT-29 cells in response to TNFα and ligation of Fas antigen. *J. Immunol.*, **155**, 4147–54.

Alimonti, J. B., Shi, L., Baijal, P. K., and Greenberg, A. H. (2001). Granzyme B induces BID-mediated cytochrome c release and mitochondrial permeability transition. *J. Biol. Chem.*, **276**, 6974–82.

Allison, J. and Strasser, A. (1998). Mechanisms of beta cell death in diabetes: a minor role for CD95. *Proc. Natl. Acad. Sci. USA*, **95**, 13818–22.

Amrani, A., Serra, P., Yamanouchi, J. *et al.* (2001). Expansion of the antigenic repertoire of a single T cell receptor upon T-cell activation. *J. Immunol.*, **167**, 655–66.

Amrani, A., Verdaguer, J., Anderson, B., Utsugi, T., and Bou, S. (1999). Perforin-independent beta cell destruction by diabetogenic CD8+ T lymphocytes in transgenic nonobese diabetic mice. *J. Clin. Invest.*, **103**, 1201–9.

Amrani, A., Verdaguer, J., Serra, P., Tafuro, S., Tan, R., and Santamaria, P. (2000a). Progression of autoimmune diabetes driven by avidity maturation of a T-cell population. *Nature*, **406**, 739–42.

Amrani, A., Verdaguer, J., Thiessen, S., Bou, S., and Santamaria, P. (2000b). IL-1α, IL-1β, and IFN-γ mark beta cells for Fas-dependent destruction by diabetogenic CD4+ T-lymphocytes. *J. Clin. Invest.*, **105**, 459–68.

Anderson, B., Park, B. J., Verdaguer, J., Amrani, A., and Santamaria, P. (1999). Prevalent CD8(+) T cell response against one peptide/MHC complex in autoimmune diabetes. *Proc. Natl. Acad. Sci. USA*, **96**, 9311–16.

Ando, K., Moriyama, T., Guidotti, L. *et al.* (1993). Mechanisms of class I restricted immunopathology: a transgenic mouse model of fulminant hepatitis. *J. Exp. Med.*, **178**, 1541.

Arnush, M., Hitmeier, M., Scarim, A., Marino, M., Manning, P., and Corbett, J. (1998a). IL-1 produced and released endogenously within human islets inhibits beta cell function. *J. Clin. Invest.*, **102**, 516–26.

Arnush, M., Scarim, A., Hitmeier, M., Kelly, C., and Corbett, J. (1998b). Potential role of resident islet macrophage activation in the initiation of autoimmune diabetes. *J. Immunol.*, **160**, 2684–91.

Arscott, P. L. and Baker, J. R. (1998). Apoptosis and thyroiditis. *Clin Immunol Immunopathol*, **87**, 207–17.

Arscott, P. L., Knapp, J., Rymaszewski, M. *et al.* (1997). Fas (APO-1, CD95)-mediated apoptosis in thyroid cells is regulated by a labile protein inhibitor. *Endocrinology*, **138**, 5019–27.

Ashany, D., Savir, A., Bhardwaj, N., and Elkon, K. (1999). Dendritic cells are resistant to apoptosis through the Fas (CD95/APO-1) pathway. *J. Immunol.*, **163**, 5303–11.

Atreya, R., Mudter, J., Finotto, S. *et al.* (2000). Blockade of interleukin 6 trans signaling suppresses T-cell resistance against apoptosis in chronic intestinal inflammation: evidence in Crohn's disease and experimental colitis in vivo. *Nat. Med.*, **6**, 583–8.

Babbe, H., Roers, A. Waisman, A. *et al.* (2000). Clonal expansions of CD8(+) T cells dominate the T cell infiltrate in active multiple sclerosis lesions as shown by micromanipulation and single cell polymerase chain reaction. *J. Exp. Med.*, **192**, 393–404.

Barry, M., Heibein, J. A., Pinkoski, M. J. *et al.* (2000). Granzyme B short-circuits the need for caspase 8 activity during granule-mediated cytotoxic T-lymphocyte killing by directly cleaving Bid. *Mol. Cell. Biol.*, **20**, 3781–94.

Blanas, E., Carbone, F., Allison, J., Miller, J., and Heath, W. (1996). Induction of autoimmune diabetes by oral administration of autoantigen. *Science*, **274**, 1707–9.

Boirivant, M., Fuss, I., Chu, A. *et al.* (1998). Oxazolone colitis: a murine model of T helper cell type 2 colitis treatable with antibodies to interleukin 4. *J. Exp. Med.*, **188**, 1929–39.

Boldin, M. P., Goncharov, T. M., Goltsev, Y. V., and Wallach, D. (1996). Involvement of MACH, a novel MORT1/FADD-interacting protease, in Fas/APO-1- and TNF receptor-induced cell death. *Cell*, **85**, 803–15.

Bottazzo, G. F., Dean, B. M., McNally, J. M., McKay, E. H., Swift, P. G. F., and Gamble, D. R. (1985). In situ characterization of autoimmune phenomenon: an expression of HLA molecules in the pancreas of diabetic insulinitis. *N. Engl. J. Med.*, **313**, 353–60.

Bouillet, P., Metcalf, D., Huang, D. *et al.* (1999). Pro-apoptotic Bcl-2 relative Bim required for certain apoptotic responses, leukocyte homeostasis, and to preclude autoimmunity. *Science*, **286**, 1735–8.

Bouillet, P., Purton, J. F., Godfrey, D. I. *et al.* (2002). BH3-only Bcl-2 family member Bim is required for apoptosis of autoreactive thymocytes. *Nature*, **415**, 922–6.

Bretz, J. D., Arscott, P. L., Myc, A., and Baker, J. R. (1999a). Inflammatory cytokine regulation of Fas-mediated apoptosis in thyroid follicular cells. *J. Biol. Chem.*, **274**, 25433–8.

Bretz, J. D., Rymaszewski, M., Arscott, P. L. *et al.* (1999b). TRAIL death pathway expression and induction in thyroid follicular cells. *J. Biol. Chem.*, **274**, 23627–32.

Brossart, P. and Bevan, M. (1996). Selective activation of Fas/Fas ligand-mediated cytotoxicity by a self peptide. *J. Exp. Med.*, **183**, 2449.

Browne, K. A., Blink, E., Sutton, V. R., Froelich, C. J., Jans, D. A., and Trapani, J. A. (1999). Cytosolic delivery of granzyme B by bacterial toxins: evidence that endosomal disruption, in addition to transmembrane pore formation, is an important function of perforin. *Mol. Cell Biol.*, **19**, 8604–15.

Cao, W., Tykodi, S., Esser, M., Braciale, V., and Braciale, T. (1995). Partial activation of CD8+ T cells by a self-derived peptide. *Nature*, **378**, 295–8.

Chan, F. and Lenardo, M. (2000). A crucial role for p80 TNF-R2 in amplifying p60 TNF-R1 apoptosis signals in T lymphocytes. *Eur. J. Immunol.*, **30**, 652–60.

Chervonsky, A., Wang, Y., Wong, F. *et al.* (1997). The role of Fas in autoimmune diabetes. *Cell*, **89**, 17–24.

Chinnaiyan, A. M., Tepper, C. G., Seldin, M. F. *et al.* (1996). FADD/MORT1 is a common mediator of CD95 (Fas/APO-1) and tumor necrosis factor receptor-induced apoptosis. *J. Biol. Chem.*, **271**, 4961–5.

Choi, C., Xu, X., Oh, J.-W. *et al.* (2001). Fas-induced expression of chemokines in human glioma cells: involvement of extracellular signal-regulated kinase 1/2 and p38 mitogen-activated protein-kinase. *Cancer Res.*, **61**, 3084–3091.

Choy, E. H. and Panayi, G. S. (2001). Mechanisms of disease: cytokine pathways and joint inflammation in rheumatoid arthritis. *N. Engl. J. Med.*, **344**, 907–16.

Christianson, S., Shultz, L., and Leiter, E. (1993). Adoptive transfer of diabetes into immunodeficient NOD-scid/scid mice. Relative contributions of CD4+ and CD8+ T-cells from diabetic versus prediabetic NOD.NON-thy-1a donors. *Diabetes*, **42**, 44–55.

Chung, I., Norris, J., and Benveniste, E. (1991). Differential tumor necrosis factor-alpha expression by astrocytes from experimental allergic encephalomyelitis-susceptible and -resistant rat strains. *J. Exp. Med.*, **173**, 801–11.

Corazza, N., Eichenberger, S., Eugster, H. P., and Mueller, C. (1999). Nonlymphocyte-derived tumor necrosis factor is required for induction of colitis in recombination activating gene (RAG)2(–/–) mice upon transfer of CD4(+)CD45RB(hi) T cells. *J. Exp. Med.*, **190**, 1479–92.

Cottet, S., Dupraz, P., Hamburger, F., Dolci, W., Jaquet, M., and Thorens, B. (2001). SOCS-1 protein prevents Janus kinase/STAT-dependent inhibition of beta cell insulin gene transcription and secretion in response to interferon-γ. *J. Biol. Chem.*, **276**, 25862–70.

Darmon, A., Nicholson, D., and Bleackley, R. (1995). Activation of the apoptotic protease CPP32 by cytotoxic T-cell-derived granzyme B. *Nature*, **377**, 446–88.

Delovitch, T. and Singh, B. (1997). The nonobese diabetic mouse as a model of autoimmune diabetes: immune disregulation gets the NOD. *Immunity*, **7**, 727–38.

Deveraux, Q., and Reed, T. (1999). IAP family proteins – suppressors of apoptosis. *Genes. Dev.*, **13**, 239–52.

DiLorenzo, T., Graser, R., Ono, T. *et al.* (1998). Major histocompatibility complex class I-restricted T cells are required for all but the end stages of diabetes development in nonobese diabetic mice and use prevalent T cell receptor α chain gene rearrangement. *Proc. Natl. Acad. Sci. USA*, **95**, 12538–43.

Dittel, B., Merchant, R., and Janeway, C. A., Jr. (1999). Evidence for Fas-dependent and Fas-independent mechanisms in the pathogenesis of experimental autoimmune encephalomyelitis. *J. Immunol.*, **162**, 6392–400.

Dressel, A., Chin, J. L., Sette, A., Gausling, R., Hollsberg, P., and Hafler, D. A. (1997). Autoantigen recognition by human CD8 T cell clones. Enhanced agonist response induced by altered peptide ligands. *J. Immunol.*, **159**, 4943–51.

Du, C., Fang, M., Li, Y., Li, L., and Wang, X. (2000). Smac, a mitochondrial protein that promotes cytochrome c-dependent caspase activation by eliminating IAP inhibition. *Cell*, **102**, 33–42.

Dupraz, P., Cottet, S., Hamburger, F., Dolci, W., Felley-Bosco, E., and Thorens, B. (2000). Dominant negative MyD88 proteins inhibit interleukin-1 beta/interferon-gamma-mediated induction of nuclear factor kappa B-dependent nitrite production and apoptosis in beta cells. *J. Biol. Chem.*, **275**, 37672–8.

Dupraz, P., Rinsch, C., Pralong, W. *et al.* (1999). Lentivirus-mediated Bcl-2 expression in beta TC-tet cells improves resistance to hypoxia and cytokine-induced apoptosis while preventing in vitro and in vivo control of insulin secretion. *Gene. Ther.*, **6**, 1160–69.

Elson, C. J. and Barker, R. N. (2000). Helper T cells in antibody-mediated, organ-specific autoimmunity. *Curr. Opin. Immunol.*, **12**, 664–9.

Enari, M., Sakahira, H., Yokoyama, H., Okawa, K., Iwamatsu, A., and Nagata, S. (1998). A caspase-activated DNase that degrades DNA during apoptosis, and its inhibitor ICAD. *Nature*, **391**, 43–50.

Esser, M., Krishnamurphy, B., and Braciale, V. (1996). Distinct T cell receptor signaling requirements for perforin- or FasL-mediated cytotoxicity. *J. Exp. Med.*, **183**, 1697–706.

Eugster, H., Frei, K., Backman, R., Bluethmann, H., Lassman, H., and Fontana, A. (1999). Severity of symptoms and demyelination in MOG-induced EAE depends on TNFR1. *Eur. J. Immunol.*, **29**, 626–32.

Feliciani, C., Toto, P., America, P. *et al.* (2000). In vitro and in vivo expression of interleukin-1-α and tumor necrosis factor-α mRNA in pemphigus vulgaris: interleukin-1a and tumor necrosis factor-a are involved in acantholysis. *J. Invest. Dermatol.*, **114**, 71–7.

Franco, A., Guidotti, L., Hobbs, M., Pasquetto, V., and Chisari, F. (1997). Pathogenetic effector function of CD4+ T-helper 1 cells in hepatitis B virus transgenic mice. *J. Immunol.*, **159**, 2001–10.

Froelich, C., Orth, K., Turbov, J. *et al.* (1996). New paradigm for lymphocyte granule-mediated cytotoxicity. Target cells bind and internalize granzyme B, but an endosomolytic agent is necessary for cytosolic delivery and apoptosis. *J. Biol. Chem.*, **271**, 29073–9.

Fujii, K., Fujii, Y., Hubscher, S., and Tanaka, Y. (2001). CD44 is the physiological trigger of Fas up-regulation on rheumatoid synovial cells. *J. Immunol.*, **167**, 1198–203.

Galon, J., Aksentijevich, I., McDermott, M., O'Shea, J., and Kastner, D. (2000). TNFRSF1A mutations and autoinflammatory syndromes. *Curr. Opin. Immunol.*, **12**, 479–86.

Garban, H. and Bonavida, B. (2001). Nitric oxide inhibits the transcription repressor yin-yang 1 binding activity at the silencer region of the Fas promoter: a pivotal role for nitric oxide in the up-regulation of Fas gene expression in human tumor cells. *J. Immunol.*, **167**, 75–81.

Giordano, C., Stassi, G., De Maria, R. *et al.* (1997). Potential involvement of Fas and its ligand in the pathogenesis of Hashimoto's thyroiditis. *Science*, **275**, 960–3.

Goverman, J. (1999). Tolerance and autoimmunity in TCR transgenic mice specific for myelin basic protein. *Immunol. Rev.*, **169**, 147–59.

Graser, R., DiLorenzo, T., Wang, F. *et al.* (1999). Identification of a CD8 T cell that can independently mediate autoimmune diabetes development in the complete absence of CD 4 T cell helper functions. *J. Immunol.*, **164**, 3913–18.

Green, E. A. and Flavell, R. A. (2000). The temporal importance of TNFalpha expression in the development of diabetes. *Immunity*, **12**, 459–69.

Grell, M., Zimmermann, G., Gottfried, E. *et al.* (1999). Induction of cell death by tumor necrosis factor (TNF) receptor 2, CD40 and CD30: a role for TNF-R1 activation by endogenous membrane- anchored TNF. *EMBO. J.*, **18**, 3034–43.

Gross, J. A., Johnston, J., Mudri, S. *et al.* (2000). TACI and BCMA are receptors for a TNF homologue implicated in B-cell autoimmune diseases. *Nature*, **404**, 995–9.

Gura, T. (1997). How TRAIL kills cancer cells, but not normal cells. *Science*, **277**, 768.

Haas, C., Ryffel, B., and Hir, M. L. (1998). INF-gamma receptor depletion prevents autoantibody production and glomerulonephritis in lupus-prone (NZB x NZW)F1 mice. *J. Immunol.*, **160**, 3173–8.

Hanninen, A., Jalkanen, S., Salmi, M., Toikkanen, S., Nikolakaros, G., and Simell, O. (1992). Macrophages, T cell receptor usage, and endothelial cell activation in the pancreas at the onset of insulin-dependent diabetes mellitus. *J. Clin. Invest.*, **90**, 1901.

Hegde, R., Srinivasula, S. M., Zhang, Z. *et al.* (2002). Identification of Omi/HtrA2 as a mitochondrial apoptotic serine protease that disrupts inhibitor of apoptosis protein-caspase interaction. *J. Biol. Chem.*, **277**, 432–8.

Heibein, J. A., Goping, I. S., Barry, M. *et al.* (2000). Granzyme B-mediated cytochrome c release is regulated by the Bcl-2 family members bid and Bax. *J. Exp. Med.*, **192**, 1391–402.

Heitmeier, M., Scarim, A., and Corbett, J. (1997). IFN-γ increases the sensitivity of islets of Langerhans for inducible nitric oxide synthase expression induced by interleukin 1. *J. Biol. Chem.*, **272**, 13697.

Heitmeier, M. R., Arnush, M., Scarim, A. L., and Corbett, J. A. (2001). Pancreatic beta-cell damage mediated by beta-cell production of interleukin-1. A novel mechanism for virus-induced diabetes. *J. Biol. Chem.*, **276**, 11151–8.

Helgason, C. D., Prendergast, J. A., Berke, G., and Bleackley, R. C. (1992). Peritoneal exudate lymphocyte and mixed lymphocyte culture hybridomas are cytolytic in the absence of cytotoxic cell proteinases and perforin. *Eur. J. Immunol.*, **22**, 3187–90.

Henkart, P. A. (1985). Mechanism of lymphocyte-mediated cytotoxicity. *Annu. Rev. Immunol.*, **3**, 31–58.

Heusel, J., Wesselschmidt, R., Shresta, S., Russell, J., and Ley, T. (1994). Cytotoxic lymphocytes require granzyme B for the rapid induction of DNA fragmentation and apoptosis in allogeneic target cells. *Cell*, **76**, 977–87.

Hildeman, D. A., Zhu, Y., Mitchell, T. C., Kappler, J., and Marrack, P. (2002). Molecular mechanisms of activated T cell death in vivo. *Curr. Opin. Immunol.*, **14**, 354–9.

Hilliard, B., Wilmen, A., Seidel, C., Liu, T. S., Goke, R., and Chen, Y. (2001). Roles of TNF-related apoptosis-inducing ligand in experimental autoimmune encephalomyelitis. *J. Immunol.*, **166**, 1314–19.

Hiromatsu, Y., Tomoaki, H., Yagita, H. *et al.* (1999). Functional fas ligand expression in thyrocytes from patients with Graves' disease. *J. Clin. Endocrinol. Metab.*, **84**, 2896–902.

Hoorens, A., Stange, G., Pavlovic, D., and Pipeleers, D. (2001). Distinction between interleukin-1-induced necrosis and apoptosis of islet cells. *Diabetes*, **50**, 551–557.

Hotta, M., Tashiro, F., Ikegami, H. *et al.* (1998). Pancreatic B cell-specific expression of thioredoxin, an antioxidative and antiapoptotic protein, prevents autoimmune and streptozotocin-induced diabetes. *J. Exp. Med.*, **188**, 1445–51.

Huseby, E. S., Ohlen, C., and Goverman, J. (1999). Cutting edge: myelin basic protein-specific cytotoxic T cell tolerance is maintained in vivo by a single dominant epitope in H-2^k mice. *J. Immunol.*, **163**, 1115–18.

Iijima, H., Takahashi, I., Kishi, D. *et al.* (1999). Alteration of interleukin 4 production results in the inhibition of T helper type 2 cell-dominated inflammatory bowel disease in T cell receptor alpha chain-deficient mice. *J. Exp. Med.*, **190**, 607–15.

Irmler, M., Thorme, M., Hahne, M. *et al.* (1997). Inhibition of death receptor signals by cellular FLIP. *Nature*, **388**, 190–5.

Itoh, N., Hanafusa, T., Miyazaki, A. *et al.* (1993). Mononuclear cell infiltration and its relation to the expression of major histocompatibility complex antigens and adhesion molecules in pancreas biopsy specimens from newly diagnosed insulin-dependent diabetes mellitus patients. *J. Clin. Invest.*, **153**, 1360–77.

Itoh, N., Imagawa, A., Hanafusa, T. *et al.* (1997). Requirement of Fas for the development of autoimmune diabetes in nonobese diabetic mice. *J. Exp. Med.*, **186**, 613–18.

Ji, H., Korganow, A. S., Mangialaio, S. *et al.* (1999). Different modes of pathogenesis in T-cell-dependent autoimmunity: clues from two TCR transgenic systems. *Immunol. Rev.*, **169**, 139–46.

Jiang, Y., Woronicz, J., Liu, W., and Goeddel, D. (1999). Prevention of constitutive TNF receptor I signaling by silencer of death domain. *Science*, **283**, 543–6.

Kagi, D., Ledermann, B., Burki, K. *et al.* (1994). Cytotoxicity mediated by T cells and natural killer cells is greatly impaired in perforin-deficient mice. *Nature*, **369**, 31–7.

Kagi, D., Odermatt, B., Ohashi, P., Zinkernagel, R., and Hengartner, H. (1996). Development of insulitis without diabetes in transgenic mice lacking perforin-dependent cytotoxicity. *J. Exp. Med.*, **183**, 2143–52.

Kagi, D., Odermatt, B., Seiler, P., Zinkernagel, R., Mak, T., and Hengartner, H. (1997). Reduced incidence and delayed onset of diabetes in perforin-deficient nonobese diabetic mice. *J. Exp. Med.*, **186**, 989–97.

Kanagawa, O., Shimizu, J., and Vaupel, B. A. (2000). Thymic and postthymic regulation of diabetogenic CD8 T cell development in TCR transgenic nonobese diabetic (NOD) mice. *J. Immunol.*, **164**, 5466–73.

Kaplan, M. J., Ray, D., Mo, R. R., Yung, R. L., and Richardson, B. C. (2000). TRAIL (Apo2 ligand) and TWEAK (Apo3 ligand) mediate CD4+ T cell killing of antigen-presenting macrophages. *J. Immunol.*, **164**, 2897–904.

Kaspar, A. A., Okada, S., Kumar, J. *et al.* (2001). A distinct pathway of cell-mediated apoptosis initiated by granulysin. *J. Immunol.*, **167**, 350–6.

Kassiotis, G. and Kollias, G. (2001). Uncoupling the proinflammatory from the immunosuppressive properties of tumor necrosis factor (TNF) at the p55 TNF receptor level: implications for pathogenesis and therapy of autoimmune demyelination. *J. Exp. Med.*, **193**, 427–34.

Kataoka, T., Budd, R. C., Holler, N. *et al.* (2000). The caspase-8 inhibitor FLIP promotes activation of NFκB and erk signaling pathways. *Curr. Biol.*, **10**, 640–8.

Katz, J., Benoist, C., and Mathis, D. (1993). Major histocompatibility complex class I molecules are required for the generation of insulitis in non-obese diabetic mice. *Eur. J. Immunol.*, **23**, 3358–60.

Kawahara, A., Yohsawa, Matsumura, H., Uchiyama, Y., and Nagata, S. (1998). Caspase-independent cell killing by Fas-associated protein with death domain. *J. Cell Biol.*, **143**, 1353–60.

Kayagaki, N., Yamaguchi, N., Nakayama, M. *et al.* (1999). Involvement of TNF-related apoptosis-inducing ligand in human CD4+ T cell-mediated cytotoxicity. *J. Immunol.*, **162**, 2639–47.

Khare, S., Sarosi, I., Xia, X. *et al.* (2000). Severe B cell hyperplasia and autoimmune disease in TALL-1 transgenic mice. *Proc. Natl. Acad. Sci. USA*, **97**, 3370–5.

Kim, S., Kim, K. A., Hwang, D. Y. *et al.* (2000). Inhibition of autoimmune diabetes by Fas ligand: the paradox is solved. *J. Immunol.*, **164**, 2931–6.

Kim, T., Zhao, Y., Barber, M., Kuharsky, D., and Yin, X. (2000). Bid-induced cytochrome c release is mediated by a pathway independent of mitochondrial permeability transition pore and Bax. *J. Biol. Chem.*, **275**, 39474–81.

Koh, D., Fung-Leung, W. P., Ho, A., Gray, D., Acha-Orbea, H., and Mak, T. W. (1992). Less mortality but more relapses in experimental allergic encephalomyelitis in CD8–/– mice. *Science*, **256**, 1210–13.

Kong, Y., Feige, U., Sarosi, I. *et al.* (1999). Activated T cells regulate bone loss and joint destruction in adjuvant arthritis through osteoprotegerin ligand. *Nature*, **402**, 304–9.

Kontoyiannis, D., Pasparakis, M., Pizzaro, T., Cominelli, F., and Kollias, G. (1999). Impaired on/off regulation of TNF biosynthesis in mice lacking TNF AU-rich elements: implications for joint and gut-associated immunopathologies. *Immunity*, **10**, 387–98.

Kovalovich, K., Li, W., DeAngelis, R., Greenbaum, L., Ciliberto, G., and Taub, R. (2001). Interleukin-6 protects against Fas-mediated death by establishing a critical level of anti-apoptotic hepatic proteins FLIP, Bcl-2 and Bcl-xL. *J. Biol. Chem.*, **276**, 26605–13.

Krammer, P. H. (2000). CD95's deadly mission in the immune system. *Nature*, **407**, 789–95.

Krensky, A. M. (2000). Granulysin: a novel antimicrobial peptide of cytolytic T lymphocytes and natural killer cells. *Biochem. Pharmacol.*, **59**, 317–20.

Kreuwel, H. T., Morgan, D. J., Krahl, T., Ko, A., Sarvetnick, N., and Sherman, L. A. (1999). Comparing the relative role of perforin/granzyme versus Fas/Fas ligand cytotoxic pathways in CD8+ T cell-mediated insulin-dependent diabetes mellitus. *J. Immunol.*, **163**, 4335–41.

Kurasawa, K., Hirose, K., Sano, H. *et al.* (2000). Increased interleukin 17 production in patients with systemic sclerosis. *Arthritis. Rheum.*, **43**, 2455–60.

Lee, S., Zhou, T., Choi, C., Wang, Z., and Benveniste, E. (2000). Differential regulation and function of Fas expression on glial cells. *J. Immunol.*, **164**, 1277–85.

Lehmann, C., Zeis, M., Schmitz, N., and Uharek, L. (2000). Impaired binding of perforin on the surface of tumor cells is a cause of target cell resistance against cytotoxic effector cells. *Blood*, **96**, 594–600.

Li, H., Zhu, H., Xu, C., and Yuan, J. (1998). Cleavage of BID by caspase 8 mediates the mitochondrial damage in the Fas pathway of apoptosis. *Cell*, **94**, 491–501.

Li, M. and Berg, A. (2000). Induction of necrotic-like cell death by tumor necrosis factor alpha and caspase inhibitors: novel mechanism for killing virus-infected cells. *J. Virol.*, **74**, 7470–7.

Liu, D., Pavlovic, D., Chen, M. C., Flodstrom, M., Sandler, S., and Eizirik, D. L. (2000). Cytokines induce apoptosis in beta-cells isolated from mice lacking the inducible isoform of nitric oxide synthase (iNOS–/–). *Diabetes*, **49**, 1116–22.

Liu, J., Marino, M., Wong, G. *et al.* (1998). TNF is a potent anti-inflammatory cytokine in autoimmune-mediated demyelination. *Nat. Med.*, **4**, 78–82.

Lovell, D., Giannini, E., Reiff, A. *et al.* (2000). Etanercept in children with polyarticular juvenile rheumatoid arthritis. *N. Engl. J. Med.*, **342**, 763–9.

Lowin, B., Beermann, F., Schmidt, A., and Tschopp, J. (1994). A null mutation in the perforin gene impairs cytolytic T lymphocyte- and natural killer cell-mediated cytotoxicity. *Proc. Natl. Acad. Sci. USA*, **91**, 11571–5.

Lubberts, E., Joosten, L., Oppers, B. *et al.* (2001). IL-1-independent role of IL-17 in synovial inflammation and joint destruction during collagen-induced arthritis. *J. Immunol.*, **167**, 1004–13.

Luo, X., Budihardjo, I., Zou, H., Slaughter, C., and Wang, X. (1998). Bid, a Bcl2 interacting protein, mediates cytochrome c release from mitochondria in response to activation of cell surface death receptors. *Cell*, **94**, 481–90.

Mackay, F., Woodcock, S., Lawton, P. *et al.* (1999). Mice transgenic for BAFF develop lymphocytic disorders along with autoimmune manifestations. *J. Exp. Med.*, **190**, 1697–710.

Mathews, C. E., Graser, R. T., Savinov, A., Serreze, D. V., and Leiter, E. H. (2001). Unusual resistance of ALR/Lt mouse beta cells to autoimmune destruction: role for beta cell-expressed resistance determinants. *Proc. Natl. Acad. Sci. USA*, **98**, 235–40.

Medana, I., Gallimore, A., Oxenius, A., Martinic, M., Wekerle, H., and Neumann, H. (2000). MHC class I-restricting killing of neurons by virus-specific CD8+ T lymphocytes is effected through the Fas/FasL, but not the perforin pathway. *Eur. J. Immunol.*, **30**, 3623–33.

Medana, I., Li, Z., Flugel, A., Tschopp, J., Wekerle, H., and Neumann, H. (2001). Fas ligand (CD95L) protects neurons against perforin-mediated T-lymphocyte cytotoxicity. *J. Immunol.*, **167**, 674–81.

Mendel, I., Katz, A., Kozak, N. *et al.* (1998). Interleukin-6 functions in autoimmune encephalomyelitis: a study in gene-targeted mice. *Eur. J. Immunol.*, **28**, 1727–37.

Mitsiades, N., Poulaki, V., Mastorakkpsi, G., Tseleni-Balafouta, S., Kotoula, V., and Koutras, A. (1999). Fas ligand expression in thyroid carcinomas: a potential mechanism for immune evasion. *J. Clin. Endocrinol. Metab.*, **84**, 2924–32.

Morgan, D., Liblan, R., Scott, B. *et al.* (1996). CD8+ T-cell-mediated spontaneous diabetes in neonatal mice. *J. Immunol.*, **157**, 978–83.

Moriwaki, M., Itoh, N., Miyagawa, J. *et al.* (1999). Fas and Fas ligand expression in inflamed islets in pancreas sections of patients with recent-onset type I diabetes mellitus. *Diabetologia*, **42**, 1332–40.

Motyka, B., Korbutt, G., Pinkoski, M. J. *et al.* (2000). Mannose 6-phosphate/insulin-like growth factor II receptor is a death receptor for granzyme B during cytotoxic T cell-induced apoptosis. *Cell*, **103**, 491–500.

Muzio, M., Chinnaiyan, A. M., Kischkel, F. C. *et al.* (1996). FLICE, a novel FADD-homologous ICE/CED-3-like protease, is recruited to the CD95 (Fas/APO-1) death-inducing signaling complex. *Cell*, **85**, 817–27.

Nagata, M., Santamaria, P., Kawamura, T., Utsugi, T., and Yoon, J.-W. (1994). Evidence for the role of CD8+ cytotoxic T cells in the destruction of pancreatic beta cells in NOD mice. *J. Immunol.*, **152**, 2042–50.

Nakamoto, Y., Guidotti, L., Pasquetto, V., Schreiber, R., and Chisari, F. (1997). Differential target cell sensitivity to CTL-activated death pathways in hepatitis B virus transgenic mice. *J. Immunol.*, **158**, 5692.

Ohashi, P., Oehen, S., Buerki, K. *et al.* (1991). Ablation of tolerance and induction of diabetes by virus infection in viral antigen transgenic mice. *Cell*, **65**, 305–17.

Ohta, A., Sekimoto, M., Sato, M. *et al.* (2000). Indispensable role for TNF-alpha and IFN-gamma at the effector phase of liver injury mediated by Th1 cells specific to hepatitis B virus surface antigen. *J. Immunol.*, **165**, 956–61.

Okuda, Y., Sakoda, S., and Bernard, C. (1998). IL-6-deficient mice are resistant to the induction of experimental encephalomyelitis provoked by myelin oligodendrocyte glycoprotein. *Int. Immunol.*, **10**, 703–8.

Ostergaard, H. L., Kane, K. P., Mescher, M. F., and Clark, W. R. (1987). Cytotoxic T lymphocyte mediated lysis without release of serine esterase. *Nature*, **330**, 71–2.

Owens, T., Wekerle, H., and Antel, J. (2001). Genetic models for CNS inflammation. *Nat. Med.*, **7**, 161–6.

Pakala, S., Chivetta, M., Kelly, C., and Katz, J. (1999). In autoimmune diabetes the transition from benign to pernicious insulitis requires an islet cell response to tumor necrosis factor alpha. *J. Exp. Med.*, **189**, 1053–62.

Peng, S., Moslehi, J., and Craft, J. (1997). Roles of interferon-gamma and interleukin-4 in murine lupus. *J. Clin. Invest.*, **99**, 1936–46.

Piguet, P., Vesin, C., Guo, J., Donati, Y., and Barazzone, C. (1998). TNF-induced enterocyte apoptosis in mice is mediated by the TNF receptor 1 and does not require p53. *Eur. J. Immunol.*, **28**, 3499–505.

Pimentel-Muinos, F. and Seed, B. (1999). Regulated commitment of TNF receptor signaling: a molecular switch for death or activation. *Immunity*, **11**, 783–93.

Pinkoski, M. J., Hobman, M., Heibein, J. A. *et al.* (1998). Entry and trafficking of granzyme B in target cells during granzyme B-perforin-mediated apoptosis. *Blood*, **92**, 1044–54.

Pinkoski, M. J., Waterhouse, N. J., Heibein, J. A. *et al.* (2001). Granzyme B-mediated apoptosis proceeds predominantly through a Bcl-2-inhibitable mitochondrial pathway. *J. Biol. Chem.*, **276**, 12060–7.

Powell, M., Mitchell, D., Lederman, J. *et al.* (1990). Lymphotoxin and tumor necrosis factor-alpha production by myelin basic protein-specific T cell clones correlates with encephalitogenicity. *Int. Immunol.*, **2**, 539–44.

Renno, T., Krakowski, M., Piccirillo, C., Lin, J.-Y., and Owens, T. (1995). TNF-α expression by resident microglia and infiltrating leukocytes in the central nervous system of mice with experimental allergic encephalomyelitis. *J. Immunol.*, **154**, 944–53.

Rescigno, M., Piguet, V., Valzasina, B. *et al.* (2000). Fas engagement induces the maturation of dendritic cells (DCs), the release of interleukin-1beta, and the production of interferon gamma in the absence of IL-12 during DC-T cell cognate interaction. A new role for Fas ligand in inflammatory responses. *J. Exp. Med.*, **192**, 1661–8.

Riminton, D., Korner, H., Strickland, D., Lemckert, F., Pollard, J., and Sedgwick, J. (1998). Challenging cytokine redundancy: inflammatory cell movement and clinical course of experimental autoimmune encephalomyelitis are normal in lymphotoxin-deficient mice, but not tumor necrosis factor-deficient mice. *J. Exp. Med.*, **187**, 1517–28.

Rouvier, E., Luciani, M. F., and Golstein, P. (1993). Fas involvement in Ca(2+)-independent T cell-mediated cytotoxicity. *J. Exp. Med.*, **177**, 195–200.

Saas, P., Boucraut, J., Quiquerez, A. *et al.* (1999). CD95 9fas/Apo-1 as a receptor governing astrocyte apoptotic or inflammatory responses: a key role in brain inflammation? *J. Immunol.*, **162**, 2326–33.

Sabelko, K., Kelly, K., Nahm, M., Cross, A., and Russell, J. (1997). Fas and Fas ligand enhance the pathogenesis of experimental allergic encephalomyelitis, but are not essential for immune privilege in the central nervous system. *J. Immunol.*, **159**, 3096–9.

Sabelko-Downes, K., Cross, A., and Russell, J. (1999). Dual role for Fas ligand in the initiation of and recovery from experimental allergic encephalomyelitis. *J. Exp. Med.*, **189**, 1195.

Sakata, K., Sakata, A., Vela-Roch, N. *et al.* (1998). Fas (CD95)-transduced signal preferentially stimulates lupus peripheral T lymphocytes. *Eur. J. Immunol.*, **28**, 2648–60.

Samoilova, E., Horton, J., Hilliard, B. *et al.* (1998). IL-6-deficient mice are resistant to experimental autoimmune encephalomyelitis: roles of OL-6 in the activation and differentiation of autoreactive T-cells. *J. Immunol.*, **161**, 6480–6.

Santamaria, P., Lewis, C., Sutherland, D., and Barbosa, J. (1992a). CD8+ T cells from isletitis of graft-recurrent type I diabetes are oligoclonal and show restricted TCR usage. *Diabetes*, **41** (Suppl. 1), 97A.

Santamaria, P., Nakhleh, R. E., Sutherland, D. E. R., and Barbosa, J. J. (1992b). Isolation and characterization of T lymphocytes infiltrating a human pancreas allograft affected by isletitis and recurrent diabetes. *Diabetes*, **41**, 53–61.

Santamaria, P., Utsugi, T., Park, B., Averill, N., Kawazu, S., and Yoon, J. (1995). Beta cell cytotoxic CD8+ T cells from non-obese diabetic mice use highly homologous T cell receptor alpha chain CDR3 sequences. *J. Immunol.*, **154**, 2494–503.

Satoh, M., Weintraub, J. P., Yoshida, H. *et al.* (2000). Fas and Fas ligand mutations inhibit autoantibody production in pristane-induced lupus. *J. Immunol.*, **165**, 1036–43.

Scaffidi, C., Fulda, S., Srinivasan, A. *et al.* (1998). Two CD95 (APO-1/Fas) signaling pathways. *EMBO. J.*, **17**, 1675–87.

Schmidt, D., Amrani, A., Verdaguer, J., Bou, S., and Santamaria, P. (1999). Autoantigen-independent deletion of diabetogenic CD4+ thymocytes by protective MHC class II molecules. *J. Immunol.*, **162**, 4627–36.

Schmidt, D., Verdaguer, J., Averill, N., and Santamaria, P. (1997). A mechanism for the major histocompatibility complex-linked resistance to autoimmunity. *J. Exp. Med.*, **186**, 1059–75.

Schneider, P., Mackay, F., Steiner, V. *et al.* (1999). BAFF, a novel ligand of the tumor necrosis factor family, stimulates B cell growth. *J. Exp. Med.*, **189**, 1747–56.

Seewaldt, S., Thomas, H. E., Ejrnaes, M. *et al.* (2000). Virus-induced autoimmune diabetes: most beta-cells die through inflammatory cytokines and not perforin from autoreactive (anti-viral) cytotoxic T-lymphocytes. *Diabetes*, **49**, 1801–9.

Segal, R., Dayan, M., Zinger, H., and Mozes, E. (2001). Suppression of experimental systemic lupus erythematosus (SLE) in mice via TNF inhibition by an anti-TNFα monoclonal antibody and by pentoxiphylline. *Lupus*, **10**, 23–31.

Serreze, D., Leiter, E., Christianson, G., Greiner, D., and Roopenian, D. (1994). Major histocompatibility complex class I-deficient NOD.β1m^{null} mice are diabetes and insulitis resistant. *Diabetes*, **43**, 505–8.

Serreze, D. V., Post, C. M., Chapman, H. D., Johnson, E. A., Lu, B., and Rothman, P. B. (2000). Interferon-gamma receptor signaling is dispensable in the development of autoimmune type 1 diabetes in NOD mice. *Diabetes*, **49**, 2007–11.

Sharif-Askari, E., Alam, A., Rheaume, E. *et al.* (2001). Direct cleavage of the human DNA fragmentation factor-45 by granzyme B induces caspase-activated DNase release and DNA fragmentation. *EMBO. J.*, **20**, 3101–13.

Shi, L., Kraut, R. P., Aebersold, R., and Greenberg, A. H. (1992). A natural killer cell granule protein that induces DNA fragmentation and apoptosis. *J. Exp. Med.*, **175**, 553–66.

Shi, L., Mai, S., Israels, S., Browne, K., Trapani, J., and Greenberg, A. (1997). Granzyme B (GraB) autonomously crosses the cell membrane and perforin initiates apoptosis and GraB nuclear localization. *J. Exp. Med.*, **185**, 855–66.

Shresta, S., Graubert, T. A., Thomas, D. A., Raptis, S. Z., and Ley, T. J. (1999). Granzyme A initiates an alternative pathway for granule-mediated apoptosis. *Immunity*, **10**, 595–605.

Sibley, R. K., Sutherland, D. E. R., Goetz, F., and Michael, A. F. (1985). Recurrent diabetes mellitus in the pancreas iso- and allograft. *Lab. Invest.*, **53**, 132–44.

Siegel, R., Chang, F., Chun, H., and Lenardo, M. (2000a). The multifaceted role of fas signaling in immune cell homeostasis and autoimmunity. *Nat. Immunol.*, **1**, 469–74.

Siegel, R. M., Chan, F. K., Chun, H. J., and Lenardo, M. J. (2000b). The multifaceted role of Fas signaling in immune cell homeostasis and autoimmunity. *Nat. Immunol.*, **1**, 469–74.

Simon, M., Hausmann, M., Tran, T. *et al.* (1997). In vitro-and ex-derived cytolytic leukocytes from granzyme A × B double knockout mice are defective in granule-mediated apoptosis but not lysis of target cells. *J. Exp. Med.*, **186**, 1781–6.

Somoza, N., Vargas, F., Roura-Mir, C. *et al.* (1994). Pancreas in recent onset insulin-dependent diabetes mellitus. *J. Immunol.*, **153**, 1360–77.

Song, K., Chen, Y., Goke, R. *et al.* (2000). Tumor necrosis factor-related apoptosis-inducing ligand (TRAIL) is an inhibitor of autoimmune inflammation and cell cycle progression. *J. Exp. Med.*, **191**, 1095–104.

Stassi, G., DeMaria, R., Trucco, G. *et al.* (1997). Nitric oxide primes pancreatic beta cells for Fas-mediated destruction in insulin-dependent diabetes mellitus. *J. Exp. Med.*, **186**, 1193.

Su, X., Hu, Q., Kristan, J. M. *et al.* (2000). Significant role for Fas in the pathogenesis of autoimmune diabetes. *J. Immunol.*, **164**, 2523–32.

Sun, D., Whitaker, J., Huang, Z. *et al.* (2001). Myelin antigen-specific CD8+ T-cells are encephalitogenic and produce severe disease in C57BL/6 mice. *J. Immunol.*, **166**, 7579–87.

Sutton, V. R., Davis, J. E., Cancilla, M. *et al.* (2000). Initiation of apoptosis by granzyme B requires direct cleavage of bid, but not direct granzyme B-mediated caspase activation. *J. Exp. Med.*, **192**, 1403–14.

Suvannavejh, G. C., Dal Canto, M. C., Matis, L. A., and Miller, S. D. (2000a). Fas-mediated apoptosis in clinical remissions of relapsing experimental autoimmune encephalomyelitis. *J. Clin. Invest.*, **105**, 223–31.

Suvannavejh, G. C., Lee, H. O., Padilla, J., Dal Canto, M. C., Barrett, T. A., and Miller, S. D. (2000b). Divergent roles for p55 and p75 tumor necrosis factor receptors in the pathogenesis of MOG(35–55)-induced experimental autoimmune encephalomyelitis. *Cell Immunol.*, **205**, 24–33.

Suzuki, H., Kundig, T., Furlonger, C. *et al.* (1995). Deregulated T cell activation and autoimmunity in mice lacking interleukin-2 receptor beta. *Science*, **268**, 1472–6.

Taylor, G., Carballo, E., Lee, D. *et al.* (1996). A pathogenetic role for TNFα in the syndrome of cachexia, arthritis and autoimmunity resulting from tristetraprolin (TTP) deficiency. *Immunity*, **4**, 445–54.

Teng, Y., Nguyen, H., Gao, X. *et al.* (2000). Functional human T cell immunity and osteoprotegerin ligand control alveolar bone destruction in periodontal infection. *J. Clin. Invest.*, **106**, R59–7.

Thomas, D. A., Scorrano, L., Putcha, G. V., Korsmeyer, S. J., and Ley, T. J. (2001). Granzyme B can cause mitochondrial depolarization and cell death in the absence of BID, BAX, and BAK. *Proc. Natl. Acad. Sci. USA*, **98**, 14985–90.

Thomas, H., Darwiche, R., Corbett, J., and Kay, T. (2002). Interleukin-1 plus γ-interferon-induced pancreatic beta cell dysfunction is mediated by beta cell nitric oxide production. *Diabetes*, **51**, 311–16.

Thomas, H. E., Darwiche, R., Corbett, J. A., and Kay, T. W. (1999). Evidence that beta cell death in the nonobese diabetic mouse is Fas independent. *J. Immunol.*, **163**, 1562–9.

Tisch, R. and McDevitt, H. (1996). Insulin-dependent diabetes mellitus. *Cell*, **85**, 291–7.

Tran, S., Holmstrom, T., Ahonen, M., Kahari, V., and Eriksson, J. (2001). MAPK/ERK overrides the apoptotic signaling from Fas, TNF and TRAIL receptors. *J. Biol. Chem.*, **276**, 16484–90.

Trautmann, A., Akdis, M., Kleemann, D. *et al.* (2000). T cell-mediated Fas-induced keratinocyte apoptosis plays a key pathogenetic role in eczematous dermatitis. *J. Clin. Invest.*, **106**, 25–35.

Tsuchida, T. (1994). Autoreactive CD8+ T-cell responses to human myelin protein-derived peptides. *Proc. Natl. Acad. Sci. USA*, **91**, 10859–63.

Vercammen, D., Brouckaert, G., Denecker, G. *et al.* (1998). Dual signaling of the Fas receptor: initiation of both apoptotic and necrotic cell death pathways. *J. Exp. Med.*, **188**, 919–30.

Verdaguer, J., Schmidt, D., Amrani, A., Anderson, B., Averill, N., and Santamaria, P. (1997). Spontaneous autoimmune diabetes in monoclonal T cell nonobese diabetic mice. *J. Exp. Med.*, **186**, 1663–76.

Verdaguer, J., Yoon, J.-W., Anderson, B. *et al.* (1996). Acceleration of spontaneous diabetes in TCRβ-transgenic nonobese diabetic mice by beta cell-cytotoxic CD8+ T cells expressing identical endogenous TCRα chains. *J. Immunol.*, **157**, 4726–35.

Verhagen, A. M., Ekert, P. G., Pakusch, M. *et al.* (2000). Identification of DIABLO, a mammalian protein that promotes apoptosis by binding to and antagonizing IAP proteins. *Cell*, **102**, 43–53.

Verma, S., Hutchings, P., Guo, J., McLachlan, S., Rapoport, B., and Cooke, A. (2000). Role of MHC class I expression and CD8(+) T cells in the evolution of iodine-induced thyroiditis in NOD-H2(h4) and NOD mice. *Eur. J. Immunol.*, **30**, 1191–202.

Villunger, A., Huang, D., Holler, N., Tschopp, J., and Strasser, A. (2000). Fas ligand-induced c-Jun kinase activation in lymphoid cells requires extensive receptor aggregation but is independent of DAXX, and Fas-mediated cell death does not involve DAXX, RIP or RAIDD. *J. Immunol.*, **165**, 1337–43.

Wahlsten, J. L., Gitchell, H. L., Chan, C. C., Wiggert, B., and Caspi, R. R. (2000). Fas and Fas ligand expressed on cells of the immune system, not on the target tissue, control induction of experimental autoimmune uveitis. *J. Immunol.*, **165**, 5480–86.

Waldner, H., Sobel, R., Howard, E., and Kuchroo, V. (1997). Fas- and FasL-deficient mice are resistant to induction of autoimmune encephalomyelitis. *J. Immunol.*, **159**, 3100–3.

Walter, U., Frantzke, A., Sarukhan, A. *et al.* (2000). Monitoring gene expression of TNFR family members by beta-cells during development of autoimmune diabetes. *Eur. J. Immunol.*, **30**, 1224–32.

Wang, B., Gonzalez, A., Benoist, C., and Mathis, D. (1996). The role of CD8+ T-cells in initiation of insulin-dependent diabetes mellitus. *Eur. J. Immunol.*, **26**, 1762–9.

Wang, H. B., Li, H., Shi, F. D., Chambers, B. J., Link, H., and Ljunggren, H. G. (2000). Tumor necrosis factor receptor-1 is critically involved in the development of experimental autoimmune myasthenia gravis. *Int. Immunol.*, **12**, 1381–8.

Wang, J., Zheng, L., Lobito, A. *et al.* (1999). Inherited human caspase 10 mutations underlie defective lymphocyte and dendritic cell apoptosis in autoimmune lymphoproliferative syndrome type II. *Cell*, **98**, 47–58.

Wicker, L., Leiter, E., Todd, J. *et al.* (1994). β2-microglobulin-deficient NOD mice do not develop insulitis or diabetes. *Diabetes*, **43**, 500–4.

Wildbaum, G., Westermann, J., Maor, G., and Karin, N. (2000). A targeted DNA vaccine encoding fas ligand defines its dual role in the regulation of experimental autoimmune encephalomyelitis. *J. Clin. Invest.*, **106**, 671–9.

Wong, F., Karttunen, J., Dumont, C. *et al.* (1999). Identification of an MHC class I-restricted autoantigen in type 1 diabetes by screening an organ-specific cDNA library. *Nat. Med.*, **9**, 1026–31.

Xia, X. Z., Treanor, J., Senaldi, G. *et al.* (2000). TACI is a TRAF-interacting receptor for TALL-1, a tumor necrosis factor family member involved in B cell regulation. *J. Exp. Med.*, **192**, 137–43.

Yamada, K., Takane-Gyotoku, N., Ichikawa, F., Inada, C., and Nokada, K. (1996). Mouse islet cell lysis mediated by interleukin-1-induced Fas. *Diabetologia*, **39**, 1306–12.

Zeine, R., Pon, R., Ladiwala, U., Antel, J., Filion, L., and Freedman, M. (1998). Mechanism of gamma delta T cell-induced human oligodendrocyte cytotoxicity: relevance to multiple sclerosis. *J. Neuroimmunol.*, **87**, 49–61.

Zhang, B., Yamamura, T., Kondo, T., Fujiwara, M., and Tabira, T. (1997). Regulation of experimental allergic encephalomyelitis by natural killer (NK) cells. *J. Exp. Med.*, **186**, 1677–87.

Zhang, D., Beresford, P. J., Greenberg, A. H., and Lieberman, J. (2001a). Granzymes A and B directly cleave lamins and disrupt the nuclear lamina during granule-mediated cytolysis. *Proc. Natl. Acad. Sci. USA*, **98**, 5746–51.

Zhang, D., Pasternack, M. S., Beresford, P. J., Wagner, L., Greenberg, A. H., and Lieberman, J. (2001b). Induction of rapid histone degradation by the cytotoxic T lymphocyte protease granzyme A. *J. Biol. Chem.*, **276**, 3683–90.

Pro- and anti-apoptotic strategies of viruses

Helmut Fickenscher[1], Bernhard Fleckenstein[2], and Armin Ensser[2]

[1]Abteilung Virologie, Hygiene-Institut, Ruprecht-Karls-Universität Heidelberg, Heidelberg, Germany
[2]Institut für Klinische und Molekulare Virologie, Friedrich-Alexander-Universität Erlangen-Nürnberg, Erlangen, Germany

7.1 Introduction

Apoptosis (Kerr *et al.*, 1972) or programmed cell death is a highly regulated and precisely coordinated program permitting the specific elimination of target cells, while neighboring cells are hardly affected. The remains of the apoptotic cells are then readily digested by phagocytes. Apoptosis is an important function in cell differentiation, embryonal development, and proliferation control. In the immune system, apoptosis is the key to the deletion of auto-reactive lymphocytes, to the regulation and restriction of immune responses, and to the elimination of cells infected by intracellular pathogens such as viruses. Consequently, defects in apoptotic pathways are associated with tumor development, autoimmune disease, immunodeficiency, and severe infections. The efficient and cautious elimination of virus-infected cells by apoptosis also degrades viral nucleic acids, even of genomically integrated proviruses. Thus, non-infectious fragments are produced, preventing the uptake of functional viral genomes by neighboring cells or phagocytes. Simultaneously, apoptotic protein material can be processed by phagocytes and other antigen-presenting cells for the presentation to helper and effector immune cells. While pro-apoptotic mechanisms are utilized by some viruses in their life cycle – e.g. for the efficient release of infectious particles – many viruses have developed anti-apoptotic functions for preventing the premature termination of their replicative cycle and for establishing latent persistence (Hay and Kannourakis, 2002). Some viruses even induce apoptosis for attacking immune cells which are directed against the virus-infected cells. Viruses utilize proteins with functional and often structural homology to cellular factors involved in apoptosis. Molecular piracy is a widespread viral strategy to collect immune modulatory or anti-apoptotic functions.

Apoptosis in Health and Disease: Clinical and Therapeutic Aspects, ed. Martin Holcik, Alex E. MacKenzie, Robert G. Korneluk, and Eric C. LaCasse. Published by Cambridge University Press. © Cambridge University Press 2004.

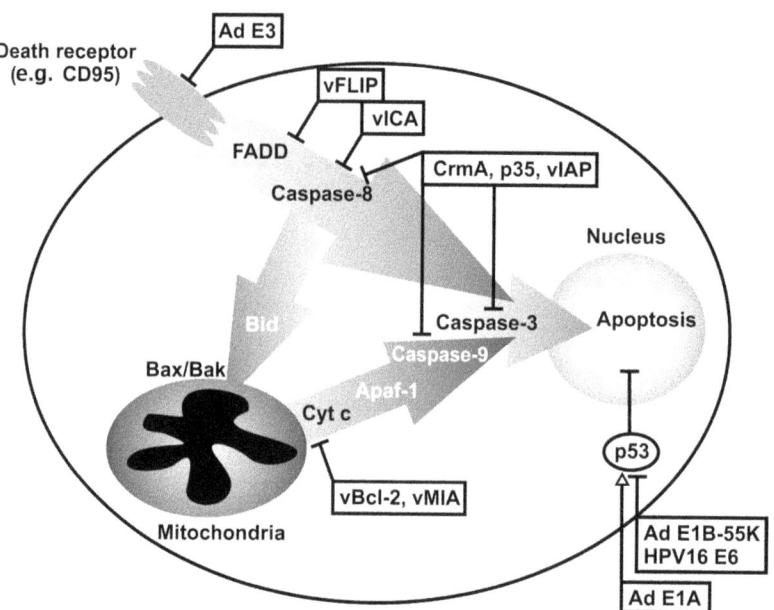

Figure 7.1   *Cellular key targets for viral factors interfering with apoptosis pathways.*

7.2   Major anti-apoptotic pathways targeted by viruses

The central and terminal events in apoptosis are mediated by cysteine-containing, aspartate-specific hydrolases termed caspases. Two distinct principal signaling pathways lead to caspase activation and cell death (Strasser *et al.*, 1995; Kaufmann and Hengartner, 2001). The first signaling pathway involves the interaction of specific soluble ligands with death receptors on the surface of eukaryotic cells, the other pathway is initiated by cell-autonomous signals within the cell and leads to the release of mitochondrial factors into the cytoplasm followed by caspase activation. Numerous viruses influence either or both of these signaling pathways in order to induce or block apoptotic cell death (Figure 7.1, Tables 7.1 and 7.2).

7.2.1   Death receptor-dependent apoptosis

The well-characterized prototype of death receptors (DR) is CD95, also known as Fas or APO-1, which is activated by its specific death ligand CD95L. The DR family (Schulze-Osthoff *et al.*, 1998) further comprises the structurally related transmembrane molecules tumor-necrosis factor receptor (TNF-R1, CD120a), DR3 (APO-3, LARD, TRAMP, WSL1), DR4 (TRAILR1, APO-2), DR5 (TRAILR2, KILLER, TRICK2), and DR6. The major functional domain of these type-I transmembrane DR-proteins is the intracytoplasmic "death domain" (DD). Receptor ligation results in homotrimerization and in homotypic interaction of the DR via its DD with

Table 7.1 *Viral anti-apoptotic effectors*

Anti-apoptotic mechanism	Viral protein	Virus	Reference
Death-receptor block	E3 RID complex	Adenovirus	Friedman and Horwitz, 2002, McNees and Gooding, 2002
	vICA, UL36	Human cytomegalovirus	Skaletskaya *et al.*, 2001
	vFLIP, E8	Equine herpesvirus type 2	Bertin *et al.*, 1997, Thome *et al.*, 1997
	vFLIP, BORFE2	Bovine herpesvirus type 4	Thome *et al.*, 1997, Wang *et al.*, 1997
	vFLIP, ORF71	Herpesvirus saimiri	Glykofrydes *et al.*, 2000, Thome *et al.*, 1997
	vFLIP, ORF71, K13	Kaposi's sarcoma-associated herpesvirus	Djerbi *et al.*, 1999, Thome *et al.*, 1997
	vFLIP, MC159L	Molluscum contagiosum virus	Bertin *et al.*, 1997, Hu *et al.*, 1997, Thome *et al.*, 97
	CrmE	Vaccinia virus	Reading *et al.*, 2002
	M-T2	Myxoma virus	Schreiber *et al.*, 1997
Caspase inhibition	p35, vIAP	Baculovirus	Clem *et al.*, 1991, Crook *et al.*, 1993
	K7, vIAP, survivin	Kaposi's sarcoma-associated herpesvirus	Wang *et al.*, 2002
	CrmA, SPI-2	Poxviruses	Tewari *et al.*, 1995, Dobbelstein and Shenk, 1996
	A224L, 4CL	African swine fever virus	Chacon *et al.*, 1995, Neilan *et al.*, 1997, Nogal *et al.*, 2001
Bcl-2 homologs	E1B 19K	Adenovirus	Han *et al.*, 1998, Perez and White, 2000
	BHRF-1	Epstein–Barr virus	Henderson *et al.*, 1993, Dumont *et al.*, 1999
	BALF-1	Epstein–Barr virus	Bellows *et al.*, 2002, Marshall *et al.*, 1999
	vBcl-2, ORF16	Herpesvirus saimiri	Derfuss *et al.*, 1998, Nava *et al.*, 1997
	vBcl-2	Kaposi's sarcoma-associated herpesvirus	Bellows *et al.*, 2000, Cheng *et al.*, 1997
	vBcl-2, M11	Murine γ-herpesvirus 68	Wang *et al.*, 1999, Gangappa *et al.*, 2002
	vBcl-2	Bovine herpesvirus type 4	Bellows *et al.*, 2000
	A179L, 5-HL	African swine fever virus	Afonso *et al.*, 1996, Neilan *et al.*, 1993

(*cont.*)

Table 7.1 (cont.)

Anti-apoptotic mechanism	Viral protein	Virus	Reference
Mitochondrial stabilization	p13II	Human T-cell leukemia virus	Ciminale *et al.*, 1999, D'Agostino *et al.*, 2002
	Tat	Human immunodeficiency virus	Kruman *et al.*, 1998, Macho *et al.*, 1999
	Nef	Human immunodeficiency virus	Geleziunas *et al.*, 2001
	E6	Human papilloma virus	Jackson *et al.*, 2000
	ICP4, US3	Herpes simplex virus type 1	Galvan and Roizman, 1998, Munger and Roizman, 2001, Zhou and Roizman, 2000
	M11L	Myxoma virus	Everett *et al.*, 2000
Mitochondrial pore modulators	HBx	Hepatitis B virus	Diao *et al.*, 2001, Rahmani *et al.*, 2000
	vMIA, UL37	Cytomegalovirus	Goldmacher *et al.*, 1999
p53 blockade	E1B-55K, E4orf6	Adenovirus	Debbas and White, 1993, Moore *et al.*, 1996
	E6	Human papilloma virus	Scheffner *et al.*, 1993
	T antigen	Simian virus 40	Yanai and Obinata, 1994
Endoplasmic reticulum blockade	M-T4	Myxoma virus	Barry *et al.*, 1997, Hnatiuk *et al.*, 1999
Protein kinase R blockade	$\sigma 3$	Reovirus	Imani and Jacobs, 1988
	NS1	Influenza virus	Bergmann *et al.*, 2000, Zhirnov *et al.*, 2002
	NS5A	Hepatitis C virus	Gale *et al.*, 1997
	Tat	Human immunodeficiency virus	Brand *et al.*, 1997
	$\gamma_1 34.5$	Herpes simplex virus type 1	Chou and Roizman, 1992
	E3L, K3L	Vaccinia virus	Lee and Esteban, 1994

cellular DD-containing adapter molecules (FADD or TRADD, respectively), thus forming the death-inducing signaling complex. The adapter molecules thereby undergo conformational changes exposing "death effector domains" (DED) which again form homotypic interactions with the DEDs of procaspases. The procaspases are then typically activated by auto-catalytic proteolysis (reviewed in Schulze-Osthoff *et al.*, 1998). DR engagement evokes the activation of two distinct

Table 7.2 *Viral pro-apoptotic effects*

Pro-apoptotic mechanism	Viral protein	Virus	Reference
Receptor engagement	gp120, gp41	Human immunodeficiency virus	Banda *et al.*, 1992, Kolesnitchenko *et al.*, 1995
	σ1	Reovirus	Tyler *et al.*, 1995, Clarke *et al.*, 2000
	E2	Sindbis virus	Ubol *et al.*, 1994, Levine *et al.*, 1993, 1996
Caspase activation	E4orf4	Adenovirus	Livne *et al.*, 2001
	E10	Equine herpesvirus type 2	Thome *et al.*, 1999, Yan *et al.*, 1999
	NS1	Parvovirus B19	Moffat *et al.*, 1998, Yaegashi *et al.*, 1999
Mitochondrial damage	Tat	Human immunodeficiency virus	Kruman *et al.*, 1998, Macho *et al.*, 1999
	Vpr	Human immunodeficiency virus	Jacotot *et al.*, 2000, Somasundaran *et al.*, 2002
	Nef	Human immunodeficiency virus	Rasola *et al.*, 2001
	PB1-F2	Influenza virus	Chen *et al.*, 2001
	Apoptin, VP3	Chicken anemia virus	Danen-van Oorschot *et al.*, 1997
	E6	Human papilloma virus 16	Brown *et al.*, 1997
p53 potentiation	E1A	Adenovirus	Debbas and White, 1993, Sabbatini *et al.*, 1995
	E2	Human papilloma virus	Webster *et al.*, 2000

response pathways leading to caspase activation (Scaffidi *et al.*, 1998). In type I cells, caspase-8 is directly and strongly activated and, in turn, activates the major effector caspase-3. In type II cells, caspase-8 activation is not sufficient for directly activating caspase-3; instead, the cytoplasmic Bid protein is cleaved which then initiates the mitochondrial pathway, resulting in the activation of caspase-9 and caspase-3 in such cells.

The DR-mediated apoptosis pathway is efficiently inhibited by various viruses, especially by large DNA viruses such as adeno-, herpes-, or pox-viruses. By using advanced biomathematical data mining methods, a new class of DED-containing viral factors was discovered. These molecules with two DEDs, but without DD, interact via their DEDs with the adapter molecule FADD and/or with procaspase-8 (FLICE), thus blocking formation of the death-inducing signaling complex.

These blocking factors were termed FLICE-inhibitory proteins (FLIPs). Such viral factors (vFLIPs) were initially found in the genomes of the γ_2-herpesviruses or rhadinoviruses equine herpesvirus type 2 (E8), bovine herpesvirus type 4 (BORF E2), herpesvirus saimiri (ORF71), and Kaposi's sarcoma-associated herpesvirus (KSHV:ORF71, K13), as well as two examples in the poxviral genome of molluscum contagiosum virus (MC159L, MC160L). The vFLIPs inhibited DR-dependent apoptosis mediated by CD95L, TNFα, TRAIL (tumor necrosis factor-related apoptosis-inducing ligand), or TRAMP. Thus, the formation of the death-inducing signaling complex was blocked, caspase-8 was not activated, and cell death did not occur (Bertin *et al.*, 1997; Hu *et al.*, 1997; Thome *et al.*, 1997; Wang *et al.*, 1997). Correspondingly, herpesvirus saimiri infection partially protected permissive cells from CD95-dependent apoptosis at a late stage of infection before the cells are lysed (Thome *et al.*, 1997). However, the anti-apoptotic activity of vFLIP of herpesvirus saimiri was not required for virus replication, T-cell transformation, and lymphoma induction (Glykofrydes *et al.*, 2000). Transgenic mice specifically expressing vFLIP/E8 in thymocytes showed very low thymocyte numbers. The thymocytes were resistant to apoptosis, even in CD95-deficient mice (Ohyama *et al.*, 2000). The vFLIP of KSHV was shown to inhibit Sindbis virus-induced apoptosis (Sarid *et al.*, 2001). Moreover, murine A20 lymphoma cells expressing the same vFLIP molecule rapidly developed into aggressive tumors in immunocompetent hosts, while the cells were shielded via vFLIP from the CD95-death signal triggered by activated T cells. Thus, apoptosis inhibitors can be regarded as tumor progression factors (Djerbi *et al.*, 1999). At least in KSHV, the vFLIP appears to be expressed in latency by Kaposi's sarcoma spindle cells (Davis *et al.*, 1997; Stürzl *et al.*, 1999) and, thus, may play an important role in blocking anti-viral cytotoxic T cell-induced apoptosis.

Besides their function in the caspase cascade, the amino-terminal DED-containing pro-domains of caspases and caspase homologs such as vFLIPs are involved in signals which inhibit apoptosis and support survival by NF-κB activation (Chaudhary *et al.*, 1999, 2000; Liu *et al.*, 2002). KSHV-infected primary effusion lymphoma cells show a constitutive NF-κB activation and IκBα phosphorylation. A chemical inhibitor of IκBα phosphorylation abrogated NF-κB activity and induced apoptosis in primary effusion lymphoma cells (Keller *et al.*, 2000). Correspondingly, KSHV vFLIP induced IκBα phosphorylation and NF-κB activation in a variety of cell lines, and was recruited to a large complex including the IκB kinases IKKα, IKKβ, NEMO/IKKγ, and RIP. Thus, vFLIP seems to mediate the NF-κB activation observed in primary effusion lymphoma cells (Liu *et al.*, 2002). Moreover, NF-κB inhibition may provide pharmacologic means for the therapy of primary effusion lymphoma in which vFLIP seems to play a crucial role (Keller *et al.*, 2000). Thus, the two major vFLIP functions – the inhibition of caspase activation and the activation of IκBα phosphorylation and NF-κB activation – have a synergic effect in the blockade of apoptosis.

On the basis of the viral FLIPs, two cellular counterparts cFLIP$_L$ and cFLIP$_S$ with two DEDs were discovered which are capable of inducing apoptosis and NF-κB activity (reviewed in Krueger *et al.*, 2001). NF-κB activation, in turn, was shown to induce cFLIP which seems to explain the anti-apoptotic function of NF-κB activation (Micheau *et al.*, 2001).

In the case of human cytomegalovirus, a functionally homologous, but structurally unrelated, viral inhibitor of caspase-8-induced apoptosis (vICA) encoded by the viral UL36 gene has been described. vICA was shown to block CD95-mediated apoptosis by binding to the pro-domain of caspase-8. Similarly to vFLIP of herpesvirus saimiri, vICA of human cytomegalovirus was shown to be dispensable for viral replication (Skaletskaya *et al.*, 2001). vFLIPs and vICA seem to block the clearance of infected cells by effector lymphocytes.

In addition to vFLIP/E8, equine herpesvirus type 2 codes for a related regulator E10 (vCarmen, CLAP, CIPER). E10 is a unique caspase-recruitment domain (CARD)-containing protein that was reported to induce both apoptosis and NF-κB activation in mammalian cells (Koseki *et al.*, 1999; Srinivasula *et al.*, 1999; Thome *et al.*, 1999; Yan *et al.*, 1999). E10 interacts with caspases and other apoptosis-regulating factors via its CARD. Moreover, E10 recruits its cellular homolog bcl-10 to the membrane and induces its phosphorylation (Thome *et al.*, 2001). NF-κB activation was reported to be mediated by the interaction of the E10 CARD domain with the carboxy-terminal region of IKKγ (Poyet *et al.*, 2001). Whereas the NF-κB activation by E10 was observed by several groups, the pro-apoptotic function of E10 is still controversial. The conserved function of diverse anti-apoptotic molecules in NF-κB activation confirms the relevance of protective NF-κB activation for the herpesviral replication.

Although adenoviruses are only distantly related to the large pox- and herpesviruses, they evolved similar functions. The E3-14.5K and E3-10.4K proteins prevented apoptosis mediated by DR TNF-R1, CD95, TRAILR1, and TRAILR2 (Gooding *et al.*, 1991; Shisler *et al.*, 1997; Benedict *et al.*, 2001; Tollefson *et al.*, 2001). Moreover, the E3 proteins induced a downregulation of the CD95 and TRAILR surface expression by inducing their degradation in lysosomes (McNees *et al.*, 2002). Therefore, the functional aggregate of the E3 proteins 14.5K, 10.4K, and 6.7K is termed the DR internalization and degradation (RID) complex. Moreover, these E3 proteins have been reported to inhibit IKK phosphorylation and NF-κB activation after TNFα and interleukin-1(IL-1) stimulation (Friedman and Horwitz, 2002). These observations might easily explain why adenoviruses are able to evade efficiently the immune response in order to establish prolonged infection (reviewed in Burgert *et al.*, 2002; McNees and Gooding, 2002).

7.2.2 Cell-autonomous apoptosis

The alternative mitochondrial pathway of apoptosis is either initiated by cell-autonomous signals such as DNA damage or it is mediated after DR-signaling by caspase-8-dependent Bid activation in class II cells. The mitochondrial pathway is regulated by the pro- and anti-apoptotic Bcl-2 family members that control the release of cytochrome c from the intermembrane space between the mitochondrial inner and outer membrane. In addition, other mitochondrial pro-apoptotic proteins such as the "apoptosis-inducing factor" (AIF), a mitochondrial oxidoreductase, the endonuclease G, or the "second mitochondrial activator of caspases" (SMAC, Diablo) are released.

Bcl-2 proteins contain short Bcl-2-homology regions BH1 to BH4, and were assigned to three groups (Adams and Cory, 1998). The anti-apoptotic group I comprises Bcl-2, Bcl-X_L, Bcl-w, and Mcl-1 which usually contain all four BH domains and a carboxy-terminal membrane anchor sequence. The pro-apoptotic representatives of group II such as Bax and Bak lack the amino-terminal BH4 domain. Translocation, conformational changes, oligomerization, and insertion of Bax or Bak into the outer mitochondrial membrane are sufficient to release cytochrome c and to induce caspase activation and apoptosis. Group III proteins including Bad, Bid, Bim, or Puma are characterized by a single BH3 domain, but are structurally rather diverse. In response to a variety of pro-apoptotic stimuli, Bid is cleaved by caspase-8, and the carboxy-terminal cleavage product of Bid can induce mitochondrial membrane insertion of Bax or Bak (Eskes *et al.*, 2000; Wei *et al.*, 2000). Two mechanisms have been discussed for the release of cytochrome c by mitochondrial membrane permeabilization: either the oligomerization of Bax or Bak directly forms pores in the outer mitochondrial membrane or, alternatively, Bax or Bak modulate the activity of the permeability transition pore complex by interaction with two pore complex components, the voltage-dependent anion channel (VDAC) and adenine nucleotide translocase (ANT) (Zamzami and Kroemer, 2001).

Thus, group III proteins integrate diverse pro-apototic signals, which are executed by pro-apoptotic group II proteins, leading to mitochondrial disintegration. In contrast, the anti-apoptotic group I proteins modulate apoptosis by inactivating or neutralizing group II effects. The viral anti-apoptotic Bcl-2 homologs resemble group II members, carry the homology domains BH1 and BH2, and are usually expressed in the lytic phase of virus replication, presumably in order to prevent premature termination of this potentially harmful process by apoptosis (Boya *et al.*, 2001).

Many γ-herpesviruses harbor a functional viral Bcl-2 homolog. For example, the product of ORF16 of the rhadinovirus herpesvirus saimiri (vBcl-2), which is expressed during lytic replication, contains the Bcl-2 homology domains BH1 and BH2 and shows an anti-apoptotic activity similar to cellular Bcl-X_L (Derfuss *et al.*,

1998). Both Bcl-X$_L$ and vBcl-2 inhibit apoptosis induced either by cell-autonomous (independent of DR) or by DR-mediated mechanisms, depending on the cell type studied. The vBcl-2 has been shown to act at the level of mitochondrial stabilization and to inhibit apoptosis induced by Sendai virus or by treatment with CD95L, dexamethasone, menadione, or irradiation (Nava *et al.*, 1997; Derfuss *et al.*, 1998). The vBcl-2 of KSHV is similarly an anti-apoptotic factor which neither homodimerizes nor heterodimerizes with other Bcl-2 family members. This suggests that it escapes any negative regulatory effects of the cellular Bax and Bak proteins (Cheng *et al.*, 1997). Thus, pairs of rhadinovirus-encoded proteins, vFLIP and vBcl-2, appear to render infected cells resistant both to DR-mediated and cell-autonomous apoptosis, resulting in enhanced virion production from permissive cells (Meinl *et al.*, 1998).

In the case of the γ_1-herpesvirus/lymphocryptovirus Epstein–Barr virus, the viral Bcl-2 homolog BHRF1 has been shown to be dispensable for virus replication and B-cell transformation (Marchini *et al.*, 1991; Lee and Yates, 1992). BHRF1 was the first Bcl-2-like anti-apoptotic viral protein to be reported (Henderson *et al.*, 1993; Tarodi *et al.*, 1994; Dumont *et al.*, 1999). Epstein–Barr virus surprisingly harbors a second vBcl-2 homolog: BALF1 conferred apoptosis resistance and was found to be associated with Bax and Bak (Marshall *et al.*, 1999). However, BALF1 did not co-localize with BHRF1. Moreover, BALF1 inhibited the anti-apoptotic activity of BHRF1, and of vBcl-2/K13 of KSHV, but not of cellular Bcl-X$_L$. Unlike cellular Bcl-2 family members, BALF1 did not show pro-apoptotic effects, not even after caspase cleavage or truncation (Bellows *et al.*, 2002). Thus, Epstein–Barr virus has an unusually finely tuned apoptosis regulatory system.

In contrast to other rhadinoviruses, the Bcl-2 homologs M11 of the murine γ-herpesvirus 68 (Wang *et al.*, 1999) and A9 of alcelaphine herpesvirus type 1 (Ensser *et al.*, 1997) are located at a different genomic position, but are similar in size and contain a conserved BH1 domain without BH2 and a hydrophobic carboxy-terminus. vBcl-2/M11 has been shown to inhibit apoptosis mediated by CD95 or TNF-R1 (Wang *et al.*, 1999). In a series of γ-herpesviral Bcl-2 family molecules, only vBcl-2/M11 of murine γ-herpesvirus 68 was susceptible to caspase digestion. However, in contrast to cellular Bcl-2, its cleavage product lacked pro-apoptotic activity. Thus, vBcl-2 molecules escape negative regulation by retaining their anti-apoptotic activities and/or by failing to be converted into pro-apoptotic proteins by caspases during programmed cell death (Bellows *et al.*, 2000).

The murine γ-herpesvirus 68 easily permits studies in vivo, in contrast to most other relevant herpesviruses. Similarly to data shown previously for the vCyclin (van Dyk *et al.*, 2000), the vBcl-2/M11 was dispensable for viral replication and pathogenicity in interferon (IFN)-γ-deficient mice, but was necessary for efficient ex vivo reactivation from latent infection, and both for persistent replication and virulence during chronic infection. While vBcl-2 and vCyclin were important for chronic infection, these genes were dispensable for viral replication in cell culture, viral replication during acute infection in vivo, establishment of latent infection, or virulence during acute infection (Gangappa *et al.*, 2002).

The release of cytochrome c and other mediators from mitochondria to the cytoplasm activates the alternative signaling cascade. In the cytoplasm, cytochrome c binds to the inactive form of the apoptotic-protease activating factor (Apaf-1), which then undergoes an energy-dependent activating conformational change and oligomerizes. Consequently, the CARD of Apaf-1 binds the CARD of procaspase-9. This tetrameric effector caspase is then activated – not by autocatalytic cleavage, but by conformational changes. Finally, the resulting apoptosome, a 700 kD complex, cleaves and activates the executing procaspases-3, -6, and -7. Although reports are still lacking, the CARD proteins viral E10 and cellular bcl-10 might also interfere with this pathway.

A protein family termed inhibitors of apoptosis (IAPs) has been discovered in baculovirus-infected insect cells (Clem *et al.*, 1991; Crook *et al.*, 1993). Later on, a number of baculoviral IAP repeat (BIR)-containing human IAP proteins were identified and termed cIAP1, cIAP2, NAIP, XIAP, survivin, and livin (reviewed in Salvesen and Duckett, 2002). Whereas XIAP seems to be the most potent factor, survivin seems not to be sufficient to inhibit caspase activity (Banks *et al.*, 2000). Cellular IAPs contain multiple zinc-binding BIR domains which form finger-like structures that directly bind caspases, while other domains of IAPs block the reactive sites of active caspases. Specific IAPs are bound and inactivated by SMAC, preventing them from inhibiting the active caspase-3, -7, and -9. Similarly to cFLIP, XIAP and other IAPs are induced by NF-κB activation (Rothe *et al.*, 1995; Stehlik *et al.*, 1998) which reflects the pronounced anti-apoptotic result of NF-κB induction. Recently, the BIR-containing glycoprotein K7 of KSHV was identified as a functional vIAP or survivin homolog (Wang *et al.*, 2002).

7.3   Specific viral strategies to induce or block apoptosis

7.3.1   RNA viruses

Many RNA viruses are characterized by a rapid replication cycle and massive virus particle production. In the case of the togavirus/alphavirus *Sindbis virus*, the viral E2 glycoprotein is involved in apoptosis induction and neurovirulence (Ubol *et al.*, 1994). Sindbis virus-induced apoptosis and encephalitis in mice can be blocked by cellular Bcl-2 which converts lytic to persistent infection in this system (Levine *et al.*, 1993, 1996). The viral attachment protein sigma 1 has been shown to be relevant for *reovirus*-induced apoptosis (Tyler *et al.*, 1995), which is mediated via the cellular DR TRAILR (Clarke *et al.*, 2000). In addition, the reovirus sigma-3 protein inhibits IFN-induced, double-stranded, RNA-activated protein kinase PKR (Imani and Jacobs, 1988). Since PKR itself is known to induce apoptosis in many cell types (Lee and Esteban, 1994), PKR inhibition by sigma-3 is expected to result in anti-apoptotic function. The *paramyxovirus* Sendai virus induces apoptosis via caspase-8 and caspase-9 activation independently of DR activation or mitochondrial damage (Bitzer *et al.*, 1999, 2002). Moreover, the short anti-genome trailer RNA of Sendai virus binds via AU-rich sequences to, and blocks, the newly described cellular pro-apoptotic protein TIAR (Iseni *et al.*, 2002). The non-structural protein NS1 of the *orthomyxovirus* influenza virus has been reported to counteract PKR effects and IFN-dependent apoptosis (Bergmann *et al.*, 2000; Zhirnov *et al.*, 2002). In contrast, the PB1-F2 protein of influenza virus targets mitochondria and induces apoptosis, presumably in attacking immune cells (Chen *et al.*, 2001).

The flavivirus *hepatitis C virus* is a major human pathogen causing chronic liver infection. However, its pathogenetic mechanisms are poorly understood. As far as apoptosis induction is concerned, two viral proteins seem to be involved. First, the hepatitis C virus non-structural protein NS5A mediates repression of the pro-apoptotic PKR, resistance to IFNs, and anti-apoptotic functions (Gale *et al.*, 1997, 1998, 1999). In addition, NS5A has been reported to interact with p53 and to inhibit p53-mediated apoptosis (Lan *et al.*, 2002). Second, the core protein of hepatitis C virus has been reported to bind to the cytoplasmic domain of TNF-R1 and to enhance TNFα-induced apoptosis (Zhu *et al.*, 1998). In contrast, hepatitis C virus core protein inhibits CD95L- and TNFα-mediated apoptosis via NFκB activation (Marusawa *et al.*, 1999). Finally, the core protein of hepatitis C viruses specifically isolated from hepatocellular carcinoma was able to activate the pro-apoptotic PKR (Delhem *et al.*, 2001); the relevance of the core protein for apoptosis regulation, however, remains controversial.

Retroviruses cause chronic infection by the integration of the reverse-transcribed viral RNA into the cellular genome. The p13II protein of the human *T-cell leukemia virus* was recently discovered as a pro-apoptotic factor which targets the inner mitochondrial membrane, disrupts mitochondrial integrity, and acts through a permeability transition pore-independent mechanism (Ciminale *et al.*, 1999; D'Agostino *et al.*, 2002). The oncogene Tax of human T-cell leukemia virus is known to induce NF-κB activity. Correspondingly, a chemical inhibitor of NF-κB activity induces apoptosis of transformed T-cell lines and primary adult T-cell leukemia cells which are infected by the virus (Mori *et al.*, 2002).

Numerous pro- and anti-apoptotic functions have been reported for the *human immunodeficiency virus* (HIV-1). Without any doubt, massive programmed cell death is responsible for immune dysfunction and progressive T-cell depletion during the course of the acquired immunodeficiency syndrome (Ameisen and Capron, 1991; Meyaard *et al.*, 1992). First, the HIV-1 envelope protein gp120 is capable of cross-linking its major receptor CD4 on T cells and, thus, of triggering activation-induced apoptosis (Banda *et al.*, 1992). The HIV-1 envelope-induced apoptosis occurs in the G2 phase of the cell cycle (Kolesnitchenko *et al.*, 1995) and engages cyclin-dependent kinase 1 and p53 (Castedo *et al.*, 2002). Second, the anti-termination protein Tat of HIV-1 can act as a substrate and inhibitor of PKR attributing an anti-apoptotic function to Tat (Brand *et al.*, 1997). In another experimental system, Tat induced apoptosis by caspase activation, calcium over-load, and oxidative stress (Kruman *et al.*, 1998). In addition, Tat primed cells to undergo apoptosis upon serum withdrawal by mitochondrial permeability tran-sition (Macho *et al.*, 1999). Vpr is a third HIV-1 protein which is involved in apoptosis pathways. Vpr induces apoptosis by a direct effect on the mitochondrial permeability transition pore (Jacotot *et al.*, 2000). The pro-apoptotic function par-allels the cytopathogenicity of specific Vpr alleles (Somasundaran *et al.*, 2002). The Nef protein is a fourth factor of HIV-1 with pro- and anti-apoptotic func-tions in addition to its involvement in many other signaling pathways (Fackler and Baur, 2002). Nef enhanced the response of CD4-positive cells to various unre-lated apoptotic agents (staurosporine, anisomycin, camptothecin, and etoposide) by DR-dependent and mitochondrial pathways (Rasola *et al.*, 2001). Nef is also known to induce CD95L expression in HIV-1-infected T cells which leads to the death of attacking non-infected cytotoxic T cells. In contrast to this DR-mediated form of fratricide, the infected cells are protected from suicide. HIV-1 Nef has been demonstrated to associate with, and to inhibit, apoptosis signal-regulating kinase 1 (ASK1), a serine/threonine kinase linking DR-signaling pathways to c-Jun amino-terminal kinase (Geleziunas *et al.*, 2001). Thus, HIV-1-infected cells are protected from DR-dependent pro-apoptotic signals through Nef-mediated ASK1 inhibition, whereas non-infected bystander cells are targeted by Nef-induced CD95L expression. A complex interplay between viral pro- and anti-apoptotic func-tions is typical for persistent viruses which actively replicate in parallel to chronic infection.

7.3.2 DNA viruses

Anti-apoptotic functions are elucidated in detail for large DNA viruses such as pox-viruses or herpesviruses. Within the herpesvirus family, the different subfamilies evolved diverse strategies (Derfuss and Meinl, 2002). For the α-*herpesviruses*, herpes simplex virus type 1 has been investigated in detail. Herpes simplex virus 1 induces and blocks apoptosis at multiple steps during infection and protects cells from exogenous inducers by caspase-independent and -dependent pathways (Galvan and Roizman, 1998; Galvan *et al.*, 1999; Zhou and Roizman, 2000). After transfection, the viral glycoproteins D and J block apoptosis induced by a virus mutant without intact glycoprotein genes (Zhou *et al.*, 2000). One of the major regulatory proteins of herpes simplex virus, ICP4, has been shown to block apoptosis induced by the virus or by hyperthermia (Leopardi and Roizman, 1996). Similarly, another regulatory protein, ICP27, was required for the prevention of apoptosis in infected human cells (Aubert and Blaho, 1999). Moreover, the US3 protein kinase of herpes simplex virus was necessary for protection from virus-induced apoptosis, mediated the post-translational modification of Bad, and prevented Bad-induced apoptosis (Leopardi *et al.*, 1997; Munger and Roizman, 2001). Finally, the $\gamma_1 34.5$ protein of herpes simplex virus 1 is able to block the total shut-off of protein synthesis which is characteristic of programmed cell death (Chou and Roizman, 1992). $\gamma_1 34.5$ is another example of a viral protein which inhibits PKR function. Although specific functions have not yet been defined, Varicella–Zoster virus has been observed to induce apoptosis in cell culture (Sadzot-Delvaux *et al.*, 1995).

In the case of human cytomegalovirus, the prototype of the β-*herpesviruses*, the immediate-early proteins IE1 and IE2 were initially reported to block apoptosis (Zhu *et al.*, 1995), a finding which was controversially discussed. Surprisingly, human cytomegalovirus harbors two additional anti-apoptotic genes. Similarly to the vBcl-2 molecules of various γ-herpesviruses, the cytomegalovirus gene UL37 codes for a mitochondrial inhibitor of apoptosis (vMIA). However, it is structurally unrelated to Bcl-2 (Goldmacher *et al.*, 1999). The neighboring UL36 encodes a viral inhibitor of caspase-8-induced apoptosis (vICA) which is able to suppress caspase-8 activation (Skaletskaya *et al.*, 2001). The binding of vICA to the pro-domain of caspase-8 blocks the activation of this enzyme. Thus, vICA is a functional, but not structural, homolog of the vFLIP proteins of various rhadino- and poxviruses. For the other human β-herpesviruses, human herpesvirus 6 and 7, apoptosis induction was observed in infected T cells, but the specific molecular mechanisms have not yet been elaborated (Inoue *et al.*, 1997; Yasukawa *et al.*, 1998). Human herpesvirus 7 induced rather necrotic lysis in productively infected cells and apoptosis in uninfected or non-productively infected cells (Secchiero *et al.*, 1997).

The anti-apoptotic proteins of the γ-*herpesviruses*, vFLIP and vBcl-2, are already described in detail in the function-specific sections above. Most γ-herpesviruses harbor a functional viral Bcl-2 homolog (Henderson *et al.*, 1993; Tarodi *et al.*, 1994; Cheng *et al.*, 1997; Ensser *et al.*, 1997; Nava *et al.*, 1997; Derfuss *et al.*, 1998; Dumont *et al.*, 1999; Wang *et al.*, 1999; Bellows *et al.* 2000). Only Epstein–Barr virus carries two Bcl-2 homologs with counteracting functions – namely, BHRF1 and BALF1 (Marshall *et al.*, 1999; Bellows *et al.*, 2002). Although the vBcl-2 genes studied so far are dispensable for virus replication (Marchini *et al.*, 1991; Lee and Yates, 1992; Gangappa *et al.*, 2002), the vBcl-2 of murine γ-herpesvirus 68 is important for a chronic infection (Gangappa *et al.*, 2002). In several rhadinoviruses (γ2-herpesviruses), viral FLIPs have been identified which efficiently inhibit caspase-8 activation by interaction with DED proteins (Bertin *et al.*, 1997;

Hu *et al.*, 1997; Thome *et al.*, 1997; Wang *et al.*, 1997; Sarid *et al.*, 2001). Since the vFLIP of KSHV is expressed in the tumor in the spindle cells (Davis *et al.*, 1997; Stürzl *et al.*, 1999), it is a candidate oncogene of this virus. In murine hosts, vFLIP behaved as a tumor progression factor in transfected tumor cells (Djerbi *et al.*, 1999). In primary effusion lymphoma cells, chemical NF-κB blockade induced apoptosis (Keller *et al.*, 2000). This could be linked to vFLIP, since it interacts with and constitutively activates the IκB kinase complex (Liu *et al.*, 2002). In addition to the vFLIP E8, a CARD-containing protein has been identified with E10 of equine herpesvirus type 2. Although its functional influence on caspases and apoptosis is controversial, the activation of NF-κB by E10 is well accepted (Koseki *et al.*, 1999; Srinivasula *et al.*, 1999; Thome *et al.*, 1999; Yan *et al.*, 1999; Poyet *et al.*, 2001; Thome *et al.*, 2001). Moreover, the LANA-1 protein (ORF73) and the LANA-2 protein (ORF10.5) of KSHV have been reported to inhibit p53 function and cell death (Friborg *et al.*, 1999; Rivas *et al.*, 2001). Recently, the glycoprotein K7 of this virus was identified as having a vIAP-blocking caspase-3 activity (Wang *et al.*, 2002). Generally, all these rhadinovirus functions are assumed to block apoptosis during replication and during pathogenesis in vivo. Whereas vFLIP may be involved in lymphomagenesis by KSHV, it is not required for T-cell transformation by herpesvirus saimiri (Glykofrydes *et al.*, 2000). Similarly to non-transformed T cells, herpesvirus saimiri-transformed human T cells can undergo activation-induced cell death in a CD95-dependent and -independent manner (Bröker *et al.*, 1997; Kraft *et al.*, 1998). Although its functional vFLIP would suggest a role in the blocking of apoptosis, infection of culture cells with bovine herpesvirus type 4 induced apoptosis and increased virus yields of lytic replication (Sciortino *et al.*, 2000).

The situation for the γ₁-herpesvirus (lympocryptovirus) *Epstein–Barr virus* and for the apoptosis of virus-transformed B cells is most complex. Besides the role of the Bcl-2 homologs, various other mechanisms have been suggested and only some of them can be mentioned. The viral oncogene EBNA2 has been reported to block Nur77-mediated apoptosis (Lee *et al.*, 2002). When the transcription of the other main viral oncogene LMP-1 was suppressed by antisense treatment, apoptosis was promoted in immortalized B cells (Kenney *et al.*, 1998). Moreover, Epstein–Barr virus RNA conferred resistance to IFNα-induced apoptosis in Burkitt's lymphoma (Nanbo *et al.*, 2002). Finally, modulation of caspase-8 and cellular FLIP expression has been suggested as a mechanism of Epstein–Barr virus tumorigenesis in Burkitt's lymphoma (Tepper and Seldin, 1999).

The interference of *adenoviruses* with the apoptosis machinery also involves a series of viral proteins (Burgert *et al.*, 2002; McNees and Gooding, 2002). Viral E1A induces p53-mediated apoptosis, which in turn is inhibited by E1B (Debbas and White, 1993; Sabbatini *et al.*, 1995). Moreover, the interaction of E1B 19K with Bax is required to block Bax-induced loss of mitochondrial membrane potential and apoptosis (Han *et al.*, 1998). E1B 19K further inhibited TNFα-mediated apoptosis through a Bid-dependent conformational change in Bax (Perez and White, 2000). The RID complex is formed by the adenoviral E3 proteins 14.5K, 10.4K, and 6.7K (Benedict *et al.*, 2001; Friedman and Horwitz, 2002; McNees *et al.*, 2002). The RID complex inhibited TNFα-induced cell death and NF-κB activation (Gooding *et al.*, 1991; Friedman and Horwitz, 2002), downregulated CD95 and TRAIL-R from the surface (Shisler *et al.*, 1997; Benedict *et al.*, 2001; Tollefson *et al.*, 2001), and protected cultured T and B cells from CD95-induced apoptosis (McNees *et al.*, 2002). In contrast to the anti-apoptotic E3 proteins, E4orf4 protein can induce DR-dependent caspase activation (Livne *et al.*, 2001). Similarly to E1B, E4orf6 protein binds and inhibits p53-dependent transcription and apoptosis (Moore *et al.*, 1996).

Among the *polyomaviruses*, apoptosis regulation has been studied mainly in simian virus 40 and in papillomaviruses. The transforming large T antigen of simian virus 40 can induce p53-dependent apoptosis (Yanai and Obinata, 1994). In human papillomavirus 16 and 18, the E6 oncoprotein functions as a ubiquitin-protein ligase in the degradation of p53 (Scheffner *et al.*, 1993; Thomas *et al.*, 1996). p53-mediated apoptosis can be abolished by human papillomavirus E6 (Thomas *et al.*, 1996). Correspondingly, E6 abrogates UV-induced and Bak-dependent apoptosis in skin cancer cells (Jackson *et al.*, 2000). In contrast, E6 has also been reported to sensitize cells to drug-induced apoptosis (Brown *et al.*, 1997). The E7 oncoprotein inhibits DR-mediated apoptosis in normal human fibroblasts and keratinocytes (Basile *et al.*, 2001; Thompson *et al.*, 2001). Blockade of E7 by aptamers induced apoptosis in cervical carcinoma cells (Nauenburg *et al.*, 2001). Moreover, the E2 protein of human papillomavirus 16 induced p53-dependent apoptosis (Webster *et al.*, 2000).

The HBx protein of *hepatitis B virus* has been linked to apoptosis control in a few diverse reports. HBx inhibited CD95-mediated apoptosis and upregulated the SAPK/JNK pathway (Diao *et al.*, 2001). Moreover, HBx co-localized to mitochondria with the voltage-dependent anion channel HVDAC3 and altered the mitochondrial transmembrane potential (Rahmani *et al.*, 2000). Finally, HBx sensitized cells to apoptotic killing by TNFα (Su and Schneider, 1997). The non-structural protein NS1 of human *parvovirus B19* induces apoptosis in erythroid lineage cells (Moffatt *et al.*, 1998; Yaegashi *et al.*, 1999). The *circovirus* chicken anemia virus encodes the pro-apoptotic apoptin (VP3) which induces apoptosis in human transformed and malignant cells, but, surprisingly, not in normal cells (Danen-Van Oorschot *et al.*, 1997).

Besides the herpesviruses, the *poxvirus* family has developed into a gold mine for the discovery of apoptosis-regulating proteins. In *vaccinia virus*, the SPI-2 (B13R) gene product protected against apoptosis (Dobbelstein and Shenk, 1996). The serpin CrmA inhibits cytotoxic T lymphocyte-mediated apoptosis (Tewari *et al.*, 1995), and CrmE encodes a soluble and cell surface-bound TNF-R that contributes to virus virulence (Reading *et al.*, 2002). In the rabbit poxvirus *myxoma virus*, M11L is a novel mitochondria-localized protein that blocks apoptosis of infected leucocytes (Everett *et al.*, 2000). M-T2 is a TNF-R homolog which mediates extracellular TNF binding and intracellular apoptosis inhibition (Schreiber *et al.*, 1997). The M-T4 RDEL-containing protein is retained within the endoplasmic reticulum and is important for the productive infection of lymphocytes. M-T4 may have a dual function in protecting infected lymphocytes from apoptosis and in modulating the inflammatory response to virus infection (Barry *et al.*, 1997; Hnatiuk *et al.*, 1999). The human *molluscum contagiosum virus* harbors two genes with DED. The MC159 vFLIP blocks CD95-induced activation of procaspases and degradation of the related MC160 protein (Shisler and Moss, 2001). However, binding of FADD and caspase-8 to MC159 vFLIP is not sufficient for its anti-apoptotic function (Garvey *et al.*, 2002). Moreover, MC159L inhibits NF-κB activation and apoptosis induced by PKR (Gil *et al.*, 2001). *African Swine fever virus*, a unique large DNA virus related to pox- and iridoviruses, harbors a huge collection of pirated cellular genes, including anti-apoptotic factors such as a vBcl-2 (Afonso *et al.*, 1996; Neilan *et al.*, 1993) and a vIAP (Chacon *et al.*, 1995; Neilan *et al.*, 1997; Nogal *et al.*, 2001). In *baculovirus*-infected insect cells, a series of IAPs (p35) have been discovered which directly bind to caspases and inhibit their function (Clem *et al.*, 1991; Crook *et al.*, 1993).

7.4 Conclusions

Apoptosis is a central function of host defense. Although the various viral pro- and anti-apoptotic strategies interfere with diverse apoptosis pathways, a few typical key targets can be discerned which are applied either in pro- or in anti-apoptotic function by diverse viruses. DR triggering, caspase activation, mitochondrial disintegration, and p53 modulation form the critical steps in apoptosis regulation which are addressed by viruses. Many viruses have even developed several anti-apoptotic mechanisms in parallel, partly by sequestering the respective homologous cellular genes or by evolving specific functional domains in regulatory or structural proteins. Study of the interaction between viruses and their hosts has revealed a variety of viral proteins that interfere with the regulation of apoptotic pathways, supporting the concept that pro- and anti-apoptotic viral and cellular functions play a central role in viral success in evolution. Some viral effectors were even discovered prior to their cellular homologs. Thus, it is not unlikely that viruses will teach us more about the fundamental mechanisms of apoptosis in the future.

ACKNOWLEDGEMENTS

Original work included in this review article was supported by the Deutsche Forschungsgemeinschaft (Sonderforschungsbereich 466), the Bundesministerium für Bildung und Forschung (IZKF Erlangen), the Bayerische Forschungsstiftung, and the Wilhelm Sander-Stiftung.

REFERENCES

Adams, J. M. and Cory, S. (1998). The Bcl-2 protein family: arbiters of cell survival. *Science*, **281**, 1322–6.

Afonso, C. L., Neilan, J. G., Kutish, G. F., and Rock, D. L. (1996). An African swine fever virus Bcl-2 homolog, 5-HL, suppresses apoptotic cell death. *J. Virol.*, **70**, 4858–63.

Ameisen, J. C. and Capron, A. (1991). Cell dysfunction and depletion in AIDS: the programmed cell death hypothesis. *Immunol. Today*, **12**, 102–5.

Aubert, M. and Blaho, J. A. (1999). The herpes simplex virus type 1 regulatory protein ICP27 is required for the prevention of apoptosis in infected human cells. *J. Virol.*, **73**, 2803–13.

Banda, N. K., Bernier, J., Kurahara, D. K. *et al.* (1992). Crosslinking CD4 by human immunodeficiency virus gp120 primes T cells for activation-induced apoptosis. *J. Exp. Med.*, **176**, 1099–106.

Banks, D. P., Plescia, J., Altieri, D. C. *et al.* (2000). Survivin does not inhibit caspase-3 activity. *Blood*, **96**, 4002–3.

Barry, M., Hnatiuk, S., Mossman, K., Lee, S. F., Boshkov, L., and McFadden, G. (1997). The myxoma virus M-T4 gene encodes a novel RDEL-containing protein that is retained within the endoplasmic reticulum and is important for the productive infection of lymphocytes. *Virology*, **239**, 360–77.

Basile, J. R., Zacny, V., and Munger, K. (2001). The cytokines tumor necrosis factor-alpha (TNF-alpha) and TNF-related apoptosis-inducing ligand differentially modulate proliferation and apoptotic pathways in human keratinocytes expressing the human papillomavirus-16 E7 oncoprotein. *J. Biol. Chem.*, **276**, 22522–8.

Bellows, D. S., Chau, B. N., Lee, P., Lazebnik, Y., Burns, W. H., and Hardwick, J. M. (2000). Anti-apoptotic herpesvirus Bcl-2 homologs escape caspase-mediated conversion to proapoptotic proteins. *J. Virol.*, **74**, 5024–31.

Bellows, D. S., Howell, M., Pearson, C., Hazlewood, S. A., and Hardwick, J. M. (2002). Epstein-Barr virus BALF1 is a BCL-2-like antagonist of the herpesvirus antiapoptotic BCL-2 proteins. *J. Virol.*, **76**, 2469–79.

Benedict, C. A., Norris, P. S., Prigozy, T. I. *et al.* (2001). Three adenovirus E3 proteins cooperate to evade apoptosis by tumor necrosis factor-related apoptosis-inducing ligand receptor-1 and -2. *J. Biol. Chem.*, **276**, 3270–8.

Bergmann, M., Garcia-Sastre, A., Carnero, E. *et al.* (2000). Influenza virus NS1 protein counteracts PKR-mediated inhibition of replication. *J. Virol.*, **74**, 6203–6.

Bertin, J., Armstrong, R. C., Ottilie, S. *et al.* (1997). Death effector domain-containing herpesvirus and poxvirus proteins inhibit both Fas- and TNFR1-induced apoptosis. *Proc. Natl. Acad. Sci. USA*, **94**, 1172–6.

Bitzer, M., Armeanu, S., Prinz, F. *et al.* (2002). Caspase-8 and Apaf-1-independent caspase-9 activation in Sendai virus-infected cells. *J. Biol. Chem.*, **277**, 29817–24.

Bitzer, M., Prinz, F., Bauer, M. *et al.* (1999). Sendai virus infection induces apoptosis through activation of caspase-8 (FLICE) and caspase-3 (CPP32). *J. Virol.*, **73**, 702–8.

Boya, P., Roques, B., and Kroemer, G. (2001). Viral and bacterial proteins regulating apoptosis at the mitochondrial level. *EMBO J.*, **20**, 4325–31.

Brand, S. R., Kobayashi, R., and Mathews, M. B. (1997). The Tat protein of human immunodeficiency virus type 1 is a substrate and inhibitor of the interferon-induced, virally activated protein kinase, PKR. *J. Biol. Chem.*, **272**, 8388–95.

Bröker, B. M., Kraft, M. S., Klauenberg, U. *et al.* (1997). Activation induces apoptosis in herpesvirus saimiri-transformed T cells independent of CD95 (Fas, APO-1). *Eur. J. Immunol.*, **27**, 2774–80.

Brown, J., Higo, H., McKalip, A., and Herman, B. (1997). Human papillomavirus (HPV) 16 E6 sensitizes cells to atractyloside-induced apoptosis: role of p53, ICE-like proteases and the mitochondrial permeability transition. *J. Cell Biochem.*, **66**, 245–55.

Burgert, H. G., Ruzsics, Z., Obermeier, S., Hilgendorf, A., Windheim, M., and Elsing, A. (2002). Subversion of host defense mechanisms by adenoviruses. *Curr. Top. Microbiol. Immunol.*, **269**, 273–318.

Castedo, M., Roumier, T., Blanco, J. *et al.* (2002). Sequential involvement of Cdk1, mTOR and p53 in apoptosis induced by the HIV-1 envelope. *EMBO J.*, **21**, 4070–80.

Chacon, M. R., Almazan, F., Nogal, M. L., Vinuela, E., and Rodriguez, J. F. (1995). The African swine fever virus IAP homolog is a late structural polypeptide. *Virology*, **214**, 670–4.

Chaudhary, P. M., Eby, M. T., Jasmin, A., Kumar, A., Liu, L., and Hood, L. (2000). Activation of the NF-kappaB pathway by caspase 8 and its homologs. *Oncogene*, **19**, 4451–60.

Chaudhary, P. M., Jasmin, A., Eby, M. T., and Hood, L. (1999). Modulation of the NF-kappa B pathway by virally encoded death effector domains-containing proteins. *Oncogene*, **18**, 5738–46.

Chen, W., Calvo, P. A., Malide, D. *et al.* (2001). A novel influenza A virus mitochondrial protein that induces cell death. *Nat. Med.*, **7**, 1306–12.

Cheng, E. H., Nicholas, J., Bellows, D. S. *et al.* (1997). A Bcl-2 homolog encoded by Kaposi sarcoma-associated virus, human herpesvirus 8, inhibits apoptosis but does not heterodimerize with Bax or Bak. *Proc. Natl. Acad. Sci. USA*, **94**, 690–4.

Chou, J. and Roizman, B. (1992). The gamma 1(34.5) gene of herpes simplex virus 1 precludes neuroblastoma cells from triggering total shutoff of protein synthesis characteristic of programmed cell death in neuronal cells. *Proc. Natl. Acad. Sci. USA*, **89**, 3266–70.

Ciminale, V., Zotti, L., D'Agostino, D. M. *et al.* (1999). Mitochondrial targeting of the p13II protein coded by the x-II ORF of human T-cell leukemia/lymphotropic virus type I (HTLV-I). *Oncogene*, **18**, 4505–14.

Clarke, P., Meintzer, S. M., Gibson, S. *et al.* (2000). Reovirus-induced apoptosis is mediated by TRAIL. *J. Virol.*, **74**, 8135–9.

Clem, R. J., Fechheimer, M., and Miller, L. K. (1991). Prevention of apoptosis by a baculovirus gene during infection of insect cells. *Science*, **254**, 1388–90.

Crook, N. E., Clem, R. J., and Miller, L. K. (1993). An apoptosis-inhibiting baculovirus gene with a zinc finger-like motif. *J. Virol.*, **67**, 2168–74.

D'Agostino, D. M., Ranzato, L., Arrigoni, G. *et al.* (2002). Mitochondrial alterations induced by the p13II protein of human T-cell leukemia virus type 1. Critical role of arginine residues. *J. Biol. Chem.*, **277**, 34424–33.

Danen-Van Oorschot, A. A., Fischer, D. F., Grimbergen, J. M. *et al.* (1997). Apoptin induces apoptosis in human transformed and malignant cells but not in normal cells. *Proc. Natl. Acad. Sci. USA*, **94**, 5843–7.

Davis, M. A., Stürzl, M. A., Blasig, C. *et al.* (1997). Expression of human herpesvirus 8-encoded cyclin D in Kaposi's sarcoma spindle cells. *J. Natl. Cancer Inst.*, **89**, 1868–74.

Debbas, M. and White, E. (1993). Wild-type p53 mediates apoptosis by E1A, which is inhibited by E1B. *Genes Dev.*, **7**, 546–54.

Delhem, N., Sabile, A., Gajardo, R. *et al.* (2001). Activation of the interferon-inducible protein kinase PKR by hepatocellular carcinoma derived-hepatitis C virus core protein. *Oncogene*, **20**, 5836–45.

Derfuss, T. and Meinl, E. (2002). Herpesviral proteins regulating apoptosis. *Curr. Top. Microbiol. Immunol.*, **269**, 7–72.

Derfuss, T., Fickenscher, H., Kraft, M. S. *et al.* (1998). Antiapoptotic activity of the herpesvirus saimiri-encoded Bcl-2 homolog: stabilization of mitochondria and inhibition of caspase-3-like activity. *J. Virol.*, **72**, 5897–904.

Diao, J., Khine, A. A., Sarangi, F. *et al.* (2001). X protein of hepatitis B virus inhibits Fas-mediated apoptosis and is associated with up-regulation of the SAPK/JNK pathway. *J. Biol. Chem.*, **276**, 8328–40.

Djerbi, M., Screpanti, V., Catrina, A. I., Bogen, B., Biberfeld, P., and Grandien, A. (1999). The inhibitor of death receptor signaling, FLICE-inhibitory protein defines a new class of tumor progression factors. *J. Exp. Med.*, **190**, 1025–32.

Dobbelstein, M. and Shenk, T. (1996). Protection against apoptosis by the vaccinia virus SPI-2 (B13R) gene product. *J. Virol.*, **70**, 6479–85.

Dumont, A., Hehner, S. P., Hofmann, T. G., Ueffing, M., Droge, W., and Schmitz, M. L. (1999). Hydrogen peroxide-induced apoptosis is CD95-independent, requires the release of mitochondria-derived reactive oxygen species and the activation of NF-kappaB. *Oncogene*, **18**, 747–57.

Ensser, A., Pflanz, R., and Fleckenstein, B. (1997). Primary structure of the alcelaphine herpes virus 1 genome. *J. Virol.*, **71**, 6517–25.

Eskes, R., Desagher, S., Antonsson, B., and Martinou, J. C. (2000). Bid induces the oligomerization and insertion of Bax into the outer mitochondrial membrane. *Mol. Cell Biol.*, **20**, 929–35.

Everett, H., Barry, M., Lee, S. F. *et al.* (2000). M11L: a novel mitochondria-localized protein of myxoma virus that blocks apoptosis of infected leukocytes. *J. Exp. Med.*, **191**, 1487–98.

Fackler, O. T. and Baur, A. S. (2002). Live and let die: Nef functions beyond HIV replication. *Immunity*, **16**, 493–7.

Friborg, J., Jr., Kong, W., Hottiger, M. O., and Nabel, G. J. (1999). p53 inhibition by the LANA protein of KSHV protects against cell death. *Nature*, **402**, 889–94.

Friedman, J. M. and Horwitz, M. S. (2002). Inhibition of tumor necrosis factor alpha-induced NF-kappa B activation by the adenovirus E3–10.4/14.5K complex. *J. Virol.*, **76**, 5515–21.

Gale, M., Jr., Blakely, C. M., Kwieciszewski, B. *et al.* (1998). Control of PKR protein kinase by hepatitis C virus nonstructural 5A protein: molecular mechanisms of kinase regulation. *Mol. Cell Biol.*, **18**, 5208–18.

Gale, M., Jr., Korth, M. J., Tang, N. M. *et al.* (1997). Evidence that hepatitis C virus resistance to interferon is mediated through repression of the PKR protein kinase by the nonstructural 5A protein. *Virology*, **230**, 217–27.

Gale, M., Jr., Kwieciszewski, B., Dossett, M., Nakao, H., and Katze, M. G. (1999). Antiapoptotic and oncogenic potentials of hepatitis C virus are linked to interferon resistance by viral repression of the PKR protein kinase. *J. Virol.*, **73**, 6506–16.

Galvan, V. and Roizman, B. (1998). Herpes simplex virus 1 induces and blocks apoptosis at multiple steps during infection and protects cells from exogenous inducers in a cell-type-dependent manner. *Proc. Natl. Acad. Sci. USA*, **95**, 3931–6.

Galvan, V., Brandimarti, R., and Roizman, B. (1999). Herpes simplex virus 1 blocks caspase-3-independent and caspase-dependent pathways to cell death. *J. Virol.*, **73**, 3219–26.

Gangappa, S., van Dyk, L. F., Jewett, T. J., Speck, S. H., and Virgin, H. W., 4th (2002). Identification of the in vivo role of a viral bcl-2. *J. Exp. Med.*, **195**, 931–40.

Garvey, T. L., Bertin, J., Siegel, R. M., Wang, G. H., Lenardo, M. J., and Cohen, J. I. (2002). Binding of FADD and caspase-8 to molluscum contagiosum virus MC159 v-FLIP is not sufficient for its antiapoptotic function. *J. Virol.*, **76**, 697–706.

Geleziunas, R., Xu, W., Takeda, K., Ichijo, H., and Greene, W. C. (2001). HIV-1 Nef inhibits ASK1-dependent death signalling providing a potential mechanism for protecting the infected host cell. *Nature*, **410**, 834–8.

Gil, J., Rullas, J., Alcami, J., and Esteban, M. (2001). MC159L protein from the poxvirus molluscum contagiosum virus inhibits NF-kappaB activation and apoptosis induced by PKR. *J. Gen. Virol.*, **82**, 3027–34.

Glykofrydes, D., Niphuis, H., Kuhn, E. M. *et al.* (2000). Herpesvirus saimiri vFLIP provides an antiapoptotic function but is not essential for viral replication, transformation, or pathogenicity. *J. Virol.*, **74**, 11919–27.

Goldmacher, V. S., Bartle, L. M., Skaletskaya, A. *et al.* (1999). A cytomegalovirus-encoded mitochondria-localized inhibitor of apoptosis structurally unrelated to Bcl-2. *Proc. Natl. Acad. Sci. USA*, **96**, 12536–41.

Gooding, L. R., Ranheim, T. S., Tollefson, A. E. *et al.* (1991). The 10,400- and 14,500-dalton proteins encoded by region E3 of adenovirus function together to protect many but not all mouse cell lines against lysis by tumor necrosis factor. *J. Virol.*, **65**, 4114–23.

Han, J., Modha, D., and White, E. (1998). Interaction of E1B 19K with Bax is required to block Bax-induced loss of mitochondrial membrane potential and apoptosis. *Oncogene*, **17**, 2993–3005.

Hay, S. and Kannourakis, G. (2002). A time to kill: viral manipulation of the cell death program. *J. Gen. Virol.*, **83**, 1547–64.

Henderson, S., Huen, D., Rowe, M., Dawson, C., Johnson, G., and Rickinson, A. (1993). Epstein-Barr virus-coded BHRF1 protein, a viral homologue of Bcl-2, protects human B cells from programmed cell death. *Proc. Natl. Acad. Sci. USA*, **90**, 8479–83.

Hnatiuk, S., Barry, M., Zeng, W. *et al.* (1999). Role of the C-terminal RDEL motif of the myxoma virus M-T4 protein in terms of apoptosis regulation and viral pathogenesis. *Virology*, **263**, 290–306.

Hu, S., Vincenz, C., Buller, M., and Dixit, V. M. (1997). A novel family of viral death effector domain-containing molecules that inhibit both CD-95- and tumor necrosis factor receptor-1-induced apoptosis. *J. Biol. Chem.*, **272**, 9621–4.

Imani, F. and Jacobs, B. L. (1988). Inhibitory activity for the interferon-induced protein kinase is associated with the reovirus serotype 1 sigma 3 protein. *Proc. Natl. Acad. Sci. USA*, **85**, 7887–91.

Inoue, Y., Yasukawa, M., and Fujita, S. (1997). Induction of T-cell apoptosis by human herpesvirus 6. *J. Virol.*, **71**, 3751–9.

Iseni, F., Garcin, D., Nishio, M., Kedersha, N., Anderson, P., and Kolakofsky, D. (2002). Sendai virus trailer RNA binds TIAR, a cellular protein involved in virus-induced apoptosis. *EMBO J.*, **21**, 5141–50.

Jackson, S., Harwood, C., Thomas, M., Banks, L., and Storey, A. (2000). Role of Bak in UV-induced apoptosis in skin cancer and abrogation by HPV E6 proteins. *Genes Dev.*, **14**, 3065–73.

Jacotot, E., Ravagnan, L., Loeffler, M. *et al.* (2000). The HIV-1 viral protein R induces apoptosis via a direct effect on the mitochondrial permeability transition pore. *J. Exp. Med.*, **191**, 33–46.

Kaufmann, S. H. and Hengartner, M. O. (2001). Programmed cell death: alive and well in the new millennium. *Trends Cell Biol.*, **11**, 526–34.

Keller, S. A., Schattner, E. J., and Cesarman, E. (2000). Inhibition of NF-kappaB induces apoptosis of KSHV-infected primary effusion lymphoma cells. *Blood*, **96**, 2537–42.

Kenney, J. L., Guinness, M. E., Curiel, T., and Lacy, J. (1998). Antisense to the Epstein-Barr virus (EBV)-encoded latent membrane protein 1 (LMP-1) suppresses LMP-1 and bcl-2 expression and promotes apoptosis in EBV-immortalized B cells. *Blood*, **92**, 1721–7.

Kerr, J. F., Wyllie, A. H., and Currie, A. R. (1972). Apoptosis: a basic biological phenomenon with wide-ranging implications in tissue kinetics. *Br. J. Cancer*, **26**, 239–57.

Kolesnitchenko, V., Wahl, L. M., Tian, H. *et al.* (1995). Human immunodeficiency virus 1 envelope-initiated G2-phase programmed cell death. *Proc. Natl. Acad. Sci. USA*, **92**, 11889–93.

Koseki, T., Inohara, N., Chen, S. *et al.* (1999). CIPER, a novel NF kappaB-activating protein containing a caspase recruitment domain with homology to herpesvirus-2 protein E10. *J. Biol. Chem.*, **274**, 9955–61.

Kraft, M. S., Henning, G., Fickenscher, H. *et al.* (1998). Herpesvirus saimiri transforms human T-cell clones to stable growth without inducing resistance to apoptosis. *J. Virol.*, **72**, 3138–45.

Krueger, A., Baumann, S., Krammer, P. H., and Kirchhoff, S. (2001). FLICE-inhibitory proteins: regulators of death receptor-mediated apoptosis. *Mol. Cell Biol.*, **21**, 8247–54.

Kruman, I. I., Nath, A., and Mattson, M. P. (1998). HIV-1 protein Tat induces apoptosis of hippocampal neurons by a mechanism involving caspase activation, calcium overload, and oxidative stress. *Exp. Neurol.*, **154**, 276–88.

Lan, K. H., Sheu, M. L., Hwang, S. J. *et al.* (2002). HCV NS5A interacts with p53 and inhibits p53-mediated apoptosis. *Oncogene*, **21**, 4801–11.

Lee, J. M., Lee, K. H., Weidner, M., Osborne, B. A., and Hayward, S. D. (2002). Epstein-Barr virus EBNA2 blocks Nur77-mediated apoptosis. *Proc. Natl. Acad. Sci. USA*, **99**, 11878–83.

Lee, M. A. and Yates, J. L. (1992). BHRF1 of Epstein-Barr virus, which is homologous to human proto-oncogene bcl2, is not essential for transformation of B cells or for virus replication in vitro. *J. Virol.*, **66**, 1899–906.

Lee, S. B. and Esteban, M. (1994). The interferon-induced double-stranded RNA-activated protein kinase induces apoptosis. *Virology*, **199**, 491–6.

Leopardi, R. and Roizman, B. (1996). The herpes simplex virus major regulatory protein ICP4 blocks apoptosis induced by the virus or by hyperthermia. *Proc. Natl. Acad. Sci. USA*, **93**, 9583–7.

Leopardi, R., van Sant, C., and Roizman, B. (1997). The herpes simplex virus 1 protein kinase US3 is required for protection from apoptosis induced by the virus. *Proc. Natl. Acad. Sci. USA*, **94**, 7891–6.

Levine, B., Goldman, J. E., Jiang, H. H., Griffin, D. E., and Hardwick, J. M. (1996). Bcl-2 protects mice against fatal alphavirus encephalitis. *Proc. Natl. Acad. Sci. USA*, **93**, 4810–15.

Levine, B., Huang, Q., Isaacs, J. T., Reed, J. C., Griffin, D. E., and Hardwick, J. M. (1993). Conversion of lytic to persistent alphavirus infection by the bcl-2 cellular oncogene. *Nature*, **361**, 739–42.

Livne, A., Shtrichman, R., and Kleinberger, T. (2001). Caspase activation by adenovirus e4orf4 protein is cell line specific and is mediated by the death receptor pathway. *J. Virol.*, **75**, 789–98.

Liu, L., Eby, M. T., Rathore, N., Sinha, S. K., Kumar, A., and Chaudhary, P. M. (2002). The human herpes virus 8-encoded viral FLICE inhibitory protein physically associates with and persistently activates the Ikappa B kinase complex. *J. Biol. Chem.*, **277**, 13745–51.

Macho, A., Calzado, M. A., Jimenez-Reina, L., Ceballos, E., Leon, J., and Munoz, E. (1999). Susceptibility of HIV-1-TAT transfected cells to undergo apoptosis: biochemical mechanisms. *Oncogene*, **18**, 7543–51.

Marchini, A., Tomkinson, B., Cohen, J. I., and Kieff, E. (1991). BHRF1, the Epstein-Barr virus gene with homology to Bcl2, is dispensable for B-lymphocyte transformation and virus replication. *J. Virol.*, **65**, 5991–6000.

Marshall, W. L., Yim, C., Gustafson, E. *et al.* (1999). Epstein-Barr virus encodes a novel homolog of the bcl-2 oncogene that inhibits apoptosis and associates with Bax and Bak. *J. Virol.*, **73**, 5181–5.

Marusawa, H., Hijikata, M., Chiba, T., and Shimotohno, K. (1999). Hepatitis C virus core protein inhibits Fas- and tumor necrosis factor alpha-mediated apoptosis via NF-kappaB activation. *J. Virol.*, **73**, 4713–20.

McNees, A. and Gooding, L. (2002). Adenoviral inhibitors of apoptotic cell death. *Virus Res.*, **88**, 87–101.

McNees, A. L., Garnett, C. T., and Gooding, L. R. (2002). The adenovirus E3 RID complex protects some cultured human T and B lymphocytes from Fas-induced apoptosis. *J. Virol.*, **76**, 9716–23.

Meinl, E., Fickenscher, H., Thome, M., Tschopp, J., and Fleckenstein, B. (1998). Anti-apoptotic strategies of lymphotropic viruses. *Immunol. Today*, **19**, 474–9.

Meyaard, L., Otto, S. A., Jonker, R. R., Mijnster, M. J., Keet, R. P., and Miedema, F. (1992). Programmed death of T cells in HIV-1 infection. *Science*, **257**, 217–19.

Micheau, O., Lens, S., Gaide, O., Alevizopoulos, K., and Tschopp, J. (2001). NF-kappaB signals induce the expression of c-FLIP. *Mol. Cell Biol.*, **21**, 5299–305.

Moffatt, S., Yaegashi, N., Tada, K., Tanaka, N., and Sugamura, K. (1998). Human parvovirus B19 nonstructural (NS1) protein induces apoptosis in erythroid lineage cells. *J. Virol.*, **72**, 3018–28.

Moore, M., Horikoshi, N., and Shenk, T. (1996). Oncogenic potential of the adenovirus E4orf6 protein. *Proc. Natl. Acad. Sci. USA*, **93**, 11295–301.

Mori, N., Yamada, Y., Ikeda, S. *et al.* (2002). Bay 11–7082 inhibits transcription factor NF-kappaB and induces apoptosis of HTLV-I-infected T-cell lines and primary adult T-cell leukemia cells. *Blood*, **100**, 1828–34.

Munger, J. and Roizman, B. (2001). The US3 protein kinase of herpes simplex virus 1 mediates the posttranslational modification of BAD and prevents BAD-induced programmed cell death in the absence of other viral proteins. *Proc. Natl. Acad. Sci. USA*, **98**, 10410–15.

Nanbo, A., Inoue, K., Adachi-Takasawa, K., and Takada, K. (2002). Epstein-Barr virus RNA confers resistance to interferon-alpha-induced apoptosis in Burkitt's lymphoma. *EMBO J.*, **21**, 954–65.

Nauenburg, S., Zwerschke, W., and Jansen-Dürr, P. (2001). Induction of apoptosis in cervical carcinoma cells by peptide aptamers that bind to the HPV-16 E7 oncoprotein. *FASEB J.*, **15**, 592–4.

Nava, V. E., Cheng, E. H., Veliuona, M. *et al.* (1997). Herpesvirus saimiri encodes a functional homolog of the human bcl-2 oncogene. *J. Virol.*, **71**, 4118–22.

Neilan, J. G., Lu, Z., Afonso, C. L., Kutish, G. F., Sussman, M. D., and Rock, D. L. (1993). An African swine fever virus gene with similarity to the proto-oncogene bcl-2 and the Epstein-Barr virus gene BHRF1. *J. Virol.*, **67**, 4391–4.

Neilan, J. G., Lu, Z., Kutish, G. F. *et al.* (1997). A BIR motif containing gene of African swine fever virus, 4CL, is nonessential for growth in vitro and viral virulence. *Virology*, **230**, 252–64.

Nogal, M. L., Gonzalez de Buitrago, G., Rodriguez, C. *et al.* (2001). African swine fever virus IAP homologue inhibits caspase activation and promotes cell survival in mammalian cells. *J. Virol.*, **75**, 2535–43.

Ohyama, T., Tsukumo, S., Yajima, N., Sakamaki, K., and Yonehara, S. (2000). Reduction of thymocyte numbers in transgenic mice expressing viral FLICE-inhibitory protein in a Fas-independent manner. *Microbiol. Immunol.*, **44**, 289–97.

Perez, D. and White, E. (2000). TNF-alpha signals apoptosis through a bid-dependent conformational change in Bax that is inhibited by E1B 19K. *Mol. Cell*, **6**, 53–63.

Poyet, J. L., Srinivasula, S. M., and Alnemri, E. S. (2001). vCLAP, a caspase-recruitment domain-containing protein of equine herpesvirus-2, persistently activates the I kappa B kinases through oligomerization of IKKgamma. *J. Biol. Chem.*, **276**, 3183–7.

Rahmani, Z., Huh, K. W., Lasher, R., and Siddiqui, A. (2000). Hepatitis B virus X protein colocalizes to mitochondria with a human voltage-dependent anion channel, HVDAC3, and alters its transmembrane potential. *J. Virol.*, **74**, 2840–6.

Rasola, A., Gramaglia, D., Boccaccio, C., and Comoglio, P. M. (2001). Apoptosis enhancement by the HIV-1 Nef protein. *J. Immunol.*, **166**, 81–8.

Reading, P. C., Khanna, A., and Smith, G. L. (2002). Vaccinia virus CrmE encodes a soluble and cell surface tumor necrosis factor receptor that contributes to virus virulence. *Virology*, **292**, 285–98.

Rivas, C., Thlick, A. E., Parravicini, C., Moore, P. S., and Chang, Y. (2001). Kaposi's sarcoma-associated herpesvirus LANA2 is a B-cell-specific latent viral protein that inhibits p53. *J. Virol.*, **75**, 429–38.

Rothe, M., Pan, M. G., Henzel, W. J., Ayres, T. M., and Goeddel, D. V. (1995). The TNFR2-TRAF signaling complex contains two novel proteins related to baculoviral inhibitor of apoptosis proteins. *Cell*, **83**, 1243–52.

Sabbatini, P., Lin, J., Levine, A. J., and White, E. (1995). Essential role for p53-mediated transcription in E1A-induced apoptosis. *Genes Dev.*, **9**, 2184–92.

Sadzot-Delvaux, C., Thonard, P., Schoonbroodt, S., Piette, J., and Rentier, B. (1995). Varicella-zoster virus induces apoptosis in cell culture. *J. Gen. Virol.*, **76**, 2875–9.

Salvesen, G. S. and Duckett, C. S. (2002). IAP proteins: blocking the road to death's door. *Nat. Rev. Mol. Cell Biol.*, **3**, 401–10.

Sarid, R., Ben-Moshe, T., Kazimirsky, G. *et al.* (2001). vFLIP protects PC-12 cells from apoptosis induced by Sindbis virus: implications for the role of TNF-alpha. *Cell Death Differ.*, **8**, 1224–31.

Scaffidi, C., Fulda, S., Srinivasan, A. *et al.* (1998). Two CD95 (APO-1/Fas) signaling pathways. *EMBO J.*, **17**, 1675–87.

Scheffner, M., Huibregtse, J. M., Vierstra, R. D., and Howley, P. M. (1993). The HPV-16 E6 and E6-AP complex functions as a ubiquitin-protein ligase in the ubiquitination of p53. *Cell*, **75**, 495–505.

Schreiber, M., Sedger, L., and McFadden, G. (1997). Distinct domains of M-T2, the myxoma virus tumor necrosis factor (TNF) receptor homolog, mediate extracellular TNF binding and intracellular apoptosis inhibition. *J. Virol.*, **71**, 2171–81.

Schulze-Osthoff, K., Ferrari, D., Los, M., Wesselborg, S., and Peter, M. E. (1998). Apoptosis signaling by death receptors. *Eur. J. Biochem.*, **254**, 439–59.

Sciortino, M. T., Perri, D., Medici, M. A., Foti, M., Orlandella, B. M., and Mastino, A. (2000). The gamma-2-herpesvirus bovine herpesvirus 4 causes apoptotic infection in permissive cell lines. *Virology*, **277**, 27–39.

Secchiero, P., Flamand, L., Gibellini, D. *et al.* (1997). Human herpesvirus 7 induces CD4(+) T-cell death by two distinct mechanisms: necrotic lysis in productively infected cells and apoptosis in uninfected or nonproductively infected cells. *Blood*, **90**, 4502–12.

Shisler, J. L. and Moss, B. (2001). Molluscum contagiosum virus inhibitors of apoptosis: the MC159 v-FLIP protein blocks Fas-induced activation of procaspases and degradation of the related MC160 protein. *Virology*, **282**, 14–25.

Shisler, J., Yang, C., Walter, B., Ware, C. F., and Gooding, L. R. (1997). The adenovirus E3–10.4K/14.5K complex mediates loss of cell surface Fas (CD95) and resistance to Fas-induced apoptosis. *J. Virol.*, **71**, 8299–306.

Skaletskaya, A., Bartle, L. M., Chittenden, T., McCormick, A. L., Mocarski, E. S., and Goldmacher, V. S. (2001). A cytomegalovirus-encoded inhibitor of apoptosis that suppresses caspase-8 activation. *Proc. Natl. Acad. Sci. USA*, **98**, 7829–34.

Somasundaran, M., Sharkey, M., Brichacek, B. *et al.* (2002). Evidence for a cytopathogenicity determinant in HIV-1 Vpr. *Proc. Natl. Acad. Sci. USA*, **99**, 9503–8.

Srinivasula, S. M., Ahmad, M., Lin, J. H. *et al.* (1999). CLAP, a novel caspase recruitment domain-containing protein in the tumor necrosis factor receptor pathway, regulates NF-kappaB activation and apoptosis. *J. Biol. Chem.*, **274**, 17946–54.

Stehlik, C., de Martin, R., Kumabashiri, I., Schmid, J. A., Binder, B. R., and Lipp, J. (1998). Nuclear factor (NF)-kappaB-regulated X-chromosome-linked IAP gene expression protects endothelial cells from tumor necrosis factor alpha-induced apoptosis. *J. Exp. Med.*, **188**, 211–16.

Strasser, A., Harris, A. W., Huang, D. C., Krammer, P. H., and Cory, S. (1995). Bcl-2 and Fas/APO-1 regulate distinct pathways to lymphocyte apoptosis. *EMBO J.*, **14**, 6136–47.

Stürzl, M., Hohenadl, C., Zietz, C. *et al.* (1999). Expression of K13/v-FLIP gene of human herpesvirus 8 and apoptosis in Kaposi's sarcoma spindle cells. *J. Natl. Cancer Inst.*, **91**, 1725–33.

Su, F. and Schneider, R. J. (1997). Hepatitis B virus HBx protein sensitizes cells to apoptotic killing by tumor necrosis factor alpha. *Proc. Natl. Acad. Sci. USA*, **94**, 8744–9.

Tarodi, B., Subramanian, T., and Chinnadurai, G. (1994). Epstein-Barr virus BHRF1 protein protects against cell death induced by DNA-damaging agents and heterologous viral infection. *Virology*, **201**, 404–7.

Tepper, C. G. and Seldin, M. F. (1999). Modulation of caspase-8 and FLICE-inhibitory protein expression as a potential mechanism of Epstein-Barr virus tumorigenesis in Burkitt's lymphoma. *Blood*, **94**, 1727–37.

Tewari, M., Telford, W. G., Miller, R. A., and Dixit, V. M. (1995). CrmA, a poxvirus-encoded serpin, inhibits cytotoxic T-lymphocyte-mediated apoptosis. *J. Biol. Chem.*, **270**, 22705–8.

Thomas, M., Matlashewski, G., Pim, D., and Banks, L. (1996). Induction of apoptosis by p53 is independent of its oligomeric state and can be abolished by HPV-18 E6 through ubiquitin mediated degradation. *Oncogene*, **13**, 265–73.

Thome, M., Gaide, O., Micheau, O. *et al.* (2001). Equine herpesvirus protein E10 induces membrane recruitment and phosphorylation of its cellular homologue, bcl-10. *J. Cell Biol.*, **152**, 1115–22.

Thome, M., Martinon, F., Hofmann, K. *et al.* (1999). Equine herpesvirus-2 E10 gene product, but not its cellular homologue, activates NF-kappaB transcription factor and c-Jun N-terminal kinase. *J. Biol. Chem.*, **274**, 9962–8.

Thome, M., Schneider, P., Hofmann, K. *et al.* (1997). Viral FLICE-inhibitory proteins (FLIPs) prevent apoptosis induced by death receptors. *Nature*, **386**, 517–21.

Thompson, D. A., Zacny, V., Belinsky, G. S. *et al.* (2001). The HPV E7 oncoprotein inhibits tumor necrosis factor alpha-mediated apoptosis in normal human fibroblasts. *Oncogene*, **20**, 3629–40.

Tollefson, A. E., Toth, K., Doronin, K. *et al.* (2001). Inhibition of TRAIL-induced apoptosis and forced internalization of TRAIL receptor 1 by adenovirus proteins. *J. Virol.*, **75**, 8875–87.

Tyler, K. L., Squier, M. K., Rodgers, S. E. *et al.* (1995). Differences in the capacity of reovirus strains to induce apoptosis are determined by the viral attachment protein sigma 1. *J. Virol.*, **69**, 6972–9.

Ubol, S., Tucker, P. C., Griffin, D. E., and Hardwick, J. M. (1994). Neurovirulent strains of alphavirus induce apoptosis in bcl-2-expressing cells: role of a single amino acid change in the E2 glycoprotein. *Proc. Natl. Acad. Sci. USA*, **91**, 5202–6.

van Dyk, L. F., Virgin, H. W., 4th, and Speck, S. H. (2000). The murine gammaherpesvirus 68 v-cyclin is a critical regulator of reactivation from latency. *J. Virol.*, **74**, 7451–61.

Wang, G. H., Bertin, J., Wang, Y. *et al.* (1997). Bovine herpesvirus 4 BORFE2 protein inhibits Fas- and tumor necrosis factor receptor 1-induced apoptosis and contains death effector domains shared with other gamma-2 herpesviruses. *J. Virol.*, **71**, 8928–32.

Wang, G. H., Garvey, T. L., and Cohen, J. I. (1999). The murine gammaherpesvirus-68 M11 protein inhibits Fas- and TNF-induced apoptosis. *J. Gen. Virol.*, **80**, 2737–40.

Wang, H. W., Sharp, T. V., Koumi, A., Koentges, G., and Boshoff, C. (2002). Characterization of an anti-apoptotic glycoprotein encoded by Kaposi's sarcoma-associated herpesvirus which resembles a spliced variant of human survivin. *EMBO J.*, **21**, 2602–15.

Webster, K., Parish, J., Pandya, M., Stern, P. L., Clarke, A. R., and Gaston, K. (2000). The human papillomavirus (HPV)16 E2 protein induces apoptosis in the absence of other HPV proteins and via a p53-dependent pathway. *J. Biol. Chem.*, **275**, 87–94.

Wei, M. C., Lindsten, T., Mootha, V. K. *et al.* (2000). tBID, a membrane-targeted death ligand, oligomerizes BAK to release cytochrome c. *Genes Dev.*, **14**, 2060–71.

Yaegashi, N., Niinuma, T., Chisaka, H. *et al.* (1999). Parvovirus B19 infection induces apoptosis of erythroid cells in vitro and in vivo. *J. Infect.*, **39**, 68–76.

Yan, M., Lee, J., Schilbach, S., Goddard, A., and Dixit, V. (1999). mE10, a novel caspase recruitment domain-containing proapoptotic molecule. *J. Biol. Chem.*, **274**, 10287–92.

Yanai, N. and Obinata, M. (1994). Apoptosis is induced at nonpermissive temperature by a transient increase in p53 in cell lines immortalized with temperature-sensitive SV40 large T-antigen gene. *Exp. Cell Res.*, **211**, 296–300.

Yasukawa, M., Inoue, Y., Ohminami, H., Terada, K., and Fujita, S. (1998). Apoptosis of CD4+ T lymphocytes in human herpesvirus-6 infection. *J. Gen. Virol.*, **79**, 143–7.

Zamzami, N. and Kroemer, G. (2001). The mitochondrion in apoptosis: how Pandora's box opens. *Nat. Rev. Mol. Cell Biol.*, **2**, 67–71.

Zhirnov, O. P., Konakova, T. E., Wolff, T., and Klenk, H. D. (2002). NS1 protein of influenza A virus down-regulates apoptosis. *J. Virol.*, **76**, 1617–25.

Zhou, G. and Roizman, B. (2000). Wild-type herpes simplex virus 1 blocks programmed cell death and release of cytochrome c but not the translocation of mitochondrial apoptosis-inducing factor to the nuclei of human embryonic lung fibroblasts. *J. Virol.*, **74**, 9048–53.

Zhou, G., Galvan, V., Campadelli-Fiume, G., and Roizman, B. (2000). Glycoprotein D or J delivered in trans blocks apoptosis in SK-N-SH cells induced by a herpes simplex virus 1 mutant lacking intact genes expressing both glycoproteins. *J. Virol.*, **74**, 11782–91.

Zhu, H., Shen, Y., and Shenk, T. (1995). Human cytomegalovirus IE1 and IE2 proteins block apoptosis. *J. Virol.*, **69**, 7960–70.

Zhu, N., Khoshnan, A., Schneider, R. *et al.* (1998). Hepatitis C virus core protein binds to the cytoplasmic domain of tumor necrosis factor (TNF) receptor 1 and enhances TNF-induced apoptosis. *J. Virol.*, **72**, 3691–7.

Index

For EU product safety concerns, contact us at Calle de José Abascal, 56–1°,
28003 Madrid, Spain or eugpsr@cambridge.org.

www.ingramcontent.com/pod-product-compliance
Ingram Content Group UK Ltd.
Pitfield, Milton Keynes, MK11 3LW, UK
UKHW060311090126
466816UK00021B/445